Information Processing in Autonomous Mobile Robots

Proceedings of the International Workshop

March 6-8, 1991

Technische Universität München

Germany

Edited by G. Schmidt

Springer-Verlag Berlin Heidelberg GmbH 1991

Organizing Chairman
Professor Dr.-Ing. Günther Schmidt
Technische Universität München
Lehrstuhl für Steuerungs- und Regelungstechnik

Sponsored by
Deutsche Forschungsgemeinschaft, Bonn

Co-sponsored by
VDI/VDE Gesellschaft Meß- und Automatisierungstechnik, Düsseldorf

ISBN 978-3-540-53964-3   ISBN 978-3-662-07896-9 (eBook)
DOI 10.1007/978-3-662-07896-9

2362/3020-543210

Preface

This volume is a collection of 22 papers presented at the International Workshop on Information Processing in Autonomous Mobile Robots, held in Munich (Germany) in March 1991.

Autonomous mobile robot technologies are generating significant interest because of their potential capabilities for future applications on the plant floor as well as in the service industry. Autonomous robots may navigate around factories and laboratories, hospitals, office-buildings, airports or similar public and semipublic places. They may deliver equipment, collect garbage and perform other such tasks. One of the major challenges for the field of autonomous mobile robot research is to develop robust and real-time systems for perception and understanding of complicated real environments as well as for intelligent decision-making with respect to proper actions.

This Workshop was set up to stimulate discussion and the exchangement of new ideas on various aspects of autonomous mobile robot methodologies and applications. The main focal points of the Workshop program were sensing and perception, navigation and control, knowledge bases and computer architechtures as well as various applications.

The papers are prepared by leading experts in theses areas from Europe, Japan, the United States and by researchers involved in the interdisciplinary research project on "Information Processing in Autonomous Mobile Robots (Sonderforschungsbereich 331)" at the Technische Universität München.

The Workshop was held under the auspices and with the financial support of the Deutsche Forschungsgemeinschaft Bonn-Bad Godesberg, which is gratefully acknowledged.

Since autonomous technologies may also become of interest for other application fields in automation the co-sponsorship of the Workshop by VDI/VDE GMA (German Society for Measurement and Automation) is highly appreciated.

As the editor, I would like to graciously thank my colleagues of the joint research project at TU München for their contributions in the preparation of the Workshop. All authors also sincerely acknowledged for the prompt availability of their manuscripts in camera-ready form. Lastly, I would like to acknowledge the invaluable assistance of my secretary, Ms. H. Will and my associate, Dr. F. Freyberger, in editing this volume.

München, May 1991   Günther Schmidt

Contents

Navigation and Control 103

Knowlege-Bases and Computer Architectures 217

*)   This research is supported by Deutsche Forschungsgemeinschaft, Bonn, as a part of
      the joint project:
          "Information Processing in Autonomous Mobile Robots (SFB 331)".

Introduction

Towards Integration of Autonomous Subsystems for Assembly and Mobility into Flexible Manufacturing

G. Schmidt

Abstract

This paper surveys scope and sample results of a long-range interdisciplinary research project established at the Technische Universität München in 1986. Its main focus is on development and evaluation of concepts, methodologies and algorithms for information processing and management in autonomous mobile robotic systems. A major goal of the project is the incorporation of basic autonomous system techniques and technologies into factory and manufacturing automation, a highly relevant industrial application area.

The paper examines the evolution of todays flexible manufacturing systems (FMS) and its conventional information and control systems architecture into a future intelligent autonomous FMS. Autonomous work stations and autonomous tranporters may play a major role in an advanced FMS. From a planning and control systems architecture viewpoint two examples for autonomous FMS subsystems, as developed in the Control Engineering Laboratory of the TU München, are discussed: a flexible autonomous robotic work station for small parts assembly and an autonomous mobile robot. Additional and more detailed reports about new information processing components and approaches for an intelligent autonomous FMS, as developed in the framework of the joint research project, are presented elsewhere in this proceeding volume.

1. Introduction

Over the past decade, intelligent autonomous systems and robots have been the focus of world-wide research and development activities. Applications such as autonomous land vehicles, underwater and space robots, multiarm-manipulators, advanced systems for nuclear power plant maintenance, diagnosis and repair, as well as agricultural, domestic, service or health-care robots have provided an additional incentive for intensive research efforts. Progress in development has been accelerated by the advent of new technologies in many fields, as for example in microelectronics, microcontrollers, parallel processor architectures and sensing devices, as well as in artificial intelligence, software development environments or real-time data bases.

- 4 -

In many instances intelligent autonomous systems are extending the capabilities of state of the art automatic engineering systems. **Automatic system** design has conventionally been based on rigorous numerical techniques for modelling and optimization, which provided automatic systems with their typical (more or less fixed) SENSE-REACT behaviour. Although tremendous improvements of performance can be achieved through reactive automatic systems, both under deterministic and stochastic conditions, their operation is often limited to adaptation to relatively small changes in fairly known and structured environments.

Intelligent autonomous systems, on the other hand, are designed to operate in less structured, uncertain, often varying or even unknown environments and therefore, require much higher levels of adaptation to unexpected external and internal events. They are also required to store, process and interprete large quantities of data generated by various types of sensors or imaging devices. Results are used for action planning and execution, monitoring and failure recovery. For these reasons, the design of intelligent autonomous systems incorporates besides well-proven automatic systems techniques, multisensory approaches and heuristic and/or symbolic tools from artificial intelligence. These tools have been conventionally developed for open-loop offline applications. They must now be integrated into the typical

PERCEPTION-PLANNING / REASONING-ACTION

closed-loop of intelligent autonomous systems and perform smoothly and accurately under stringent real-time and real-world conditions.

2. Goals of Autonomous Systems Research

2.1 Joint Research Project

Todate, studies of intelligent autonomous system techniques with respect to industrial applications are relatively scarce. Nevertheless, autonomous system research and development may have a major impact on most relevant industrial fields such as factory automation and flexible manufacturing. Although todays flexible manufacturing systems (FMS) prove to be highly automated systems, their overall capabilities and performance could be extended and improved through development and application of specific autonomous system technologies. As a result of this idea, a long-range interdisciplinary research project on "Concepts, Methodologies and Algorithms of Information Processing and Management in Autonomous, Mobile Robots" was established at the Technische Universität München in 1986. This project, known as Sonderforschungsbereich 331, is funded by the Deutsche Forschungsgemeinschaft and the Bavarian State Ministery for Science and Arts. A total of seven teams from Electrical Engineering, Mechanical Engineering and Computer Science are engaged in cooperative studies, research and development work. The scope of this project is strongly influenced by experience with todays manufacturing automation systems and state-of-the-art CIM techniques. Its main objective is to contribute to the improvement of efficiency and productivity through incorporation of flexible and autonomous system techniques into future manufacturing systems.

The first project phase (1986-1989) was concerned with the development of selected concepts of information processing and management, the design of specific hardware/software tools and components, both for autonomous robotic manipulation and mobility tasks. Current research efforts (1990-1993) are directed to integration of newly developed approaches and subsystems into a realistic FMS and CIM environment, accompanied by preliminary tests and evaluation. A large-scale demonstration of enhanced autonomy features in an intelligent FMS scenario will be the final goal of the joint research project (1994-1997).

2.2 Test and Demonstration Environment

In addition to the usual experimental and demonstration facilities available in the laboratories involved in the project, one of the groups operates a medium-sized experimental FMS factory within the Technische Universität München. This FMS test-bed comprises CNC-lathes, a machining and sawing center, robotic work cells, material and tool storages, as well as loading/unloading stations. An upgraded conventional automatic guided vehicle (AGV) is directed at material transportation jobs, as well as at moving a full-size manipulator to different work stations and locations. At its destination, the mobile manipulator loads and unloads parts and tools.

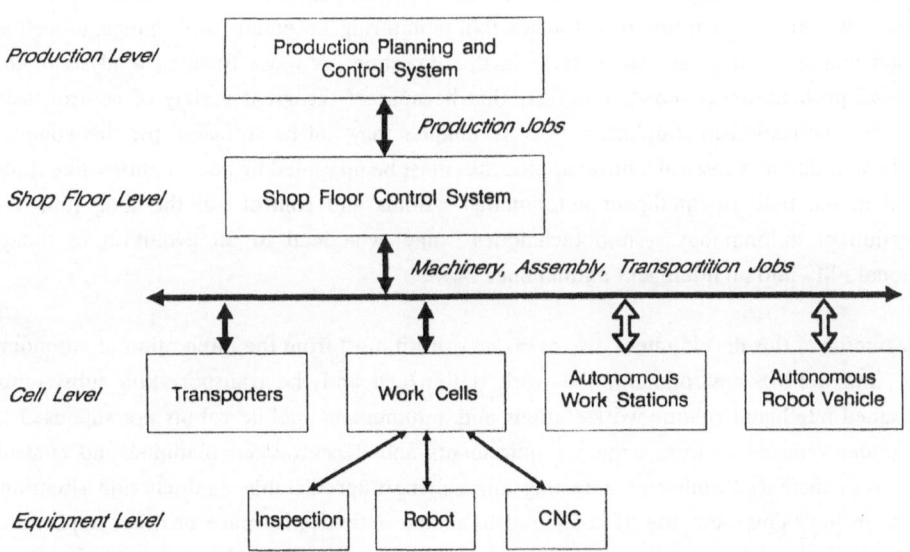

Fig. 1: Planning and control system of experimental FMS factory

The FMS's information and control systems are embedded in a CIM environment, Fig. 1.

Planning, dispatching, supervision and control of the manufacturing system are performed through a hierarchically structured, networked computer system comprising the

* production planning and control system (PPC),
* shop floor control system,
* work cell, work station and transporter controllers,
* equipment/machine controllers.

Production jobs created by the PPC system are transferred to the shop floor control system. They are expanded into individual tasks for execution by the work stations, machining and transportation resources. A knowledge base contains all information required on this level for scheduling control and error recovery purposes. Work station and machine controllers perform typical low-level CNC or robot control operations during parts processing and assembly.

2.3 Intelligent autonomous FMS

From a control engineering viewpoint the operation of todays FMS is still more or less centralized, planned and open-loop. This fact may be one of the reasons that FMS operations are often prone to degradation or malfunction, causing even partial standstill of resources and requiring frequent manual intervention for repair and recovery. To overcome some of these shortcomings, there exists a trend for introducing feedback and closed-loop control techniques into FMS subsystems and operations. The incorporation of automatic tool monitoring, automatic tool change, as well as closed-loop quality control, are some steps in this direction. A more detailed analysis of the various FMS problem areas shows, however, that because of the great variety of control tasks involved, the application of standard control techniques may not be sufficient for this complex industrial environment. Classical control approaches must be upgraded by new qualities, like those developed in the field of intelligent autonomous systems and control. In the long run, the incorporation of autonomous system techniques may even lead to an evolution of todays conventional FMS into an **intelligent autonomous FMS**.

In the beginning of this development, two areas may profit most from the integration of autonomy functions into a FMS environment: the work station/cell and the transportation subsystems. Self-contained intelligent robotic work stations and autommomous mobile robots are supposed to operate under reduced communication requirements and decentralized planning and control, however, with increased ability of detecting and solving unpredictable conflicts and situations coming up in their environments. If compared to similar activities in space or other application fields, the levels of autonomy required for a semi-structured and reasonably predictable industrial application area, like parts manufacturing, are not extremely high. Nevertheless, detailed scientific and practical experience gained through systematic research and development work in an economically relevant real-world environment, may generate useful results for future more ambitious autonomous systems projects and applications.

It is immediately obvious that together with the introduction of autonomous subsystems on the work station and transportation level of a FMS (see Fig. 1), a few other problems need to be considered simultaneously. For closing the PERCEPTION-PLANNING-ACTION loop, FMS specific techniques, components and algorithms for perception, planning and simulation, execution of operations or failure diagnosis and recovery need to be developed. Furthermore, autonomous operations on the lower levels of a FMS control hierarchy have to be supported by appropriate activities on the higher levels. This means, that in the long run the successful incorporation of autonomous system techniques may lead to major changes in the traditional FMS information and control system architecture and its functionalities. These phenomena must be studied very carefully in an early phase.

2.4 Research Topics

Consequently, the development and evaluation activities in the joint research project are focusing on three major areas:

* autonomous system techniques for robotic assembly and manipulation
* autonomous system techniques for robotic mobility
* techniques for integration of autonomous subsystems and operations into the overall information and control systems architecture of an intelligent autonomous FMS.

In the proceeding volume, the survey is accompanied by five technical papers prepared by members of the joint research project, highlighting related research topics. The paper prepared by Prof. SIEGERT's group is concerned with upper-level integration aspects and an object-oriented knowledge base for a CIM environment [1]. Integration aspects, from a FMS task-level programming viewpoint, are outlined in a paper prepared by Prof. MILBERG's group [2].

Two specific examples of autonomous subsystems for assembly and mobility, as developed by the Robotics Group of the Control Engineering Laboratory, will be examined next

* the planning and control architecture of an autonomous robotic work station for flexible assembly and testing of electro-mechanical components,
* the information processing, decision and control structure of the autonomous robot vehicle MACROBE and some of its intelligent subsystems.

3. Planning and control architecture of an autonomous work station

3.1 System Overview

The development of an integrated planning and control structure for a flexible autonomous work

station is focusing on three major areas:
* task level programming techniques
* automatic assembly planning techniques
* various forms of failure tolerant behaviour.

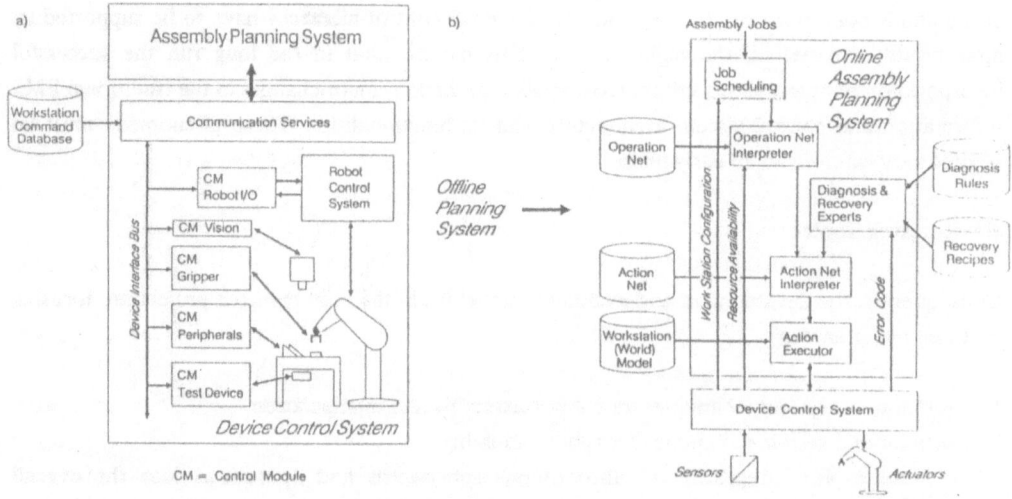

Fig. 2: *Work station planning and control system*
   a) device control system, b) assembly planning system

We will show how each of these topics contributes to the enhancement of system autonomy.
The flexible robotic work station under consideration is designed for small lot-size assembly of a
family of electro-mechanical relais. It is used in our laboratory as a test-bed for evaluation of
newly developed theoretical approaches and components. The station comprises the following
basic resources: a programmable industrial robot as a universal manipulation facility, a smart
two-fingered gripper with additional axis of fine motion for performing rapid and precise assembly
operations through hybrid position/force control, sensors, part feeders, a pneumatically driven
fixture for subassemblies and a test device. All components and devices on this level are
interconnected through intelligent peripheral controller modules and a serial semi-standardized
bus, Fig. 2a.

3.2 Flexibility and Autonomy

In the context of the work station, **flexibility** is understood as the capacity of the station to be
applied to a variety of assembly tasks (within a certain scope) solely by means of reprogramming
and not by redesign. The industrial requirement for enhanced system flexibility and
programmability usually implies higher levels of system autonomy. **Autonomy** is understood here
as the capacity of the work station and its resources to interact with a partly unpredictable,

uncertain or varying environment without manual assistance.

Variations in the assembly environment may result from various types of failures. We will consider here

* transient and permanent **assembly system failures** as for example caused by electromagnetic interference or breakdown of a tool or feeder
* **assembly process failures** as for example, caused by an erroneous part in a feeder, a jammed component or by incorrect geometrical orientation of a part in the gripper-fingers.

For an industrial application, like assembly, three levels of autonomy may be distinguished:

Level 1 autonomy is achieved by a work station control system, which can detect system and/or process failures and initiate a safe shutdown, if required, in order to prevent potential damage. The risk of unsupervised operation is made sufficiently small through error confinement measures. Failure detection requires sufficient sensor information in order to check actual system behaviour against expected system behaviour in real-time.

Level 2 autonomy refers to the additional ability of a control system to adapt operations to typical environmental situations and to react appropriately to unexpected but basically known events. This capability requires recognition of the relevant process conditions through sensor information, modification of excecutional parameters, selection of suitable plans and strategies for failure recovery. It leads to a failure tolerant system behaviour.

Level 3 autonomy allows a control system to react to unknown events through generation of new plans and strategies. Implementation of this type of autonomy seems to be questionable in a semi-structured industrial environment. Unconstraint plan generation may cause a system getting out of user's control and may even lead to unforeseeable dangers. However, a reduced form of this highest level of autonomy may prove to be useful in the work station environment if implemented as some sort of advisory or expert system assisting the user or operator in the solution of critical or complex situations.

An aspect closely related to an industrial application area, such as FMS, is the fact that assembly and related operations should always have a deterministic outcome and can, thus, be preplanned to a certain degree. As a consequence, autonomy can be increased by assisting the planning process through automatic planners and by making planners responsive to the actual state of the assembly process.

The development of the autonomous assembly sequence planning and control system is constrained to level 1 and 2 autonomy features. However, integration of autonomy features into a work station control system requires a number of additional features in its hardware and software architecture.

3.3 Planning and Control System

The work station task planning and control system, as developed in our laboratory [3], comprises four major subsystems

* offline assembly planning system
* online assembly planning system including
  a failure detection, diagnosis and recovery system
* device control system
* integrated sensory and mechanical components.

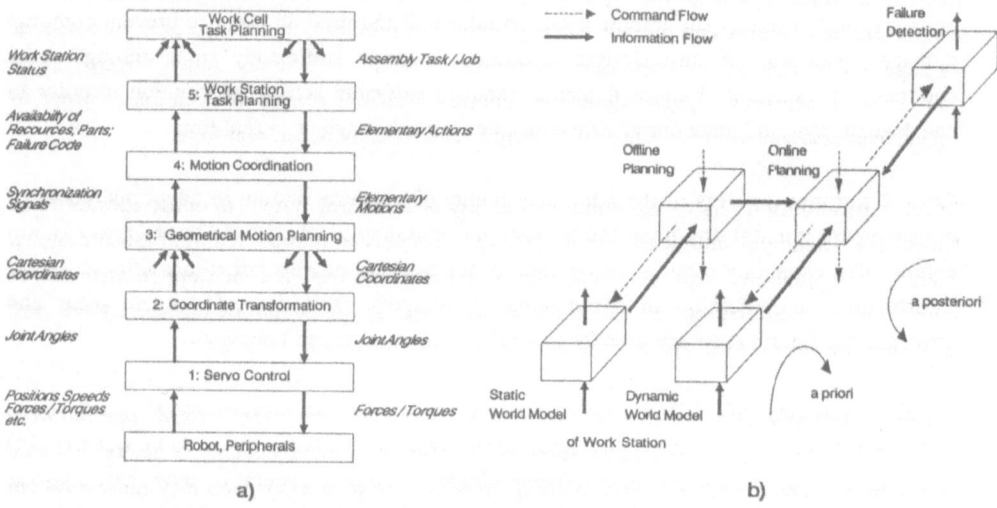

Fig. 3: *Reference model for assembly work station control system architecture*
   a) model layers, b) activites within and between layers

An abstract reference model for a general robotic work cell control system architecture forms the base of the designed assembly planning and control system, Fig. 3a. The hierachically structured multiple layered model shows similarities to other recently published architectures, as for example the NASA/NTS Standard Reference Model (NASREM) for telerobot control systems [4]. The reference model proves to be an ideal framework for accomodating advanced concepts of task planning, monitoring and autonomous system control techniques. Basically, three types of interrelated fundamental operations are performed on each model layer:

* temporal and spatial task decomposition, i.e. planning and excecution of higher level goals
  and operations into lower level actions

* world model up-dating, i.e. bookkeeping of the state of objects, resources and subsystems within the work station
* failure detection, diagnosis and recovery based on comprehensive sensory information.

Information and command flow between and within the layers are shown in Fig. 3b. With the objective of reducing real-time planning effort, planning operations are decomposed into a preparatory or offline part and an incremental or online part. Offline planning depends on static world model aspects, while online planning is intimately related to adequately processed sensory information and dynamic world model aspects.

In the autonomous work station, the planning and control system under consideration layer 1 and 2 operations are incorporated within the device control system, which proves to be a hardware dependent subsystem. The assembly planning subsystems, on the other hand, are based on an abstract architecture and comprise layer 3 to 5 activities. A high-level robot motion programming language forms the interface between planning and control layers.

3.4 Three Types of Autonomy

Three types of autonomy are implemented within the framework of the work station planning and control system. Autonomy based on

* reactive system behaviour
* adaptive system behaviour
* online planning, failure detection and failure recovery.

Autonomy functions through **reactive behaviour** are provided by means of intelligent feedback control loops located on layer 1 of the reference model. A good example are gripper-fingers with active compliance [5]. Force vector sensors measure reaction forces applied to each finger separately. Without external contact, gripping forces are equal in both fingers. When an external force is applied, finger position is controlled, such that the difference of the finger forces is minimized.

Through this type of low-level autonomy, small positional errors of the part or the robot are compensated during the course of an activity (a posteriori). Upper control layers are not involved and shielded from the effect of these disturbances. As a further consequence, requirements for positional accuracy of robot motion may be reduced to about 1 mm instead of 0.1 mm.

Autonomy through **adaptive behaviour** is related to layer 3 of the reference model and thus included in the assembly planning system. For example, robot motion can be adapted to changes in the environment, when gripping from an inaccurately positioned pallet. In this case, a vision sensor is employed to determine actual part position and orientation. The sensor identifies also the type of part to be grasped, in order to <u>detect</u> faulty or wrong components. Both aspects contribute to the autonomy of the system <u>before an activity is started</u> (a priori).

Planned behaviour combined with adequate failure recovery techniques are related to layer 4 and 5 of the reference model and provide high level autonomy functions. Without entering into a detailed discussion of the model- and rule-based activities of the offline planning system [6] (Fig. 2b), let us just mention that an "assembly job" is automatically decomposed into the set of product related sequences of "operations" (like INSERTION, TESTING, etc.), represented by an "operation net". Operations are in turn transformed into a set of sequences of "elementary actions" represented by an "action net".

An action net can be considered a state transition graph representing all <u>relevant</u> system states and <u>permissible</u> transitions, Fig. 4b. The system state defines, for example, the actual location of the robot arm and of the part under consideration, plus restrictions which influence the permissible action sequences. Arcs between nodes represent the elementary actions. They are control procedures, in terms of device commands, effecting state transitions.

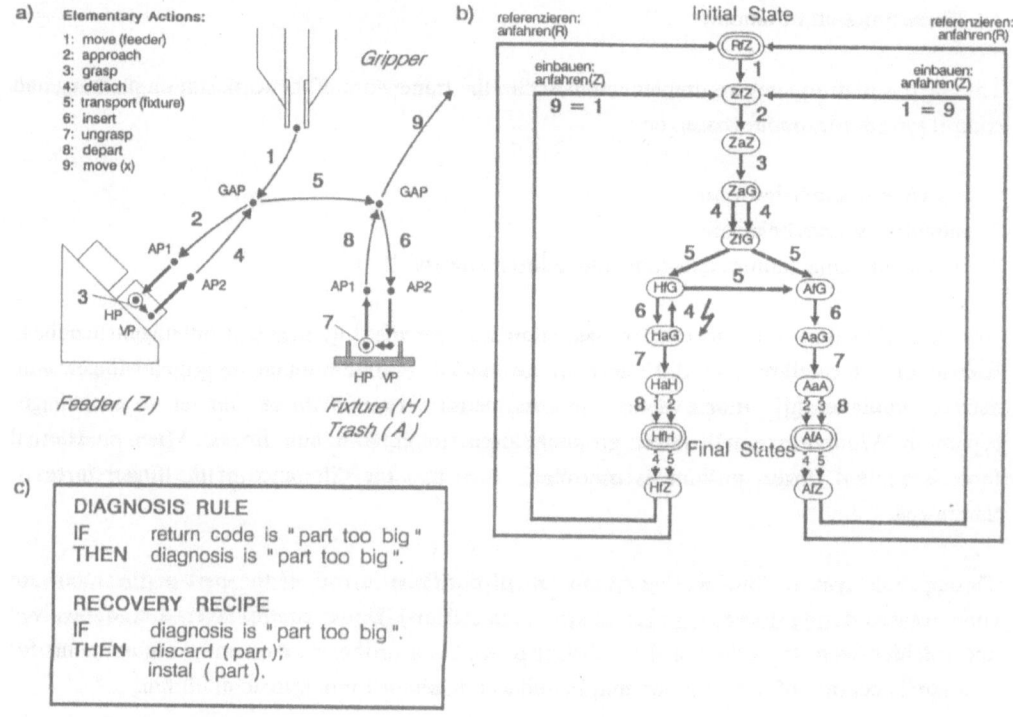

Fig. 4: Operation "Insertion"
a) insertion task, b) action net, c) diagnosis and recovery rules

For both, nominal or exception procedures, the action sequence from the actual state to any of the goal states is always unique. For example, starting from initial state RfZ (Fig. 4b) the assembly "operation" INSERTION is defined by the stereotype sequence of actions: *move, approach, grasp, detach, transport, insert, ungrasp, depart* to final state HfH.

If action "insert" fails, due to a faulty part at state HaG, the sensor equipped gripper detects that the expected force value is exceeded. The regular action sequence is aborted and control is transferred to the diagnosis and recovery experts. Depending on the contents of the diagnosis rule-base and the recovery recipes, recovery operations are invoked. For a simple scenario as discussed here (Fig. 4c), the action net interpreter selects an action sequence, leading to the new goal state AfA, where the faulty part is discarded. As its last activity the recovery expert reinvokes the earlier interrupted action, and normal operation is resumed (install (part)). It is clear that this type of planned behaviour can be extended to more complex scenarios, thus contributing to system autonomy in case of assembly failures.

It is noteworthy that the use of an action net enforces the command of permissible action seqences only. This is a major difference to conventional expert system applications, where minor changes in a rule-base can cause unforeseen side-effects. The worst thing that could happen here is that the action net interpreter would not be able to excecute a given operation command.

Similar autonomy features result from application of the operation net. It is a representation of the set of all permissible operation sequences leading to a completely assembled and tested product. In a given assembly state, the next operation is selected according to current part and resource availability. Thus, the planning system responds again autonomously to changing environmental conditions, Fig. 2b.

This discussion has shown how efficient autonomous system operations can be integrated into different layers of an advanced planning and control system for an assembly work station. The planning algorithms and implementation details are reported in more detail in [3].

Similar autonomous system techniques are employed in the information processing system of the mobile robot MACROBE.

4. Information Processing Structure of an Autonomous Robot Vehicle

4.1 System Overview

MACROBE is an experimental robot vehicle developed by the Robotics Group of our laboratory for autonomous operations in semi-structured real-world environments. The vehicle serves as a test-bed for evaluation of information processing methodologies and techniques needed for performing typical motion tasks on the laboratory or factory floor and between various building floors. Missions may include simple manipulation and inspection tasks or exploration of partly unknown environments. MACROBE has a three-wheel kinematic; its motion speed is up to 1m/sec.

A structural overview of the vehicle's perception, planning, action and communication subsystems is shown in Fig. 5. The underlying computer architecture will not be discussed here. Prof. FÄRBER outlines architectural features of computer systems for autonomous mobile robots in his paper [7].

In an FMS environment, MACROBE's operation is as follows: A transportation job is transmitted via radio link from the shop floor control system to the vehicle. The **global planner** decodes motion-relevant contents from a command, such as

TRANSPORT_PALLET (.) TO WORK_STATION (.).

By use of an internal topological map of the FMS environment, the planner generates an appropriate route from the start to the goal location. The route is embedded in a free motion space corridor.

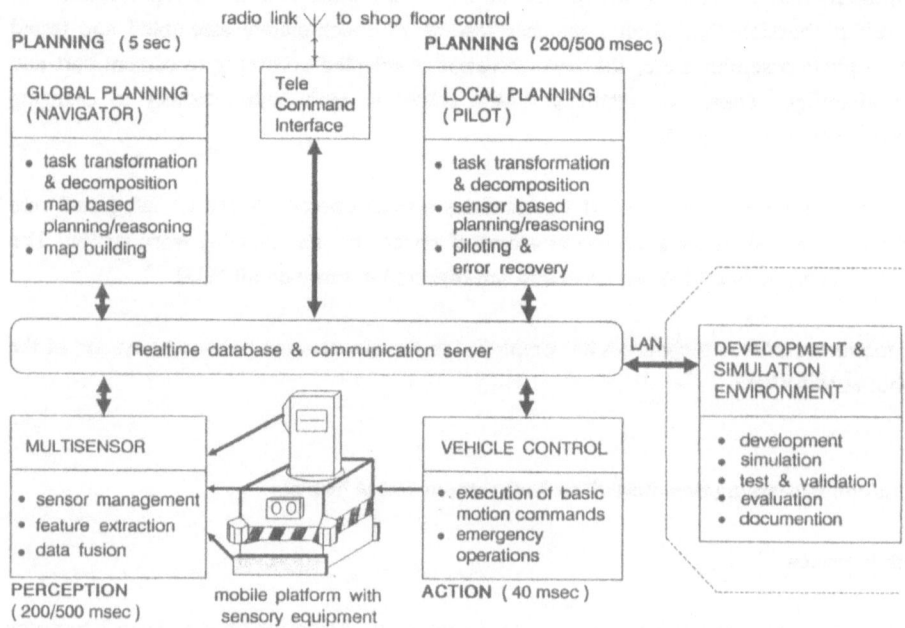

Fig. 5:   Mobile robot MACROBE, system overview

Details of map-based route planning, sensor-based map construction and up-dating are discussed in Peter KAMPMANN's paper [8].

A **multisensor system** provides motion-relevant information of the robot's current 3-D environment. To date, geometrical and motion information is acquired by a 3-D laser camera, a 2-D multi sonar system and usual odometry, including a fiber optical gyro for heading measurement. Environmental information coming from a microwave range imaging device [9] developed by Prof. DETLEFSEN's group and from interpreted CCD camera images will be incorporated in the near future. Prof. RADIG discusses in his paper [10] results of his work on interpretation of camera images from the FMS environment. Christoph FRÖHLICH, on the other hand, reports about issues of sensor data reduction, fusion and feature extraction by software-based virtual sensors within the multisensor system [11].

The **local planner** generates from route information, the required sequence of basic cartesian motion commands executed by the **vehicle control system**. Furthermore, the motion command sequence is expanded by so-called "autonomy functions" into an "expanded command sequence". Sensor-information based autonomy functions support the solution of unexpected conflicts during vehicle motion, as for example caused by driving errors, transient or permanent obstacles in the vehicle's motion path, and discrepancies between real-world and the internal map.

4.3 Three Types of Autonomy

Again three types of autonomy are incorporated in MACROBE's planning and control system. Autonomy achieved through

* reactive behaviour,
* adaptive behaviour,
* global planning.

With respect to the vehicle's specific operational environment, the selected autonomy functions are restricted to standard situation patterns. If **reactive behaviour** is not successful in solving a local conflict by use of internal feedback mechanisms, control is transferred to the **adaptive** layer. Error recovery mechanisms modify a preplanned expanded motion sequence and return control to the reactive layer. Error detection and recovery are based on error reports of the failed functional units, on actual multisensor information and a set of recovery rules. In case of a severe conflict, **route planning** on the highest autonomy layer is invoked and route replanning is initiated.

While autonomy through planning and reactive behaviour is already implemented in MACROBE's information processing system, application of adaptive autonomy features will be a topic for future research.

With respect to typical conflict situations in an FMS scenario the following basic requirements for autonomous vehicle operation were identified:

* collision avoidance with preview capability,
* evasion in case of simple obstacle configurations,
* precise maneuvering in narrow corridors or through gates,
* docking at load and unload stations,
* reorientation during execution of extended motion tasks.

These requirements can be fulfilled to some degree with the introduction of the following basic **autonomy functions** within the reactive layer (Table 1).

Table 1:   Set of autonomy functions of the reactive layer

---

speed adaption:	ENABLE/DISABLE SPEED_ADAPTION (D)
	the vehicle is always kept in a minimum longitudinal distance D to an obstacle within a speed dependent safety zone,
local evasion:	ENABLE/DISABLE LOCAL_EVASION (A)
	in case of an obstacle partially blocking the motion corridor, a standardized evasion maneuver in known free space and parallel to the preplanned route is performed with a maximum lateral distance < A,
contour following:	ENABLE/DISABLE CONTOUR_FOLLOWING (S,D,F)
	the vehicle follows a standard contour S with a lateral distance D up to frame F; smooth vehicle motion is achieved through appropriate filtering of the sensed contour,
gate traversing:	ENABLE/DISABLE GATE_TRAVERSING (R,F)
	a narrow gate with the geometrical representation R is sensed as a natural landmark and traversed through frame F,
docking:	ACTIVATE DOCKING (N,F)
	a standardized docking maneuver of type N into the load/unload station frame F is performed.

4.5 Autonomous Operations

The effect of some of the autonomy functions can be exemplified through the following transportation task "TRANSPORT_PALLET (P1) TO WORK_STATION (P7)", performed in the FMS facility, Fig. 6 and Table 2 presents both, the related basic and expanded command sequences as generated by the global and local planner. To date, generation of the expanded command sequence is performed manually.

The long aisle between P1 and P2 is traversed by activating the autonomy function CONTOUR FOLLOWING (S,D2,P2) thus counteracting accumulation of lateral route deviations due to odmetry errors. Through the command sequence ENABLE GATE_TRAVERSING (R,P5) and MOVE_ABS (P5), a possible discrepancy between the actual environment and the map is compensated at P5. By resetting the dead-reckoning system to vehicle position P5, the command SET POSITION (P5) thus upgrades this command sequence to a complex reorientation procedure.SPEED_ADAPTION is enabled during the complete transportation task. This means that the 3-D laser camera previews permanent or transient obstacles within the motion corridor and vehicle speed is smoothly adapted instead of emergency braking as known from conventional AGVs.

Table 2:   *Basic and expanded motion command sequence for task:*
         TRANSPORT_PALLET(P1) TO WORK_STATION(P7)

Motion Command Sequence	Expanded Command Sequence	Comment
	ENABLE SPEED_ADAPTION (D1)	collision avoidance
SET SPEED (V)	SET SPEED (V)	
	ENABLE CONTOUR_FOLLOWING (S,D2,P2)	along work station
MOVE_ABS (P2)		to P2
	DISABLE CONTOUR_FOLLOWING ()	
MOVE_ABS(P3)	MOVE_ABS (P3)	by dead-reckoning
MOVE_ABS(P4)	MOVE_ABS (P4)	to P4
	ENABLE LOCAL_EVASION (D3)	evasion allowed in wide aisle
	ENABLE GATE_TRAVERSING (R,P5)	
	MOVE_ABS (P5)	vehicle in frame P5
	SET POSITION (P5)	reset dead-reckoning system to P5
	DISABLE LOCAL_EVADING ()	
	DISABLE GATE_TRAVERSING ()	
MOVE_ABS (P6)	MOVE_ABS (P6)	
SET SPEED (0)		stop vehicle
	ROTATE LASER_SENSOR (PHI)	rotate sensor in order
SET SPEED (V)		to prepare for docking
MOVE_ABS (P7)	ACTIVATE DOCKING (N, P7)	docking maneuver N
SET SPEED (0)		stop at load position

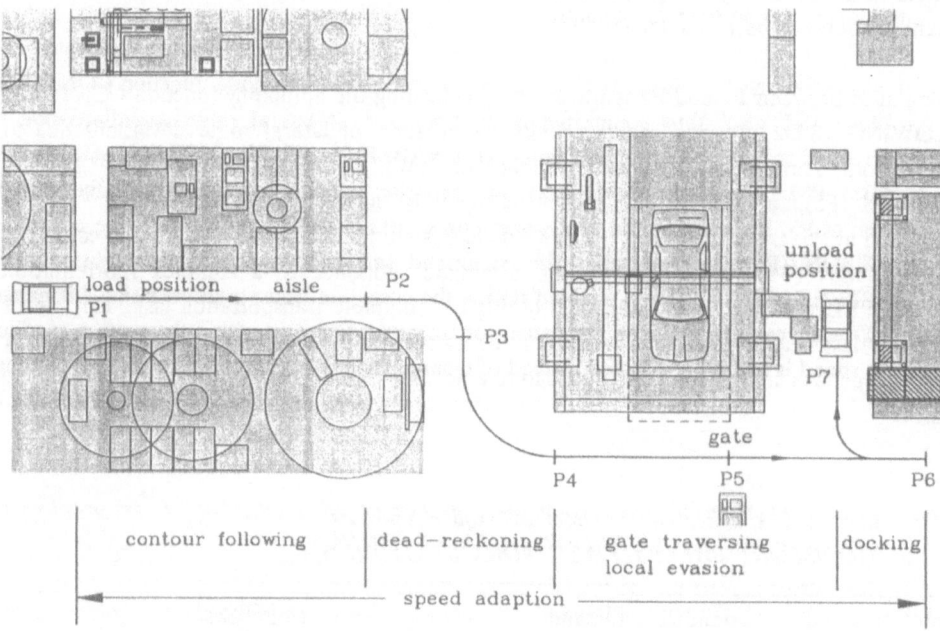

Fig. 6:   *Example of autonomous mobile robot operation in FMS environment*

The preliminary experiments performed with MACROBE in a FMS factory demonstrate the effectiveness of the implemented sensor-based autonomy functions in a variety of conflict

Fig. 7:   *Mobile robot MACROBE operating in a FMS factory*

situations. The chosen approach of expanded motion command sequences opens a path for future rule-based generation and modification of these sequences. This capability will provide MACROBEs planning and control system with the still missing adaptive autonomy features.

5. Conclusions

Todays FMSs may steadily develop into intelligent semi-autonomous or autonomous FMSs. Information processing and management will play a major role in this development. Related issues are topics of a joint interdisciplinary research project at the TU München.

The examples of autonomous system research surveyed in this paper are embedded in the broader research and development activities of the joint project. The approaches chosen for incorporation of autonomous system techniques into subsystems of a FMS show certain similarities. Integration of multisensor information and application of appropriate reactive, adaptive and planning operations lead to a balanced design for the planning and control systems of an semi-autonomous robotic work station and mobile robot.

The integration of autonomous subsystems needs adequate autonomous system techniques on the higher levels of a FMS information processing system. Related approaches are discussed by members of our joint research project in the prodeeding volume.

6. References

[1]   Schweiger, J.; Siegert, H.-J.: An object-oriented knowledge base for CIM environments. In: Proc. International Workshop "Information Processing in Autonomous Mobile Robots", March 6-8, 1991. Ed. G. Schmidt. Berlin, Springer Verlag, 1991, pp. 231 - 246.

[2]   Fischer, K.; Kupec, T.; Milberg, J.; Siegert, H.-J.; Stetter, R.: Layers of task planning for intelligent autonomous systems in a flexible manufacturing environment. In: Proc. International Workshop "Information Processing in Autonomous Mobile Robots", March 6-8, 1991. Ed. G. Schmidt, Berlin, Springer Verlag, 1991, pp. 331 - 348.

[3]   v. Dungern, O.: Planungs- und Autonomiefunktionen zur Steuerung flexibler Montagezellen. Fortschritt-Berichte VDI, Reihe 8, Nr. 235, Düsseldorf, VDI Verlag, 1991.

[4]   Albus, J. S.; McCain, H.G.; Lumia, R.: NASA/NBS Standard Reference Model for Telerobot Control System Architecture (NASREM), NIST Technical Note 1235, U.S. Dept. of Commerce, Gaithersburg, MD, 1989.

[5]   v. Dungern, O.; Freyberger, F.; Schmidt, G.: Ein systemfähiger Robotergreifer mit lokalen Korrekturfreiheitsgraden für die flexible Kleinmontage. atp 32 (1990) 11, 541-544; atp 32 (1990) 12, 585-589.

[6] v. Dungern, O.; Schmidt, G.: Vorbereitende und begleitende Ablaufplanung für flexible Montagezellen in industrieller Umgebung. Robotersysteme 6 (1990) 225-235.

[7] Färber, G.; Helling, S.; Ruß, A.: Architectural features of computer systems for autonomous mobile robot applications. In: Proc. International Workshop "Information Processing in Autonomous Mobile Robots", March 6-8, 1991. Ed. G. Schmidt, Berlin, Springer Verlag, 1991, pp. 247 - 264.

[8] Kampmann, P.; Schmidt, G.: Indoor navigation of mobile robots by use of learned maps. In: Proc. International Workshop "Information Processing in Autonomous Mobile Robots", March 6-8, 1991. Ed. G. Schmidt, Berlin, Springer Verlag, 1991, pp. 151 - 170.

[9] Detlefsen, J.; Rožman, M.; Lange, M.: Contributions of a microwave radar sensor to a multisensor system used for autonomous vehicles. In: Proc. International Workshop "Information Processing in Autonomous Mobile Robots", March 6-8, 1991. Ed. G. Schmidt, Berlin, Springer Verlag, 1991, pp. 93 - 102.

[10] Levi, P.; Munkelt, O.; Radig, B.; Sattler, R.: An application of image processing and image interpretation in manufacturing environments. In: Proc. International Workshop "Information Processing in Autonomous Mobile Robots", March 6-8, 1991. Ed. G. Schmidt, Berlin, Springer Verlag, 1991, pp. 35 - 44.

[11] Fröhlich, C.; Freyberger, F.; Karl, G.; Schmidt, G.: Multisensor system for an autonomous robot vehicle. In: Proc. International Workshop "Information Processing in Autonomous Mobile Robots", March 6-8, 1991. Ed. G. Schmidt, Berlin, Springer Verlag, 1991, pp. 61 - 76.

Sensing and Perception

Temporal Integration of Multiple Sensor Observations for Dynamic World Modeling: A Multiple Hypothesis Approach

I. J. Cox and J. J. Leonard

Abstract

We propose a multiple hypothesis approach for building and maintaining a world model for an autonomous robot vehicle. Dynamic world modeling requires the integration of multiple sensor observations obtained from multiple vehicle locations at different times. A crucial problem in this interpretation task is the presence of uncertainty in the *origins* of measurements (data association uncertainty) as well as in the *values* of measurements (noise uncertainty). The extended Kalman filter (EKF) has seen widespread use in robotics for dealing with the latter problem. The multiple hypothesis filter, first proposed by Reid[13] in the context of multitarget tracking, combines the basic machinery of the EKF with a rigorous Bayesian framework in which to address temporal and spatial data association uncertainty. For dynamic world modeling, the approach results in multiple world models at a given time step, each one representing a possible interpretation of all past and current measurements and each having an associated probability. A single unified world model can be constructed by integrating all of the hypotheses to form a single hypothesis. This framework is being implemented using infrared and ultrasonic range data; experimental results will be presented.

1 Introduction

Autonomous vehicles need to *automatically* build and maintain a map of their environment. A coherent interpretation of a robot's external environment is necessary in order for it to reason about such things as its position, motion planning, and obstacle avoidance. Maintenance of the map is needed because even if a map is provided *a priori*, few robot environments are completely static. Consequently, as the robot's environment alters, so must its world model. Map building is a dynamic process that involves providing an interpretation of observed sensor information in terms of physical features in the environment. The prediction of expected events relies on the fact that a correct interpretation has been found for geometric features that are repeatedly observed. Further, geometric features that are initiated to explain unexpected events are an attempt to develop such an interpretation over time. Thus the map is built up as new events are observed and explained, and is refined by reobserving events that have been correctly interpreted.

Constructing a map (or world model) for a mobile robot is problematic because of the difficulty in interpreting sensor observations. Sensor measurements are corrupted by noise; the physical operating conditions may lead to spurious measurements, e.g. specular reflections associated with ultrasonic ranging; or we may get false or missing readings due to poor performance of feature extractors. Further, dynamic perception, with its goal of integrating information from successive measurement scans, requires the evaluation of temporal (as well as spatial) correspondence, i.e. data association across multiple (time steps, images, scans.) The correspondence problem (or data association problem) is what makes perception and sensor-based control different from traditional estimation and control problems. *The fundamental problem is that we have both uncertainty in the* **origins** *of measurements as well as in the* **values** *of the measurements.*

To address this problem, an earlier paper[10] suggested two independent measures of uncertainty. The first represents the uncertainty in the value of a measurement due to noise and is adequately modeled by standard covariance matrices. The second "credibility" measure represented our uncertainty in the origin of the measurement. Did the measurement originate from a valid target, specular reflection etc.? Section (2) discusses this earlier work and contrasts it with our current approach.

We believe that dynamic world modeling for a mobile robot can be viewed as a problem in target tracking[11]. Usually one is concerned with a stationary observer and moving targets, but one can just as well have a moving observer and stationary targets (a mobile robot and a static environment) or both a moving observer and stationary and moving targets (a mobile robot in a dynamic environment)[7]. In Section (3) we illustrate how the "multiple hypothesis tracking" (MHT) filter[13] can be used to build and maintain a dynamic model of a robot's environment. MHT techniques provide a powerful formalism for addressing the correspondence problem and ambiguities that may arise at any time in the interpretation of a sensor measurement. The appeal of MHT techniques is that they can refrain from making irreversible decisions too early in the interpretation process:

> The multiple hypothesis approach makes 'soft' data association decisions when the available data are not of high enough quality. Instead of making firm commitments of associating measurements to tracks, it forms multiple data association hypotheses corresponding to alternate explanations of the origins of the measurements. These hypotheses are evaluated with respect to their probabilities of being true using the target and sensor models[3].

Section (4) discusses the implementation of the algorithm using ultrasonic and infrared range data. Finally, Section (5) concludes with a discussion of the advantages and disadvantages of the multiple hypothesis approach and some areas of possible future work.

2 Earlier Work

Previous attempts at map building for autonomous robot vehicles have not discriminated between spatial uncertainty due to noise and uncertainty in the origin of sensor measurements[5]. Significant progress has been made towards dealing with noisy measurements and the corresponding spatial uncertainty that is introduced. Impressive results are reported by Ayache and Faugeras[1] for a mobile vehicle using trinocular stereo to determine depth. Kriegman *et al* [8] also describe how a map can be built using binocular stereo. Crowley[6] describes a similar approach using ultrasonic sensing. Common to all these approaches is the use of the Kalman filter to model and propagate uncertainty in both the position of the robot and the geometric features (of the world model) using covariance matrices. The Kalman filter is a powerful tool for dealing with noise, but is of little help in modeling the uncertainty in the origin of sensor measurements. The experimental results presented by these groups are all for static scenes, i.e. the environment is unchanging. It is unlikely that these approaches would be successful in a dynamic environment without extension, perhaps along the lines of the work described here.

In an earlier paper[10], we examined the problem of constructing and maintaining a map, i.e. a world model, from which the vehicle's position could be accurately computed. This paper addressed the central issue of uncertainty management, pointing out the need for two independent measures of uncertainty. Figure (1) illustrates our original concept for map building and maintenance for purposes of mobile robot position estimation. In this paper, we expand on this earlier work in two ways. First, we provide the capability to model dynamic as well as static features in our model. Second, we replace the heuristic based "credibility" measure with a rigorous Bayesian formalism that estimates the probability of a hypothesized world model.

Because of the unreliable nature of sensor observations, the interpretation of past events and current sensor measurements is likely to contain ambiguities. Thus, at any one time, we propose to have numerous world models, each one representing different assumptions regarding the origin of sensor measurements. The proposed algorithm generates multiple hypotheses, assigning a probability to each. Figure (2) shows the structure of our current proposal.

Note that in our earlier algorithm the credibility measure was assigned to an individual feature in the *single* world model of the robot. In contrast, the algorithm described here generates multiple world models, i.e. multiple interpretation hypotheses, and assigns a probability to each. Within each world model, we do not explicitly represent the probability of a model feature. However, probabilities for individual model features can be determined. One method for doing so is to use the sum of the probabilities of the hypotheses which include the feature[12].

3 Multiple Hypothesis Framework for Dynamic World Modeling

The multiple hypothesis filter[13] was originally developed for the purpose of target tracking[2], in particular, for the tracking of multiple targets in a clutter, i.e. noise, false alarms, missing measurements etc. In this section we describe the multiple hypothesis filter and demonstrate how such a framework is applicable to the modeling of dynamic environments for autonomous robot vehicles.

We envision a world model consisting of simple geometric primitives such as line or arc segments, together with associated information describing the temporal dynamics of each

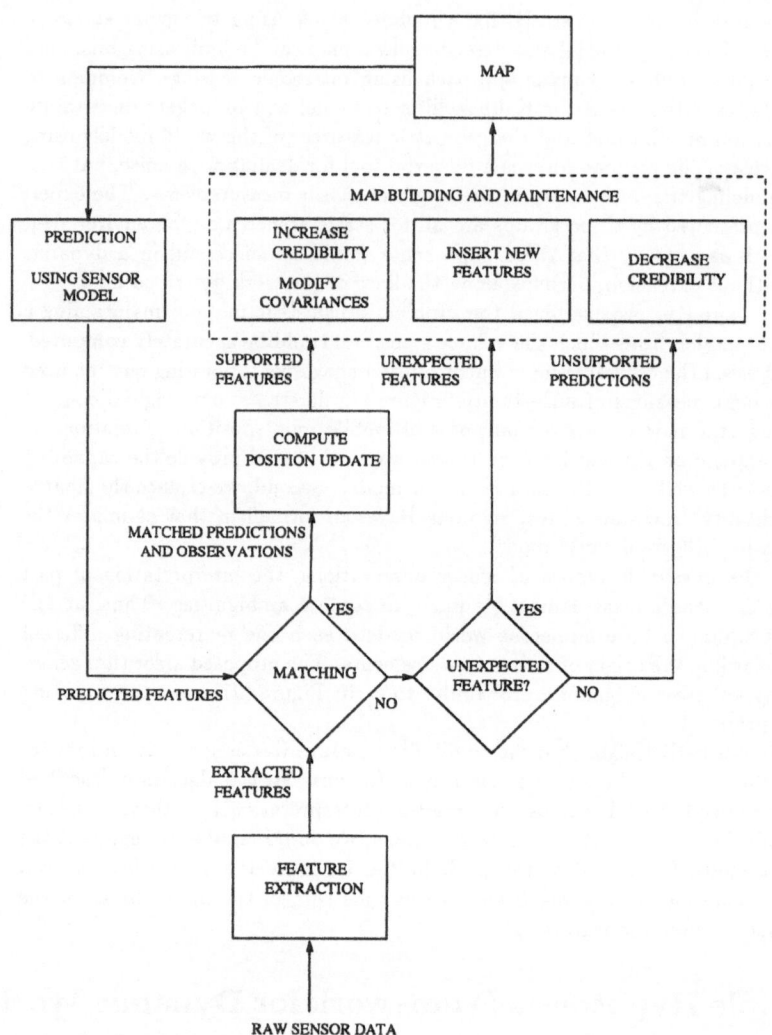

Figure 1: Model-based localization incorporating map building and maintenance.

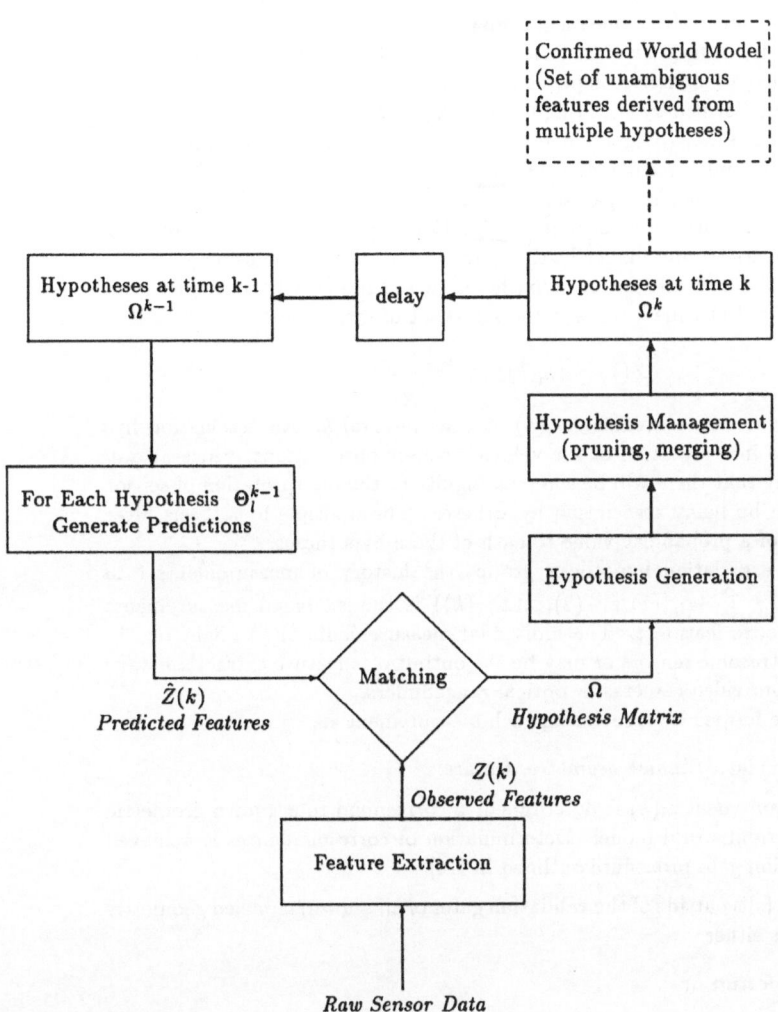

Figure 2: A Multiple Hypothesis Framework for Dynamic World Modeling.

feature, e.g. velocity, and acceleration. For those readers with familiarity of the tracking literature we then have that a target is a geometric feature, extracted from sensory measurements; a track is a sequence from one measurement scan to the next, of corresponding geometric features. A track may be static, representing a stationary geometric feature such as a wall or corner, or dynamic, representing a moving geometric feature, e.g. a sequence of line segments formed by a moving person. A global hypothesis is one possible world model of the robot's environment together with an associated probability of the hypothesis given all past and current measurements. A world model consists of a set of geometric features, each of which has spatial (position) and dynamic (velocity) attributes with corresponding uncertainties represented by traditional covariances. The grouping together of geometric features into higher level objects is not currently considered [1].

At time k we have a set of association hypotheses (world models) Ω^k obtained from the set of hypotheses Ω^{k-1} at time $k-1$ and the latest set of measurements

$$Z(k) \triangleq \{\mathbf{z}_i(k)\}_{i=1}^{m_k} \tag{1}$$

where m_k is the number of measurements $\mathbf{z}_i(k)$ at time interval k. An association hypothesis is one possible interpretation of the vehicle's sensor observations. Since at any one time it is very likely that there will be some ambiguity to the interpretation of sensor observations, there can be many association hypotheses. The multiple hypothesis filter allows the assignment of a probability value to each of these hypotheses.

More formally, an association hypothesis groups the history of measurements into partitions such that $Z^{k,g} \triangleq \{\mathbf{z}_{i_{1,t}}(1), \mathbf{z}_{i_{2,t}}(2), ..., \mathbf{z}_{i_{k,t}}(k)\}$ is the set of all measurements originating from geometric feature t. The individual measurements $\mathbf{z}_i(k)$ might be obtained directly from ultrasonic sensors or may be the output of a feature extraction stage applied to raw data from video cameras or optical rangefinders.

New hypotheses are formed by associating each measurement as

1. *belonging to a previously known geometric feature.*

   That is, the measurement $\mathbf{z}_i(k)$ is determined to correspond to a known geometric feature in our current world model. Determination of correspondences is achieved using the validation gate procedure outlined in [10].

   If a measurement falls outside of the validation gates of all known/modeled geometric features then it is either

2. *a new geometric feature* or

3. *a false alarm.*

   In addition, for geometric features that are not assigned measurements, we also have the possibility of

4. *deletion of geometric feature*

   This situation may arise when a learned feature such as a stationary desk, is moved to a new position.

---

[1]In fact, in the current context, the terms geometric feature and object are interchangeable.

We define a particular global hypothesis at time k by Θ_m^k. Let $\Theta_{l(m)}^{k-1}$ denote the parent hypothesis from which Θ_m^k is derived, and $\theta_m(k)$ denote the hypothesis that indicates the specific status of all geometric features postulated by $\Theta_{l(m)}^{k-1}$ at time k and the specific origin of all measurements received at time k.

Let T denote the total number of geometric features postulated by the parent hypothesis $\Theta_{l(m)}^{k-1}$. Next, we define the event $\theta_m(k)$ based on the current measurements to consist of:

τ measurements from known geometric features

ν measurements from new geometric features

ϕ false alarms and

χ deleted (or obsolete) geometric features from the parent hypothesis

For all the current measurements, $z_i(k)$, $i = 1, ..., m_k$ we define the indicator variables

$$\tau_i \triangleq \begin{cases} 1 & z_i(k) \text{ came from a known geometric feature} \\ 0 & \text{otherwise} \end{cases} \quad (2)$$

$$\nu_i \triangleq \begin{cases} 1 & z_i(k) \text{ is a new geometric feature} \\ 0 & \text{otherwise} \end{cases} \quad (3)$$

$$\delta_t \triangleq \begin{cases} 1 & \text{if geometric feature } t \text{ (in } \Theta_{l(m)}^{k-1}) \text{ is detected at time } k \\ 0 & \text{otherwise} \end{cases} \quad (4)$$

$$\chi_t \triangleq \begin{cases} 1 & \text{if geometric feature } t \text{ (in } \Theta_{l(m)}^{k-1}) \text{ is deleted at time } k \\ 0 & \text{otherwise} \end{cases} \quad (5)$$

From these definitions it holds that the number of measurements in the current event originating from existing geometric features is

$$\tau = \sum_{i=1}^{m_k} \tau_i \quad (6)$$

the number of new geometric features is

$$\nu = \sum_{i=1}^{m_k} \nu_i \quad (7)$$

and the number of false alarms is

$$\phi = m_k - \tau - \nu \quad (8)$$

The number of deleted or obsolete geometric features is

$$\chi = \sum_{t=1}^{T} \chi_t \quad (9)$$

We can construct a set of events $\theta_m(k)$ by first creating a hypothesis matrix in which known geometric features are represented by the columns of the matrix and the current

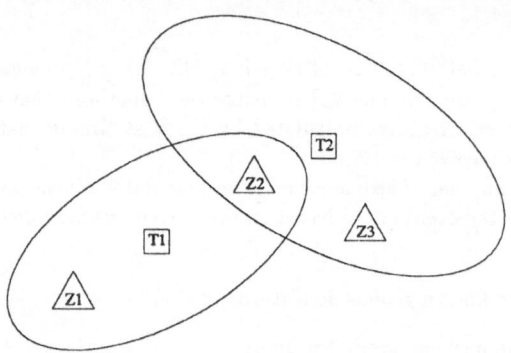

Figure 3: Predicted target locations and elliptical validation regions for a situation with two known geometric features (T_1 and T_2) and three new measurements ($\mathbf{z}_1(k)$, $\mathbf{z}_2(k)$ and $\mathbf{z}_3(k)$).

measurements by the rows. A non-zero element at matrix position $c_{i,j}$ denotes that measurement $\mathbf{z}_i(k)$ is contained in the validation region of geometric feature t_j. In addition to the T known geometric features in the world model the hypothesis matrix has appended to it a column 0 denoting false alarms and a column $N+1$ denoting new geometric feature. Figure (3) depicts a situation in which we have two known geometric features (T_1 and T_2) and three new measurements ($\mathbf{z}_1(k)$, $\mathbf{z}_2(k)$ and $\mathbf{z}_3(k)$). This situation is represented by the hypothesis matrix

$$\Omega = \begin{array}{cccc} T_F & T_1 & T_2 & T_N \\ \left(\begin{array}{cccc} 1 & 1 & 0 & 1 \\ 1 & 1 & 1 & 1 \\ 1 & 0 & 1 & 1 \end{array}\right) & & & \begin{array}{c} \mathbf{z}_1(k) \\ \mathbf{z}_2(k) \\ \mathbf{z}_3(k) \end{array} \end{array}$$

Hypothesis generation is then performed by picking one unit per row and one unit per column except for columns T_F and T_N, where the number of false alarms and new targets is not restricted. This procedure assumes that the following two conditions are met (the so-called no-split/no-merge assumption[12]).

1. a measurement has only one source and

2. a geometric feature has at most one associated measurement per scan

We may relax these conditions in the future.

3.1 Probability calculations

The new hypothesis at time k, Θ_m^k is made up of the current event and a previous hypothesis based on measurements up to and including time $k-1$, i.e.

$$\Theta_m^k = \left\{\Theta_{l(m)}^{k-1}, \theta_m(k)\right\} \tag{10}$$

We need to calculate the probability of such a hypothesis, i.e.

$$P\left\{\Theta_m^k | Z^k\right\} \tag{11}$$

where Z^k denotes all measurements up to and including time k. Using Bayes' rule we have

$$P\{\Theta_m^k|Z^k\} = P\{\theta_m(k), \Theta_{l(m)}^{k-1}|Z(k), Z^{k-1}\}$$

$$= \frac{1}{c}p[Z(k)|\theta_m(k), \Theta_{l(m)}^{k-1}, Z^{k-1}]P\{\theta_m(k)|\Theta_{l(m)}^{k-1}, Z^{k-1}\}P\{\Theta_{l(m)}^{k-1}|Z^{k-1}\} \quad (12)$$

The last term of this equation, $P\{\Theta_{l(m)}^{k-1}|Z^{k-1}\}$, represents the probability of the parent global hypothesis and is therefore available from the previous scan. The remaining two terms may be evaluated as follows.

The second factor of Equation (12) is obtained by combining results from [2] and [9] to yield

$$P\{\theta_m(k)|\Theta_{l(m)}^{k-1}, Z^{k-1}\} = \frac{\phi!\nu!}{m_k!}\mu_F(\phi)\mu_N(\nu)\prod_t \left(P_D^t\right)^{\delta_t}\left(1 - P_D^t\right)^{1-\delta_t}\left(P_\chi^t\right)^{\chi_t}\left(1 - P_\chi^t\right)^{1-\chi_t}$$

$$(13)$$

where c is a normalization constant, $\mu_F(\phi)$ ans $\mu_N(\nu)$ are the prior PMFs of the number of false measurements and new geometric features, and P_D^t and P_χ^t are the probabilities of detection and termination (deletion) of track t.

To determine the first term on the right hand side of Equation (12) we assume that a measurement $\mathbf{z}_i(k)$ has a Gaussian probability density function (PDF)

$$N_{t_i} = N[\mathbf{z}_i(k)] \triangleq N[\mathbf{z}_i(k); \hat{\mathbf{z}}_i(k|k-1), \mathbf{S}^{t_i}(k)]$$

$$= \left|2\pi\mathbf{S}^{t_i}(k)\right|^{-\frac{1}{2}}e^{-\frac{1}{2}\left\{(\mathbf{z}(k)-\hat{\mathbf{z}}(k|k-1))^T\{\mathbf{S}^{t_i}(k)\}^{-1}(\mathbf{z}(k)-\hat{\mathbf{z}}(k|k-1))\right\}}$$

if it is associated with geometric feature t_i, where $\hat{\mathbf{z}}_i(k|k-1)$ denotes the predicted measurement for geometric feature t_i and $S^{t_i}(k)$ is the associated innovation covariance. The prediction $\hat{\mathbf{z}}_i(k|k-1)$ and innovation covariance S^{t_i} are precisely what is calculated using the (extended) Kalman filter. Thus, a Kalman filter is associated with every geometric feature; each state vector represents the position and velocity of the associated feature. If the measurement is a false alarm, then its PDF is assumed uniform in the observation volume, i.e. V^{-1}. The probability of a new geometric feature is also taken to be uniform[2] with PDF V^{-1}. Under these assumptions, we have that

$$p[Z(k)|\theta_m(k), \Theta_{l(m)}^{k-1}, Z^{k-1}] = \prod_{i=1}^{m_k}[N_{t_i}[\mathbf{z}_i(k)]]^{\tau_i}V^{-(1-\tau_i)} = V^{-\phi-\nu}\prod_{i=1}^{m_k}[N_{t_i}[\mathbf{z}_i(k)]]^{\tau_i} \quad (14)$$

Substituting Equations (14) and (13) into Equation (12) yields the final expression for the conditional probability of an association hypothesis

$$P\{\Theta_m^k|Z^k\} = \frac{1}{c}\frac{\phi!\nu!}{m_k}\mu_F(\phi)\mu_N(\nu)V^{-\phi-\nu}\prod_{i=1}^{m_k}[N_{t_i}[\mathbf{z}_i(k)]]^{\tau_i}$$

$$\times \left\{\prod_t (P_D^t)^{\delta_t}(1 - P_D^t)^{1-\delta_t}\left(P_\chi^t\right)^{\chi_t}\left(1 - P_\chi^t\right)^{1-\chi_t}\right\}P\{\Theta_{l(m)}^{k-1}|Z^{k-1}\} \quad (15)$$

---

[2]Intuitively, the choice of uniform PDFs for false alarms and new features seems less justifiable for robotic applications than for traditional radar and underwater sonar tracking applications. The impact of these assumptions is a topic we intend to investigate in our experimental work.

4   Implementation

We have implemented all of the major components of the MHT filter: creation of the hypothesis matrix, hypothesis generation, calculation of their respective probabilities, the Kalman filters associated with each geometric feature and hypothesis management strategies.

In practice, hypothesis management plays a crucial role due to the potentially overwhelming growth in the number of hypotheses. We eliminate unlikely hypotheses by a combination of both screening (prior to the generation of hypotheses) and pruning (after the generation of hypotheses). One simple screening strategy that is very effective is to not consider the possibility of false alarms or new targets if the corresponding measurement is matched to only one target and this target has no other measurements matched to it. Similarly, we only consider the possibility of track termination if the corresponding track has no measurements associated with it at any given time step.

Pruning is based on "N-scan-back" procedure[9]. If the leaves of the tree are at the kth level we look at the parent at the $(k - N)$th node. For each of its say, m, branches, we sum the probabilities of the leaves associated with each of the m branches. Only that branch with the greatest probability is kept, all other branches are pruned. This pruning procedure is based on the assumption that any ambiguities at $k - N$ are effectively resolved N sets of measurements further on. As a result of this procedure, only a single (most likely) hypothesis is eventually generated, all other hypotheses being pruned away. This may not always be desirable and alternative pruning procedures are likely to be considered.

One further action being taken to reduce the number of hypotheses was proposed by Kurien:

> The factor that contributes most to the computational complexity...is the multitude of global hypotheses that may be formed. The number of global hypotheses is a combinatorial function of the number of track hypotheses (geometric features). Because the purpose of forming global hypotheses is to select the most likely assignment from the multiple assignments of reports (measurements) to targets (features), it is not necessary to form global hypotheses for tracks that do not have any common report assignments. This motivates the idea of partitioning targets (features) into separate *clusters* for the purpose of forming global hypotheses. Targets (features) within each cluster share common reports (measurements), whereas targets in different clusters do not share any common reports. By forming such clusters, we can form global hypotheses and select the most likely report-to-target assignment independently for each cluster; consequently, the combinatorial problem associated with forming global hypotheses is reduced significantly[9].

Of course, with each new set of measurements, we must check whether a measurement is shared (falls in the validation region) between two or more clusters. If so, these clusters must be merged. Similarly, a cluster containing two or more geometric features that do not share common measurements may be split.

The implementation of the above framework is currently in progress using infra-red[4] and ultrasonic[11] range data.

5 Discussion

To build and maintain a world model, an autonomous robot vehicle needs to integrate spatial information over successive temporal intervals. This requires confronting the dual uncertainties associated with the values and origins of sensor measurements. We propose to achieve this within the Bayesian framework of a multiple hypothesis (tracking) filter. The primary advantage of the approach is that it postpones making irrevocable decisions. Instead, when ambiguities arise, each possible interpretation or hypothesis is generated and its associated probability calculated. Subsequent measurements are then applied with each hypothesis. Temporal integration of measurements resolves many (hopefully all) past ambiguities.

Consequently, at any time we have multiple world models for an autonomous robot vehicle, each one representing a possible interpretation of all past and current measurements and each having an associated probability. If desired, a single unified world model can be constructed by integrating all of the hypotheses to form a single hypothesis. Geometric features within this single world model then have associated probabilities. This approach may quickly become infeasible if the hypothesis tree is not adequately pruned. However, we are confident that this will not be the case in practice.

References

[1] N. Ayache and O. Faugeras. Maintaining representations of the environment of a mobile robot. *IEEE Trans. Robotics and Automation*, 5(6):804–819, 1989.

[2] Y. Bar-Shalom and T. E. Fortmann. *Tracking and Data Association*. Academic Press, 1988.

[3] C. Chong, S. Mori, and K. Chang. Distributed multitarget multisensor tracking. In *Multitarget-Multisensor Tracking: Advanced Applications*, pages 247–295. Artech House, 1990.

[4] I. J. Cox. Blanche: Position estimation for an autonomous robot vehicle. In I. J. Cox and G. T. Wilfong, editors, *Autonomous Robot Vehicles*. Springer-Verlag, 1990.

[5] I. J. Cox and G. T. Wilfong. *Autonomous Robot Vehicles*. Springer-Verlag, 1990.

[6] J. L. Crowley. World modeling and position estimation for a mobile robot using ultrasonic ranging. In *Proc. IEEE Int. Conf. Robotics and Automation*, pages 674–681, 1989.

[7] J. Hallam. *Intelligent Automatic Interpretation of Active Marine Sonar*. PhD thesis, University of Edinburgh, 1984.

[8] D. Kriegman, E. Triendl, and T. Binford. Stereo vision and navigation in buildings for mobile robots. *IEEE Trans. Robotics and Automation*, 5(6), December 1989.

[9] T. Kurien. Issues in the design of practical multitarget tracking algorithms. In *Multitarget-Multisensor Tracking: Advanced Applications*, pages 43–83. Artech House, 1990.

[10] J. J. Leonard, I. J. Cox, and H. F. Durrant-Whyte. Dynamic map building for an autonomous mobile robot. In *Proc. IEEE Int. Workshop on Intelligent Robots and Systems*, 1990.

[11] J. J. Leonard and H. F. Durrant-Whyte. Application of multi-target tracking to sonar-based mobile robot navigation. In *29th IEEE Int. Conference on Decision and Control*, 1990.

[12] S. Mori, C. Chong, E. Tse, and R. Wishner. Tracking and classifying multiple targets without a priori identification. *IEEE Transactions on Automatic Control*, AC-31(5), May 1986.

[13] D. B. Reid. An algorithm for tracking multiple targets. *IEEE Transactions on Automatic Control*, AC-24(6), December 1979.

An Application of Image Processing and Image Interpretation in Manufacturing Environments

P. Levi, O. Munkelt, B. Radig, and R. Sattler

Abstract

This contribution presents a computer vision approach to help an autonomous, mobile system to determine its own position and orientation in general manufacturing environment without special optical landmarks. Furthermore, the detection and identification of unexpected obstacles is supported.

We model objects which are of interest for this task as CAD-Models. Each is transformed off-line, automatically into a boundary representation which in turn serves as a basis to compute a two dimensional representation of relevant aspects. Images are segmented into region and lines. The resulting representation is matched with the aspects to obtain the identity and the relative orientation of the vehicle with respect to an object.

To perform the matching fast, a network of decision rules is generated from the views which will in most cases identify objects in very few steps.

The overall architecture of the system follows the blackboard paradigm. Its modularity guarantees that the system will be incrementally extendable by adding knowledge sources as well as information from other subsystems of the vehicle.

1. Introduction

Computer Vision has numerous applications in the area of robotics and autonomous, mobile systems. In the joint research project "Information Processing in Autonomous Mobile Robots" of the Technical University of Munich, computer vision is employed to monitor action of an manipulator [1] and a calibrated high resolution camera is developed for high precision measurements [2].

We describe an approach to utilize a video camera to support the self orientation of an autonomous, mobile system in an industrial environment. See Fig. 1 (a) for a picture taken from our experimental manufacturing environment. To the left of the corridor there are load stands and to the right there are toolboxes. There is an obstacle in the corridor, namely a tool carriage.

Fig. 1: (a) example of an industrial environment   (b) model of a tool carriage

The task of self-orientation has to be solved without the use of optical land marks. Moreover, the illumination of the environment will not be adjusted specially. Both are challenging requirements for image interpretation.

For object detection, which is a prerequisite for position determination, we designed CAD-models of all the objects in our industrial environment. Fig. 1 (b) shows a view of the CAD-Model of the tool carriage.

---

[1]   Project B4 of the SFB 331, research group of Prof. Dr.-Ing. Milberg

[2]   Project B2 of the SFB 331, research group of Prof. Dr.-Ing. Marko

Besides these object models, we have the following knowledge:
- a layout of the industrial environment,
- odometric data giving the approximate position of the mobile system,
- knowledge on the applicability and effects of image interpretation techniques, and
- a camera model.

This knowledge has to be adapted to the respective situations /Lie89/.

The following chapter describes the automatic creation of aspects from CAD-models. In chapter 3, we will introduce control structures which will later be used for object recognition and position determination. Chapter 4 describes the software architecture of the system. In Chapter 5 an example is given which explains how the system perform an identification task. Chapter 6 gives some remarks on the implementation.

2. Models

All the objects occurring in the environment of our autonomous, mobile system have been modelled in the CAD-system Euclid. These models will automatically be transformed in 3 steps off-line.

1. Generation of the boundary-representaion
The boundary-representation (b-rep) is a 3D-model describing the visible surfaces of an object. Since there are, at present, no standardized interfaces for CAD-systems commercially available, we have used the programming interface of the Euclid-system for the implementation of algorithms generating the boundary representation, using some of the results of the Esprit-Project CAD-I /Sch89/.

2. Generation of views
Object views relevant and necessary for a later object recognition will be generated as follows: an imaginary sphere round an object will be subdivided into triangles of the same size. For each triangle a 2D-view is computed which represents the whole set of 3D-viewpoints within this triangle. For a typical triangulation about 240 views are generated for each object. For each of these 2D-views, a number of features will be generated, like number of faces, size of faces and number of edges. Besides these features, a data structure of a view will also consist of a description of its edges, faces and components, as well as the coordinates of its triangle.

3. Generation of aspects
Equivalent views of an object are grouped after certain criteria into the so-called aspects. This will be based on the features of the views. One criterion is, for example, that the number of larger faces will have to be constant. The data structure of an aspect and of a view are the same, the description of the views are combined to yield the description of the aspect. The integration of a simple camera model will now make it possible to reduce the number of aspects. Height, angle of inclination and possible

positions of the camera rule out a large number of aspects, e.g. bird and frog-perspectives of objects are as impossible as perspectives which cannot be seen by the camera mounted on the vehicle. In this way we usually obtain about 10 aspects of an object.

The result of the aspect generation is the aspect graph. The aspect graph is hierarchical with respect to one feature, e.g. the number of faces. In this case, the top level collects all aspects where one face is visible. The next level is the level where two faces are visible, etc. As a heuristic to guide the selection of an aspect for matching it with the image representation, the aspects are ordered with respect to the likelihood of their occurrence.

Besides this description of objects by a relatively low number of aspects, we can often describe objects by a few characteristic features. These serve as a key for the creation of object hypotheses (see Chapter 5).

 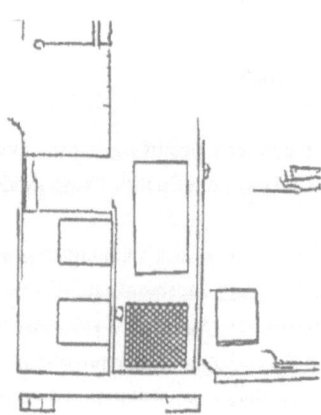

Fig. 2 (a): characteristic feature of our mobile robot   (b) detection of this feature

The airing bar (Fig. 2 (a)) is the characteristic feature of our mobile robot. Characteristic features can be generated off-line out of views. At present they are, however, being generated manually. They describe each object by the triple object, component and sequence of image interpretation operations necessary for the recognition of that component. This sequence is at present being determined experimentally and will be represented as image interpretation knowledge-source in the system (see Chapter 4). See Fig. 2 (b) for the result of the application of such a knowledge-source.

The off-line modelling for each object results in an aspect graph and the characteristic features for the hypotheses generation.

3. Control Structures

Strategies and tactics with which an object is first recognized and with which thereafter its position relative to our mobile robot will be determined, will be considered as control knowledge. The common data structure of control knowledge is an And/Or-graph. The nodes of this graph are rules. They determine which features are characteristic for a position and how these will really be extracted out of the image. The position determination is a qualitative one (such as "diagonal to the right and front"). Since rules resemble object features in a certain position, they are called visibility-rules, and the And/Or-graph visibility graph /MLR89/. Geometric descriptions as well as components are object features .

Views which are sufficient to determine the rough orientation from e.g. 8 directions are used to build the And/Or-graph.

Fig. 3: (a) Image of the tool carriage   (b) simplified CAD-representation

The example of a tool carriage, see Fig. 3 (a), explains the approach. Its corresponding view is shown in Fig. 3 (b). In this view characteristic object features are the shape of its major faces, the drawers and the three wheels.

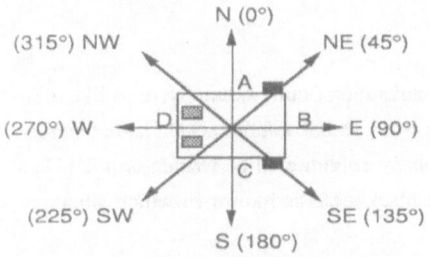

To describe the rough relative position, we use the 8 directions given in the object centred compass card of Fig. 4.

Fig. 4: Compass card for the specification of relative positions

A part of the visibility graph, related to the tool carriage of Fig. 3 (b), is shown in Fig. 5. Rule 2 is expanded to demonstrate the final determination of the relative position. Additional information in the rules refer to the operators which are able to extract the desired features from the image and to group image segments to object components.

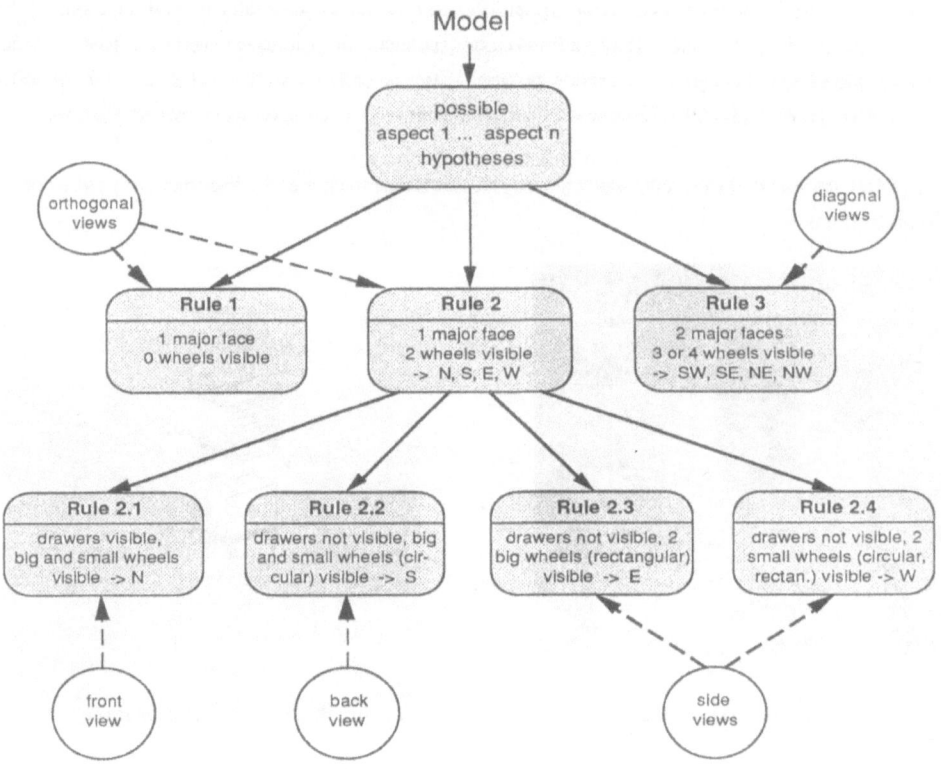

Fig. 5: Part of the visibility graph for the tool carriage

4. System Structure

We have chosen a blackboard architecture /Eng89/ for the realization of our system. Systems like these consist of a blackboard and a number of knowledge sources. All partial solutions which have been generated by the knowledge sources during the process of problem solving will be written on the blackboard. The knowledge sources will observe the state of the blackboard and know in which situations they can contribute to the problem solving.

Communicating only by way of entries in the blackboard, the knowledge sources are more or less independent from each other. This makes blackboard architectures very modular and facilitates the extendibility and parallelizability of such systems.

The principle of a blackboard system is a very simple one. As soon as a new entry has been created on the blackboard, a series of knowledge sources will be applicable. As a general rule, one new entry on the blackboard will cause the applicability of several knowledge sources. One of them will be selected by the scheduler. Executing the chosen knowledge source, new entries will be created on the blackboard. This process will be repeated until a problem solution has been found.

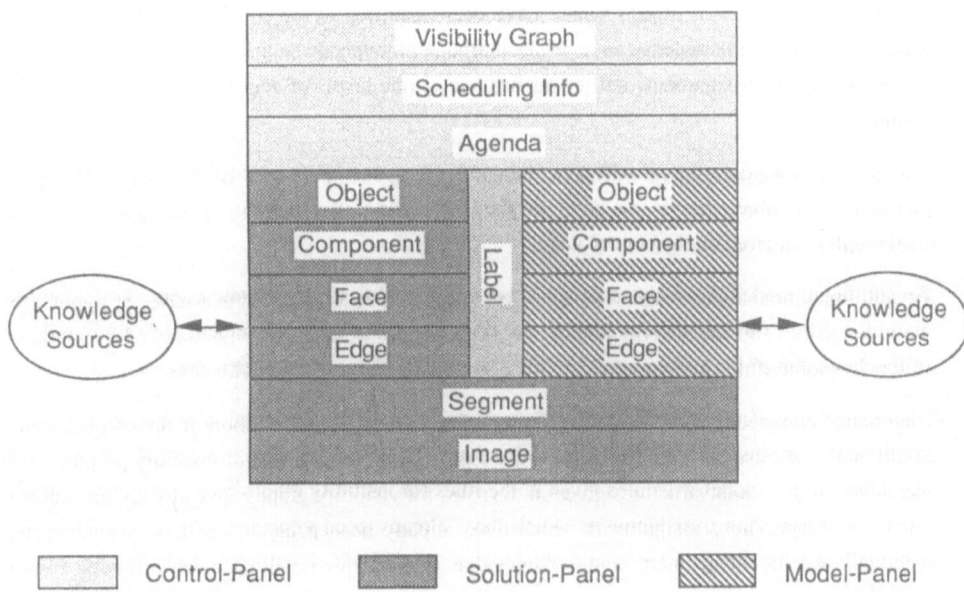

Fig. 6: Blackboard-Architecture

The blackboard of our system (see Fig. 6) has three panels:

- The solution panel stores detected image structures and matching results. It contains levels for segments, edges, faces, components and objects.

- The model panel will save one or more aspects of an object. These will be consulted during the matching process. The model field, too, consists of levels for segments, edges, faces, components and objects.

- In the control panel we will store data needed for system control. We have introduced this panel in order to have an explicit representation of those control structures which in other blackboard systems are an integral part of the scheduler.

The control blackboard is divided into three fields. One field will save the visibility graphs. The agenda holds all knowledge sources applicable in a certain situation. Further information on the knowledge source next to be used and their priorities are stored in the third field.

The knowledge sources of our system can be classified into three categories as follows:

- The image interpretation knowledge sources contain operations for preprocessing, segmentation, grouping and detection of characteristic features. Besides standardized operations (like filters, morphologically operations, etc.) there are highly specialized operators for detection of those characteristic features of objects which have been specified by object models. The category of image interpretation knowledge sources also harbours knowledge sources for matching image and model structures. Assignments will be determined on the levels of edges, faces, components and objects.

- The model knowledge sources will generate and load aspects to be used for matching. As soon as an aspect of an object has been entered on the model panel, the visibility graph belonging to the model will be entered on the control panel.

  An additional model knowledge source has been designed so as to generate a series of hypotheses about the visible objects, thereby making use of the layout of the environment and of a hypothesis of the viewpoint of the autonomous, mobile system gained out of odometric data.

- The control knowledge sources realize the scheduler using the information of the control panel. Additionally, we also have control knowledge sources for interpretation of visibility graphs. They check back if the model structures given in the rules for visibility graphs have already been detected in the image.Those assignments which have already been generated will be evaluated and eventually execute image interpretation knowledge sources which will try to detect model structures in the image.

5. Example

The working order of the system when recognizing an object will be shown by the example of the tool carriage seen in Fig. 3 (a). The image interpretation process will start by loading an image onto the blackboard. This will cause the triggering of a number of knowledge sources for image preprocessing (filter operations). Based on the features of the image (illumination, grey level histogram, etc.), a preprocessing operation will be chosen. The preprocessed image will trigger a number of operations for segmentation. See Fig. 7 (a) for the results of the application of such an image processing operation.

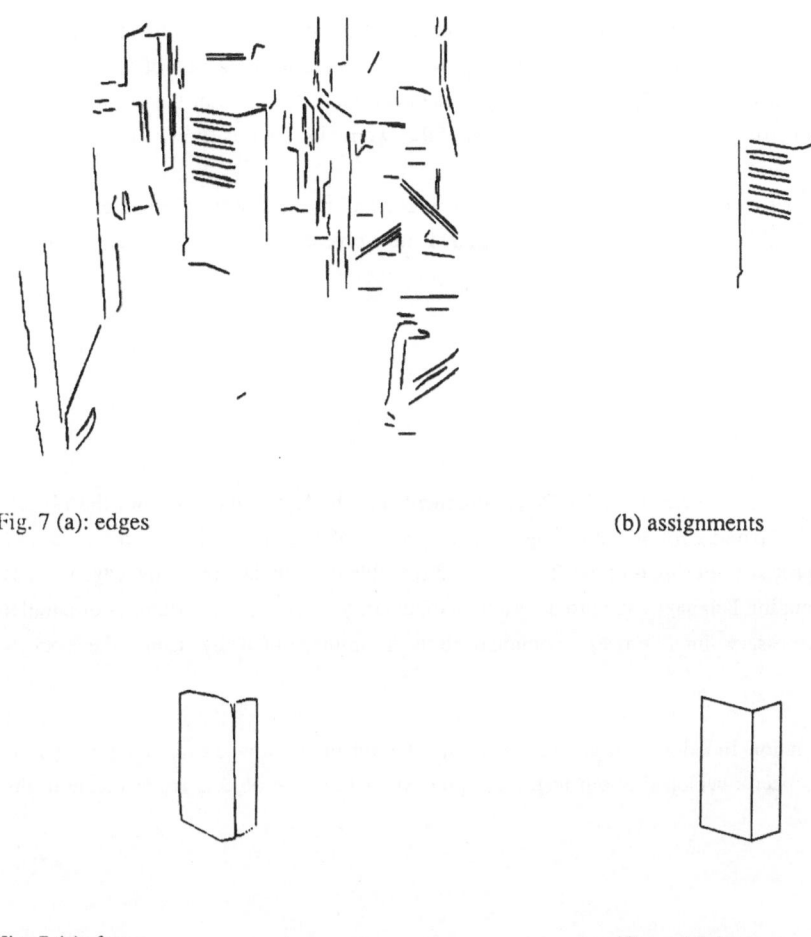

Fig. 7 (a): edges   (b) assignments

Fig. 7 (c): faces   (d) model faces

Fig. 7: (a) - (d) Steps in the process of recognition of the tool carriage

Parallel to the application of these knowledge sources, the system generate object hypotheses. For the orientation relative to mobile objects, those image interpretation knowledge sources are applied which are able to detect characteristic features. For the tool carriage, the drawers are characteristic features. In case of orientation relative to non-moving objects, those model knowledge sources will be executed which generate object hypotheses with the aid of odometric and layout data.

An object hypotheses will be entered on the model blackboard, in our example the tool carriage. In addition to this, the respective visibility graph will be loaded in the control blackboard. To verify the object hypothesis, the visibility graph will be interpreted by a knowledge source. Starting at the root, the rules of the graph will be applied, checking if the requested model structures have already been assigned to image structures. See Fig. 7 (b) for the assignments of our example. If the model structures have not yet been detected the image interpretation knowledge source for detection of the respective model structure will be executed. Fig. 7 (c) shows the results of applying rule 3 of the visibility-graph (Fig. 5) for the tool carriage. The corresponding model of the two major faces is shown in Fig. 7 (d).

If an object hypothesis cannot be verified it will be dismissed and the next object hypothesis not yet considered will be loaded onto the model blackboard. This new object hypothesis will then be treated in the same way.

6. Implementation

The presented system is realized in an UNIX environment, the implementation follows the object-oriented programming paradigm. The knowledge sources and the blackboard itself have been realized as objects and are processes during runtime. This makes it possible to formulate the knowledge sources in different programming languages and provides a high modularity and offers opportunities of parallel execution. The necessary inter process communication is implemented by using the socket-mechanisms.

The image interpretation functions are programmed with the aid of the image interpretation system HORUS which has been developed at our institute. The CAD data of the objects are available in the CAD-system Euclid.

References

/Eng88/:   Engelmore, R; Morgan, T. (eds.)
        Blackboard Systems
        Addison Wesley Publishing Comp., Wokingham, 1988
/Lie89/:   Liedtke, C.E.; Ender, M.
        Wissensbasierte Bildverarbeitung
        Springer-Verlag, Berlin, 1989
/Maj89/:   Majumdar, J.; Levi, P.; Rembold, U.
        3-D Model Based Robot Vision by Matching Scene Description with the Object Model from a CAD Modeller
        Robotics and Autonomous Systems 5, North Holland, 1989, pp. 69 - 83
/Sch89/:   Schlechtendahl, E.G. (Hrsg.),
        CAD Data Transfer for Solid Models
        Research Reports ESPRIT, Project 322, CAD Interfaces (CAD*I), Volume 3
        Springer-Verlag, Berlin, 1989

Dynamic Sensing and Multi-Task Performance with a Mobile Manipulator System

W. F. Carriker, P. K. Khosla and B. H. Krogh

Abstract

We consider the problem of multiple task execution by a mobile robot system in an unstructured, possibly changing environment. The run-time scheduling problem is to choose the sequence of robot sensing and motion operations to accomplish the objectives of the multiple tasks in an efficient manner. In this paper we present a set of representations for the run-time scheduler that facilitate the solution of the on-line decision problem. Task specifications, robot operations, and knowledge of the robot state and environment are represented with independent, interconnected structures that enable the run-time scheduler to identify sequences of system operations that progress toward completion of the tasks most effectively. Operations are represented using sense/think/act primitives to build operation macros which allow multiple task steps to be mapped into single robot functions. The representations and their application to run-time scheduling for dynamic sensing and multi-task execution are illustrated with an example robot system for laboratory maintenance.

1. Introduction

In a variety of future applications, such as planetary exploration and hazardous environment operations, mission-oriented robotic systems will be required to perform multiple tasks in unstructured environments [1, 2, 3]. Upon entering the mission environment, the robot will have to acquire the information necessary to execute its assigned tasks. To be effective, dynamic, task-dependent sensing strategies will be required to focus sensory resources on the relevant features of the environment. Moreover, to achieve efficient performance the sequence of task operations will have to be determined on-line as information is accumulated, taking into account multiple task objectives and system operating constraints such as available time and energy. The robot will also have to be able to accommodate changes in the environment, as well as changes to task descriptions and additional task plans received from a remote supervisor.

To address these problems, the robot must have a task executor which coordinates the acquisition and interpretation of sensory data, and orders the robot actions to carry out the specified tasks. In the context of the many architectures proposed for mobile robotic systems, these problems of dynamic sensing and multi-task execution must be solved at a high level in the on-line control system. In this paper, we use the term *run-time scheduler* (RTS) to identify that part of the control system which determines the sequence of sensory and motion operations for the system. To carry out the types of complex tasks described above, an RTS as described in this paper will be necessary in any control architecture, whether it be centralized and hierarchical [1, 4, 5], or layered [6, 7, 8].

There are two major components to the design of the RTS for a mobile robot: the decision algorithm or *scheduling strategy*, and the *representations* used for the elements of the on-line scheduling problem. An RTS selects the next operation to be performed as a function of the current state of the robot, the knowledge of the environment, and the status of the various tasks to be performed. This is a difficult decision problem that is not amenable to standard planning or optimization algorithms due to the lack of *a priori* information about the environment, multiple task objectives, and the potential for changes in the environment and task definitions. Moreover, before a scheduling strategy can be developed, representations of the functional capabilities of the system, knowledge about the environment, and definitions of the tasks to be performed must be specified. Our objective, in this paper, is to develop a comprehensive set of representations for the RTS which permit the development of scheduling strategies that can take advantage of the rich number of alternative operating sequences available to a mobile robot system executing multiple tasks.

Our approach involves the decoupling of the representations of *tasks*, *operations* (functionality), and *knowledge* so that tasks can be specified independently and then coupled dynamically at run-time. In the remainder of the paper we define and illustrate the proposed representations using as an example a mobile manipulator system for simple laboratory maintenance applications. The operation sequences for each task and the current state of the task execution are represented using a Petri-net-like structure called *task nets*. Operations which the robot can perform are defined by *operation macros* which are constructed from generic *sense/think/act primitives*. Finally, the knowledge base is divided into robot state information and environmental information, with pointers to information sources and links to task conditions.

The remainder of the paper is organized as follows. Section 2 introduces a laboratory maintenance example used for illustrations in the remainder of the paper and describes the mobile manipulator system being developed in our laboratory. The three major data structures for dynamic sensing and multi-task execution are defined and illustrated in section 3. In section 4 we illustrate how the representations introduced in this paper enable the scheduler to achieve efficiency and flexibility in acquiring sensory data and executing multiple tasks. The concluding section discusses our current research into scheduling strategies based on the representations introduced in this paper.

2. Example Scenario: Laboratory Maintenance

Consider a robot designed to assist humans working in a laboratory environment. Such a system could be expected to collect experimental samples in the workspace and deposit them in specified receptacles. It might also be responsible for monitoring and making necessary adjustments to other automatic equipment such as fume hoods, filter systems, stationary manipulators, or even other mobile robot systems. As an example, we consider the environment shown in Figure 1, and assign our robot four tasks. The first of these consists of regularly checking an equipment monitor for a filter system and changing the filter when signalled to do so. The remaining tasks consist of locating and retrieving three samples in the workspace and placing them in a depository.

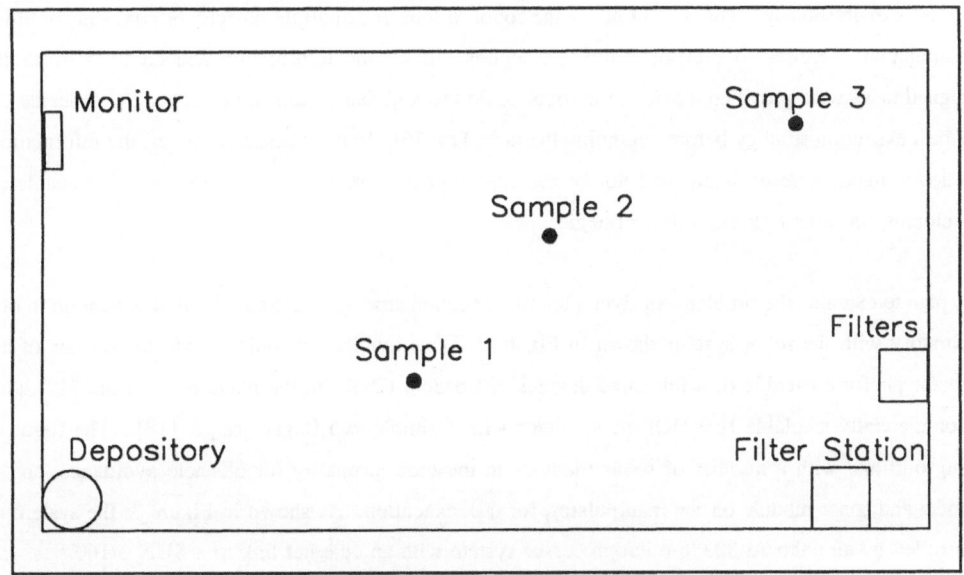

Figure 1: Laboratory floor plan

This scenario allows us to examine a number of issues. At one level we are faced with problems of sensor processing, sensing strategies, and gripping strategies [9, 10, 11, 12]. At a higher level, we are faced with problems of task planning for the individual jobs assigned to the robot [13, 14]. Time limits, such as requiring that the filter be changed within several minutes of the signal appearing on the monitor, create additional constraints on the execution of specific tasks. Finally, there are the problems that arise when the robot must schedule its activities at run-time to acquire information about the environment and perform multiple tasks in the most efficient manner. It is to these issues of dynamic sensing and multi-task performance that we turn our attention.

In the example above, there are seven objects which the robot must locate to carry out all of the prescribed tasks. The locations of some or possibly all of these objects may not be known in advance, and thus they must be discovered by the robot. If each task is considered independently, then the robot will have to perform up to seven "locate object" operations, possibly ignoring nearby objects in one step and then, having moved away in search of something else, searching for the remaining objects in future steps. Or, consider the case where two objects are situated close together, such as the monitor and one of the samples. Without proper consideration, the robot will not be able to take advantage of this arrangement, and could waste time by moving to the monitor, observing a signal, moving to the opposite end of the lab to change the filter, and then traversing the lab again to get back to the sample.

Both of these examples indicate a need for some type of parallel task execution strategy that allows the robot to schedule its sensing and motion activities dynamically, taking all of its assigned tasks into account simultaneously. This would allow the robot to look for multiple objects, or consider multiple manipulation exercises to prevent redundant actions. If all the information needed to perform its assigned tasks were provided *a priori*, the robot could use a global optimization procedure to decide on the best execution strategy before beginning the tasks [15, 16]. In most cases, however, the information needed to make these decisions will not be available *a priori*, and thus the system must be capable of developing run-time task execution strategies.

We plan to examine the problems of dynamic task execution strategies and parallel task execution in our laboratory with the robot system shown in Figure 2. This mobile manipulator system consists of an Ilonator platform capable of a full three degrees-of-freedom (DOF) in the plane of the floor [17], and a commercially available five DOF manipulator with a simple two finger gripper [18]. The robot is being outfitted with a number of sonar modules in the base, primarily for obstacle avoidance, and a camera and sonar module on the manipulator, for task execution. As shown in Figure 3, the system is controlled by an onboard 80286 microprocessor system with an ethernet link to a SUN 3/160.

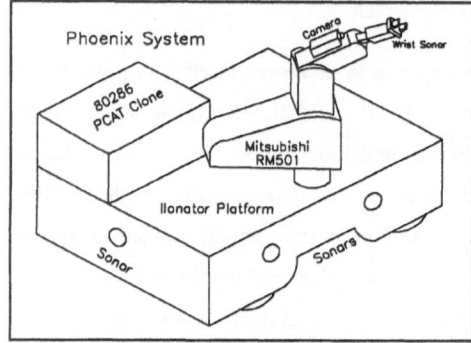

Figure 2: Experimental robot

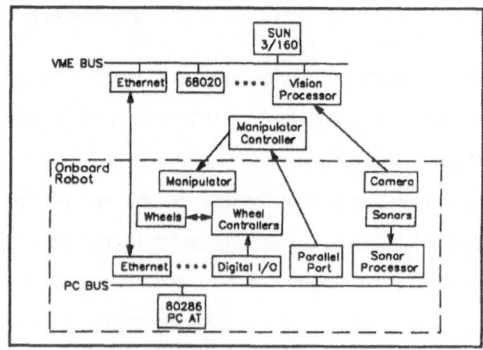

Figure 3: Hardware architecture

3. A Representation for Dynamic, Multi-Task Execution

To implement multi-task execution, controlled by an RTS, we have developed the set of interconnected representations shown in Figure 4, consisting of three primary structures: a collection of generic operations that the robot can perform, a set of task nets that describe the individual tasks assigned to the robot, and an assortment of data representing the knowledge about the world. These structures are interconnected through *information links* and *control lines* that regulate the use of the robot's actuators, sensors, and data processors, and the flow of data during task execution.

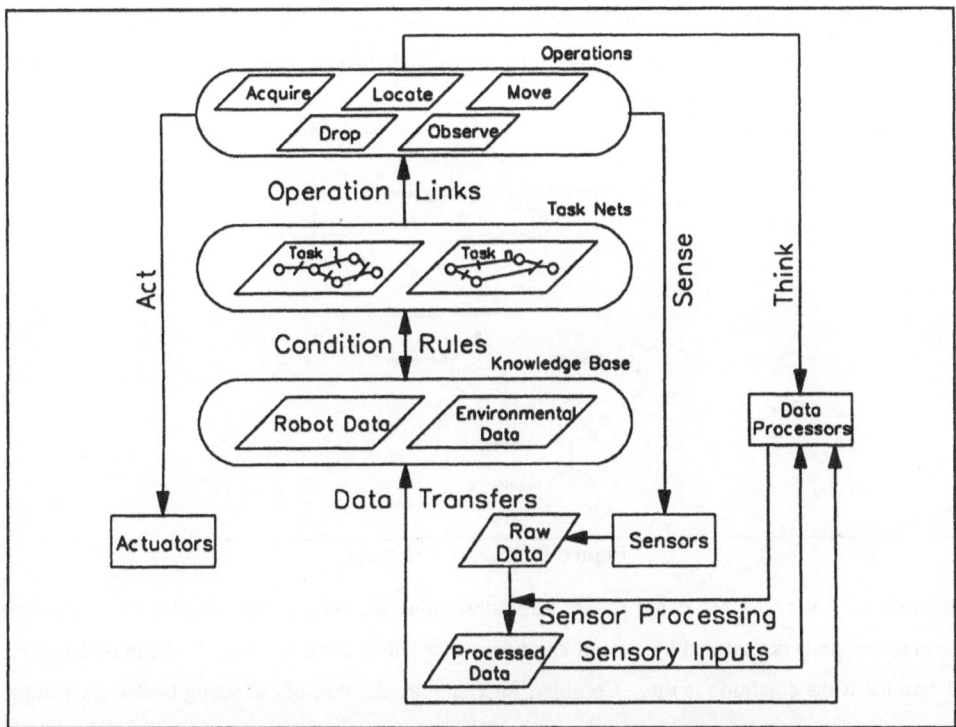

Figure 4: Interconnected operation/task/knowledge representations

As is often noted in the artificial intelligence (AI) and robotics literature, representation is important in determining the capabilities of a system [19, 14, 20, 21, 22]. Our decomposition of the task representation is designed to permit a great deal of flexibility during task execution, while providing a concise picture of the state of each task at each moment. Separating the task descriptions from both the underlying database and the operational capabilities of the system allows each task net to be generated independently of other system tasks. At the same time, the linking of the operation, task, and knowledge representations allows interactions between tasks to be considered at run-time. To illustrate how this can be accomplished, we next describe each of the structures in some detail, and then briefly discuss one possible run-time scheduling strategy.

3.1 Task Representation

Each task to be performed by the robot is characterized by a task net consisting of *conditions* and *operations* as shown in Figure 5. Task nets for the tasks from our example in section 2 are shown in Figures 6 and 7. In the graphical representation of task nets, conditions are indicated by circles and operations are represented by bars. Conditions represent preconditions that must be satisfied before an operation can be executed. The set of conditions required for executing a particular operation is indicated by arcs in the task net. An input condition for an operation can be negated as shown with the small, filled circle on OPERATION 4 in Figure 5. For example, in Figures 6 and 7, the negated condition ROBOT_FULL is used to prevent acquire operations, rather than using the (positive) condition ROBOT_NOT_FULL to allow the same operation.

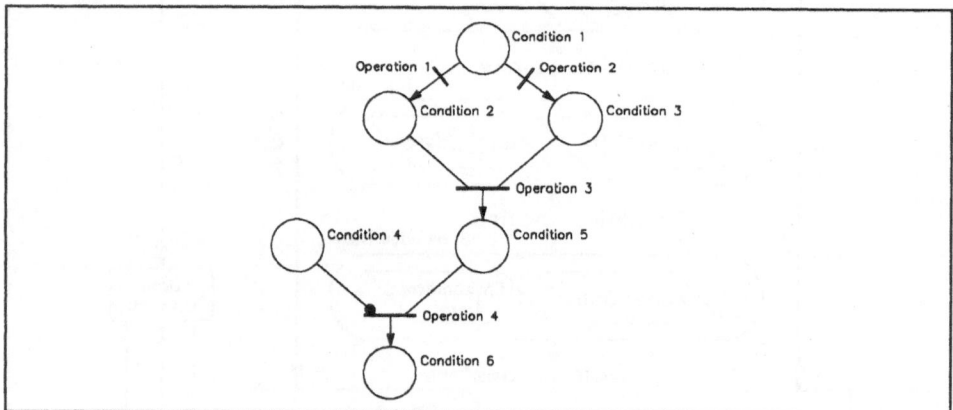

Figure 5: A general task net

Conditions are binary, valued either true or false, describing the current state of each task. The value of each condition is determined by a set of *condition rules* which use data stored in the knowledge base to determine if the condition is true. Consider, for example, our task of collecting laboratory samples, which can be represented by the net shown in Figure 6. Suppose the system is interested in the condition, SAMPLE_N_ACQUIRED. Our robot has two onboard storage bins for samples plus the end-effector gripper which can hold a sample. Thus, the condition rule for the SAMPLE_N_ACQUIRED condition can be stated IF (STORAGE_1_CONTENTS = SAMPLE_N) OR (STORAGE_2_CONTENTS = SAMPLE_N) OR (GRIPPER_CONTENTS = SAMPLE_N) THEN TRUE. If the predicates of the condition rule are not satisfied, the condition defaults to false.

Operations describe the actions that the robot must perform to progress through a task. As described above, each operation has a set of input conditions that must be true before it can be attempted. For example, in Figure 7 we can see that the three conditions, FILTER_ACQUIRED, SIGNAL_OBSERVED, and NEAR_STATION must all be satisfied before the robot can attempt the CHANGE_FILTER operation. Further, each operation has one explicit output condition expected to become true when the operation is executed.

Graphically, task nets are similar to Petri nets, which have been proposed for modelling task execution progress in robotic systems [23, 4, 24, 25]. In contrast to Petri nets, however there is no concept of a token "moving" in a task net when an operation is attempted, for an operation may fail, leaving the output condition unchanged. Also, for some operations, such as ACQUIRE_OBJECT, there are outcomes that are not the result of a single execution of the operation. An example of this is shown by the ROBOT_FULL condition. This is not an outcome of any particular operation (because we do not know *a priori* the order of task execution) but may occur after any ACQUIRE_OBJECT operation. Finally, some operations may be required by more than one task, such as LOCATE_DEPOSITORY, and the robot should recognize that doing this once affects conditions in more than one task. Linking the conditions to the underlying knowledge base provides a solution for each of these problems, since the value of each condition is determined by its condition rules, and not the execution of individual operations.

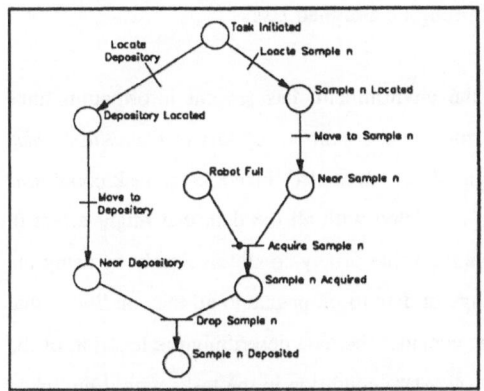 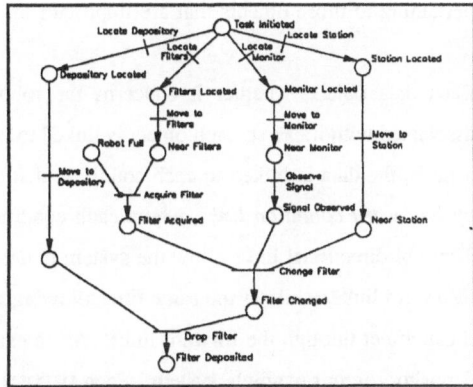

Figure 6: Sample collection tasks   **Figure 7**: Filter system maintenance

Task operations represent specific instantiations of the generic actions that the robot can perform. During task execution, task operations can be grouped into generic, system operations through *operation links*, allowing the robot to consider parallel task execution possibilities. This not only allows the system to perform an operation used in multiple tasks, such as the LOCATE_DEPOSITORY operation in Figures 6 and 7, only once, but also allows it to perform similar actions such as LOCATE_SAMPLE and LOCATE_FILTERS concurrently. Operations in a single task can also be performed simultaneously. An example of this can be seen in the MOVE_TO_FILTERS and the MOVE_TO_STATION operations in Figure 7. As shown in Figure 1, these objects are close to one another, allowing these two operations to be grouped together provided the robot has already determined their positions.

3.2 Task-Oriented Knowledge Representation

The knowledge base structure contains two primary sets of data: data about the robot itself, and data about objects of interest in the environment. Data about the robot can be further divided into position and orientation information, and information describing the capabilities and "health" of the robot. By way of example, consider our experimental robot. The robot knowledge base should contain data describing the base position and orientation with respect to a global frame of reference, and the manipulator joint parameters. From this information, the system can obtain the position and orientation of the end-effector with respect to either the global frame or the base frame of reference. We would also maintain the state of the gripper, open or closed, and its contents, if any. The capabilities of the robot can be described by such information as the location and contents of onboard storage bins, maximum speeds of the base and manipulator, and payload capacity of the manipulator. For some systems it might also be necessary to keep track of fuel levels or battery capacity. Finally, the robot knowledge base might maintain a list of equipment believed to be malfunctioning. The environmental data which is maintained depends entirely on the individual tasks assigned to the system. Thus, while there may be a tremendous amount of data concerning the robot's environment, the robot is only interested in data pertaining to those objects that are important for performing its assigned tasks.

Each data object, whether it concerns the robot or the environment, has several information links associated with it. First, each object is linked to one or more task conditions by *forward condition links*. That is, the data is linked to each condition that it might affect. It is also linked to the task conditions by *backward condition links*, where each condition is associated with all the data that might affect it. These bi-directional links allow the system to determine the value of any condition simply by using the backward links, while at the same time allowing a change of data to propagate to all the conditions that it can affect through the forward links. As an example, consider the data describing the location of the depository in our example system. Each DEPOSITORY_LOCATED condition in the tasks would be linked to this information, and likewise, this information would contain pointers to each of these conditions. The RTS can determine at any time whether a depository located condition is true by following a link to this data, or any time this data changes, the system can update the associated conditions.

Each element of the knowledge base is also associated with a *source link* that describes how the robot might obtain the information. For example, the location of an object might be identified by a particular vision image and wrist sonar reading. Alternatively, this information could be supplied to the robot through a supervisory system. This information is used during task execution by the data processors to determine what types of sensory processing to do and what to search for in the processed data to obtain the desired data for the knowledge base. Again, we consider our example problem to illustrate the source links. Suppose the robot has chosen to locate the filter system's monitor. The robot may search part of the workspace, comparing sensor inputs to the information it has stored about the monitor. If it does not locate the monitor in a few minutes, it may decide that the operation has failed. It could then attempt another search, or ask the supervisor for information concerning the monitor's location.

Finally, the robot needs to keep track of whether the data in the knowledge base are valid or not. For example, if the knowledge base contains a description of SAMPLE N's location with the respect to the current position of the base, this information will become invalid when the robot moves. To monitor data status, each data structure has a *time-stamp* and a set of *dependency links*. Thus, the sample position information receives a time-stamp indicating when it was obtained, with a dependency link to the position of the robot. When the robot's position is changed, the robot position data will receive a new time-stamp. If the robot then examines the position data for the sample, it will be able to determine whether it is still valid based on the order of the time-stamps of the pieces of data indicated by the dependency links. Whenever data becomes invalid, the RTS distributes this information to the affected conditions, and decides what operations would now be possible.

3.3 Operations: Sense/Think/Act

The primitive operations that can be performed by our example robot or any robotic system can be broadly classified into three basic functions. The robot can *sense* the environment through various sensors, it can analyze information, or *think*, using data processors, or it can *act*, moving about or interacting with its environment. This decomposition is shown in Figure 4 through the sense, think, and act links from the operations structure to the sensors, data processors, and actuators. Each operation that the robot performs is defined by a macro built from these three sets of primitives. Furthermore, we note that the operations themselves can be grouped into sensing, thinking, or acting categories, and even at the highest level of task description, such as MONITOR THE FILTER STATION or COLLECT SAMPLES, tasks can be categorized into one of these three functions. This type of functional classification has been observed by other authors [26].

This "recursive" decomposition of operations can be exploited during the production of the *operation macros* that the robot uses to perform requested operations. Typically, these macros represent general system operations such as LOCATE_OBJECTS {X1, X2, ... XN}. The specific information about the number of targets and their identifying characteristics come from the individual task operations that are combined to form the system operation. These macros can be quite complex when several goals are attempted at once. This can be seen in the complexity of even a simple operation like DROP_SAMPLE_N. As seen in Figure 6, two conditions must be true before this operation can be attempted: the robot must be near the depository, and it must be carrying the sample. Given these conditions, there are still some variables to be considered, however, such as whether the sample is in the gripper or a storage location. Moreover, there can be some uncertainty in the location of the depository and the robot itself. As shown in Figure 8, even a simple operation may contain many sense/think/act levels.

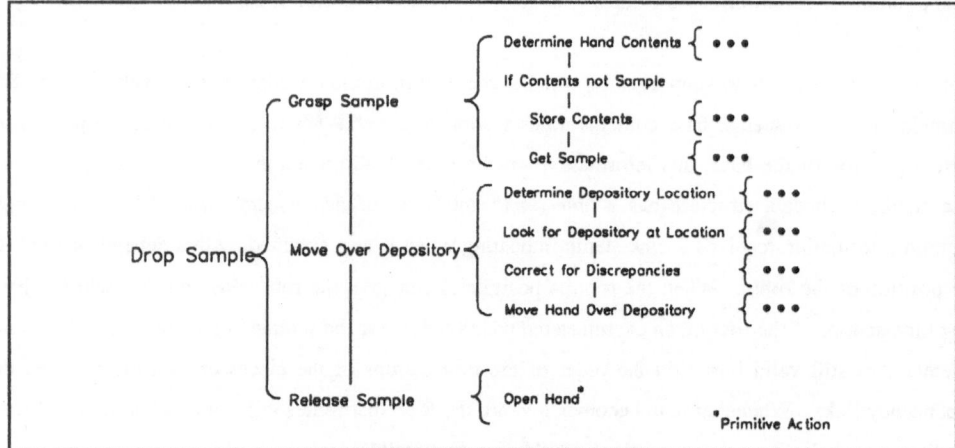

Figure 8: Operation decomposition

Eventually, however, all system operations can be expanded to an ordered set of primitive operations that directly represent the robot's basic sense/think/act capabilities. Each of these primitive operations, whether they involve sensing the environment, processing information, or acting on the robot's surroundings, requires some time, and has some possibility of failure. These attributes combine to produce the overall behavior of the operation.

3.4 Task Execution

To illustrate the use of the representations introduced above, we consider an RTS that uses a myopic scheduling strategy in which task execution proceeds in four basic steps. First the system determines the current state of each task, setting the value of each condition based on the condition rules and the underlying data structures. To minimize data checking, currently executing tasks would be maintained by propagating the effects of modified data structures through the task nets, while new tasks, added to the currently running set, would be initialized by checking each condition against all the underlying data. The values of all the conditions in the tasks would then provide a snapshot of the current state of the robot's tasks.

Evaluation of the task net conditions indicates which task operations can be attempted next. The RTS then groups these operations, through the operation links, into generic system operations. This grouping process eliminates duplicate task operations, and combines similar operations, reducing the total number of options to consider, and allowing for parallel task execution. Once this step has been completed, the RTS then selects which system operation to attempt. This decision would be based on the results of an optimization procedure that would weigh the expected time and energy expenditures against the expected progress made toward task completion. The problem of quantifying operation costs and payoffs is a topic of current research as discussed in section 5. Finally, the RTS instructs the robot to attempt the chosen operation.

When the robot executes an operation, there are two possible outcomes. The robot might succeed in its attempt, modifying the world and knowledge base in the expected manner. In this case, the robot will make progress in at least one task, with possible cross-over affects in other tasks. This would be noted by updating the appropriate conditions based on the modified data in the knowledge base, and the RTS would repeat the cycle. The second possibility is that the robot might fail in its attempt. In this case, the robot might perform some automatic correcting action, moving away from an intervening obstacle, for example. The RTS will then have to decide how to handle the failure. One possibility would be to immediately repeat the operation. This could be effective when the robot has different initial conditions for the second attempt. Another possibility would be for the RTS to ignore the problem and schedule a new operation. If all possible operations result in failure, the RTS could request assistance from the supervisor. Once the failure has been resolved, the RTS would start the cycle again.

4. Representation, Information Processing, and Planning Flexibility

An autonomous system asked to perform several simultaneous tasks is faced with a number of decisions. If the robot is to have any flexibility in its decisions, it will require a flexible representation scheme and the ability to process information at different levels. We believe the representations we have described in the previously provides this flexibility. In this section, we briefly present two illustrations from our example problem to demonstrate the ways that this flexibility can be achieved.

As shown in Figures 6 and 7, the first step for the robot in each of its tasks is to find the objects of interest in the workspace. If we consider each task individually, the robot will have to perform a total of seven searches if no information is supplied, or three if only the sample locations are unknown. Consider for the moment, the situation where the robot is searching for the samples, starting in the situation shown in Figure 9. If the robot now attempts to locate sample 1, it will have to leave its current location and search for the desired sample. Then, once it has located the sample, the system would begin its search for sample 2, leading to another extensive exploration of the workspace.

If we consider the problem under our representation, however, the search operations for the two tasks could be combined into a single operation, resulting in a savings of time and energy. If the system has been told the locations of the monitor, depository, filters, and filter station, then each of the conditions related to these data would be set true. The next available operations would then be LOCATE_SAMPLE_1, LOCATE_SAMPLE_2, LOCATE_SAMPLE_3, MOVE_TO_DEPOSITORY (in each task), MOVE_TO_FILTERS, MOVE_TO_MONITOR, and MOVE_TO_STATION. The robot would pass these options through the operation links and consider a set of generic operations. In the case of the three locate sample operations, they would become lumped into one LOCATE_OBJECTS {SAMPLE_1, SAMPLE_2, SAMPLE_3} operation. During the execution of this operation, sample 2 would be located before the robot moved off in search of other objects, eliminating unnecessary backtracking later.

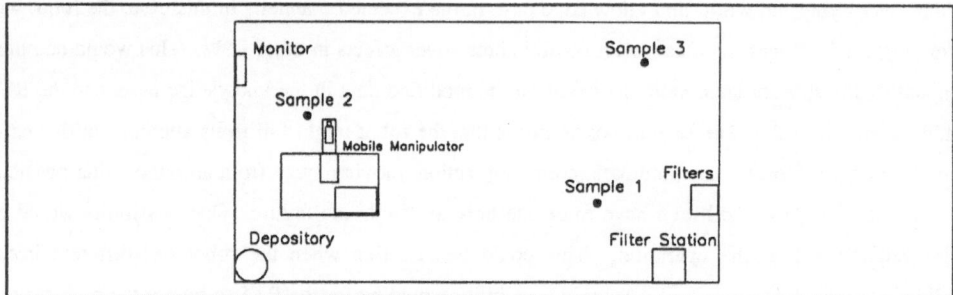

Figure 9: Multiple sensing operations

Another example of the potential for parallel task execution can be seen in Figure 10. The placement of the filters in close proximity to the filter station would allow the robot to position itself once such that it could perform two task steps. If the tasks are considered separately, however, this would likely not be done, and the robot might acquire a filter and then, not realizing that it has already satisfied the preconditions for changing the filter, move off in search of one of the samples. The robot would then waste more time and energy returning to the filter station later.

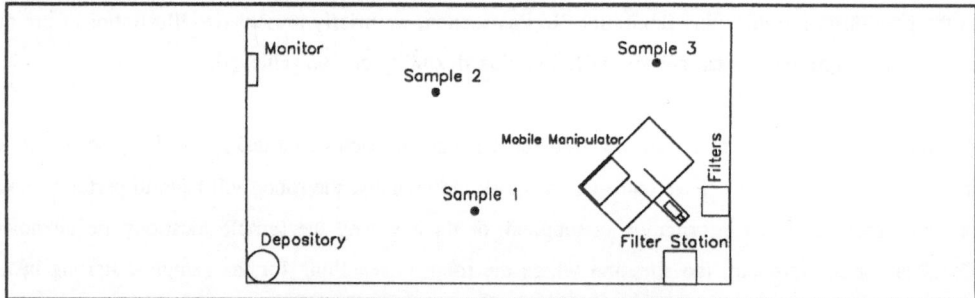

Figure 10: Concurrent task performance

If we consider this set of tasks using our representation, however, the robot can consider the MOVE_TO_FILTERS and MOVE_TO_STATION operations as a single step. In this case, the robot will be able to move to a centralized point, acquiring a filter and then placing it into the system. Of course, a simple myopic scheduling strategy might not produce these types of actions, but that is a result of the strategy, not the underlying representation.

5. Current Research

This paper introduces a set of interconnected data representations for tasks, operations, and knowledge for an RTS for a mobile robotic system. The objective is to create representations of the elements of the on-line scheduling problem to enable the system to perform dynamic sensing for executing multiple tasks concurrently and efficiently. The use of the data structures is illustrated with a simple myopic scheduling strategy.

We are currently investigating several issues related to the representation and execution of tasks for mobile manipulator systems. For the robot in our laboratory, we have created a set of task nets for a particular case of the laboratory maintenance example in section 2. This is a prototype application with which we will investigate alternative scheduling strategies. Using the representations introduced in this paper, we are developing a set of basic sense/think/act operation primitives for the mobile manipulator in our laboratory. From these primitives, we are defining higher level operation macros that take advantage of the functionality of the robot system in a way that permits general operation definitions to be applied to specific task operations. Data objects are also being defined for the knowledge base to represent the robot state and environmental data efficiently for our application.

Having developed an appropriate representation for the RTS, different scheduling strategies will be investigated. To quantify the decision alternatives, a method for assigning costs to individual operations and operation sequences will be developed. Costs and rewards will include factors related to energy, time, and likelihood of successful completion. For myopic scheduling, local cost functions representing the short-term pay off for next-step decisions will be used to generate run-time schedules. Multiple task operations will be combined when possible into single system operations, using operation macros. The analytical issue to be investigated in this case is termination; that is, will the robot be able to determine if a solution exists to the multi-task execution problem, and if a solution exists, can it be guaranteed that a solution will be found?

Alternatives to strictly myopic scheduling will also be investigated. Since the scheduling decisions are based on information acquired in real time, rigorous optimization algorithms such as dynamic programming are computationally intractable for the run-time scheduling problem. As an alternative, we will consider on-line versions of search techniques used in (AI), such as the A* search [22]. In contrast to a standard AI planning application where the results of decisions are evaluated by a computation, our search involves the execution of robot operations. Problems of recoverability and execution of alternative decisions when an operation fails will have to be addressed in the context of the run-time scheduler. A further extension to be investigated is the inclusion of "learning" in the scheduler algorithm so that decisions can be influenced as more experience is acquired for repeated tasks in a given environment.

References

1. M. Beckerman, D. Barnett, J. Einstein, J. Jones, P. Spelt, and C. Weisbin, "Robust Performance of Multiple Tasks by an Autonomous Robot", *Workshop on Integration of AI and Robotic Systems*, 1989.

2. W. Hamel, S. Babcock, M. Hall, C. Jorgensen, S. Killough, and C. Weisbin, "Autonomous Robots for Hazardous and Unstructured Environments", *Proceedings of the Robots-10 Conference*, 1986.

3. D. Wettergreen, H. Thomas, and C. Thorpe, "Planning Strategies for the Ambler Walking Robot", *Proceedings of the IEEE International Conference on Systems Engineering*, 1990, pp. 198-203.

4. F. Wang, K. Kyriakopoulos, A. Tsolkas, and G. Saridis, "A Petri-net Coordination Model for an Intelligent Mobile Robot", *Proceedings of the 29th Conference on Decision and Control*, December 1990, pp. 1531-1536.

5. U. Rembold, "The Karlsruhe Autonomous Mobile Assembly Robot", *1988 IEEE International Conference on Robotics and Automation*, 1988, pp. 598-603.

6. J. Connell, "A Behavior-Based Arm Controller", *IEEE Transactions on Robotics and Automation*, Vol. 5, No. 6, December 1989, pp. 784-791.

7. R. Brooks, "A Robust Layered Control System For a Mobile Robot", *IEEE Journal of Robotics and Automation*, No. 1, March 1986, pp. 14-23.

8. E. Badreddin, "Recursive Nested Behavior Control Structure for Mobile Robots", *Intelligent Autonomous Systems*, December 1989.

9. D. Kriegman, E. Triendl, and T. Binford, "Stereo Vision and Navigation in Buildings for Mobile Robots", *IEEE Transactions on Robotics and Automation*, Vol. 5, No. 6, December 1989, pp. 792-803.

10. Y. Zheng, "Integration of Multiple Sensors into a Robotic System and its Performance Evaluation", *IEEE Transactions on Robotics and Automation*, Vol. 5, No. 5, October 1989, pp. 658-669.

11. A. Elfes and L. Matthies, "Sensor Integration for Robot Navigation: Combining Sonar and Stereo Range Data in a Grid-Based Representation", *Proceedings of IEEE Conference on Decision and Control*, 1987, pp. 1802-1807.

12. E. Ammeen and H. Stephanou, "Intelligent Grasp Planning", *Proceedings of the IEEE International Conference on Systems Engineering*, 1990, pp. 52-55.

13. R. Fikes and N. Nilsson, "STRIPS: A New Approach to the Application of Theorem Proving to Problem Solving", *Artificial Intelligence*, 1971, pp. 189-208.

14. M. Ginsberg and D. Smith, "Reasoning about Action I: A Possible Worlds Approach", *Artificial Intelligence*, 1988, pp. 165-195.

15. W. Carriker, P. Khosla, and B. Krogh, "An Approach for Coordinating Mobility and Manipulation", *IEEE International Conference on Systems Engineering*, 1989.

16. W. Carriker, P. Khosla, and B. Krogh, "The Use of Simulated Annealing to Solve the Mobile Manipulator Path Planning Problem", *1990 IEEE International Conference on Robotics and Automation*, 1990, pp. 204-209.

17. D. Feng, M. Friedman, and B. Krogh, "The Servo-Control System for an Omnidirectional Mobile Robot", *Proceedings of the IEEE International Conference on Robotics and Automation*, 1989.

18. Move Master II Model RM-501 Instruction Manual, Mitsubishi Industrial Micro Robot.

19. E. Palma-Villalon and P. Dauchez, "World representation and path planning for a mobile robot", *Robotica*, Vol. 6, 1988, pp. 35-40.

20. S. Amarel, "On Representations of Problems of Reasoning about Actions", *Machine Intelligence*, 1968.

21. D. Payton, J. Rosenblatt, and D. Keirsey, "Plan Guided Reaction", *IEEE Transactions on Systems, Man, and Cybernetics*, Vol. 20, No. 6, November/December 1990, pp. 1370-1382.

22. E. Rich and K. Knight, *Artificial Intelligence*, McGraw-Hill, Inc., 1991.

23. F. Noreils, "Integrating Error Recovery in a Mobile Robot Control System", *1990 IEEE International Conference on Robotics and Automation*, May 1990, pp. 396-401.

24. J. Peterson, *Petri Net Theory and the Modeling of Systems*, Prentice-Hall, Cliffs, N.J., 1981.

25. R. Valette, M. Courvoisier, H. Demmou, J. Bigou, C. Desclaux, "Putting Petri Nets to Work for Controlling Flexible Manufacturing Systems", *Proceedings of ISCAS 85*, 1985, pp. 929-932.

26. A. Sood, M. Herman, M. Trivedi, and H. Wechsler, "Computational Perspective on Perception, Planning, Action and Systems Integration", *IEEE Transactions on Systems, Man, and Cybernetics*, Vol. 20, No. 6, November/December 1990, pp. 1241-1244.

Multisensor System for an Autonomous Robot Vehicle

C. Fröhlich, F. Freyberger, G. Karl and G. Schmidt

ABSTRACT

Autonomous vehicles are supposed to perform motion tasks without artificial guiding references or other external support. For operation in indoor environments, precise and up-to-date sensory information of a robot's 3-D surrounding proves to be most important. As a key subsystem within the planning and control system of the full scale autonomous mobile robot MACROBE, a multisensor system with real-time capability was developed.

This paper focuses on the architecture of MACROBE's multisensor system, including various physical sensor devices and software based virtual sensors. A medium-range, wide-angle, eyesafe 3-D laser range camera serves as a primary geometry sensor together with a 2-D multisonar system. The paper discusses the application of virtual sensors for purposes of motion-oriented reduction of range image data and of sensor data fusion. Experimental results performed with the multisensor system as part of the mobile robot MACROBE are reported.

1. INTRODUCTION

Basic features of an autonomous vehicle are its capability to perform motion tasks without artificial guiding references or external support and to react appropriately in case of unexpected situations and events. For this purpose, dead-reckoning based operation must be accompanied by sensor-assisted navigation with respect to natural landmarks like walls, doors or gates.

This paper discusses a multisensor system for indoor navigation as designed for the experimental mobile robot MACROBE. This full scale autonomous mobile robot is developed by the Robotics Group of the Laboratory for Automatic Control Engineering, Technical University of Munich. It plans and executes indoor operations through a motion planning and control system. Typical autonomous maneuvers considered, are guarded motion, following of path's center line, traversing of narrow passages, following or evading moving objects, docking as well as reorientation and exploration. Varying motion tasks and environmental conditions require, however various kinds of sensor assistance in real-time. The multisensor system integrates multiple physical sensor devices which provide basic geometrical data of the vehicle's surroundings. Furthermore, it organizes algorithms for the extraction of motion relevant features out of sensor data by means of so-called virtual sensors [1]. An architectural overview of MACROBE's perception, planning, reasoning, action and communication capabilities is shown in Fig. 1.

This paper focuses on MACROBE's multisensor system. The architecture of the multisensor system, including physical sensor devices and virtual sensors is presented in Section 2. The 3-D geometric range images of the environment are provided by an eyesafe laser range camera. The other physical sensor devices within the multisensor system, are a 2-D short to medium range imaging multisonar system at the vehicle's frontend, six side looking ultrasonic range finders, two shaft encoders mounted at separate measuring wheels for odometric purposes and an optical fibre gyro for heading measurement.

As a basic virtual sensor, Section 3 introduces 2-D top view profiles as obtained from motion relevant reduction of 3-D range images. Experimental results from sensor-assisted navigation in a semi-structured environment using a network of virtual sensors are presented in Section 4. To increase accuracy and reliability of sensor information, sensor data of complementary and competitive physical sensor devices are fused. Section 5 introduces sensor data fusion by means of virtual sensors and discusses the application to the execution of guarded motion. Section 6 concludes the paper with an outlook toward future work in the mobile robot project MACROBE and its multisensor system.

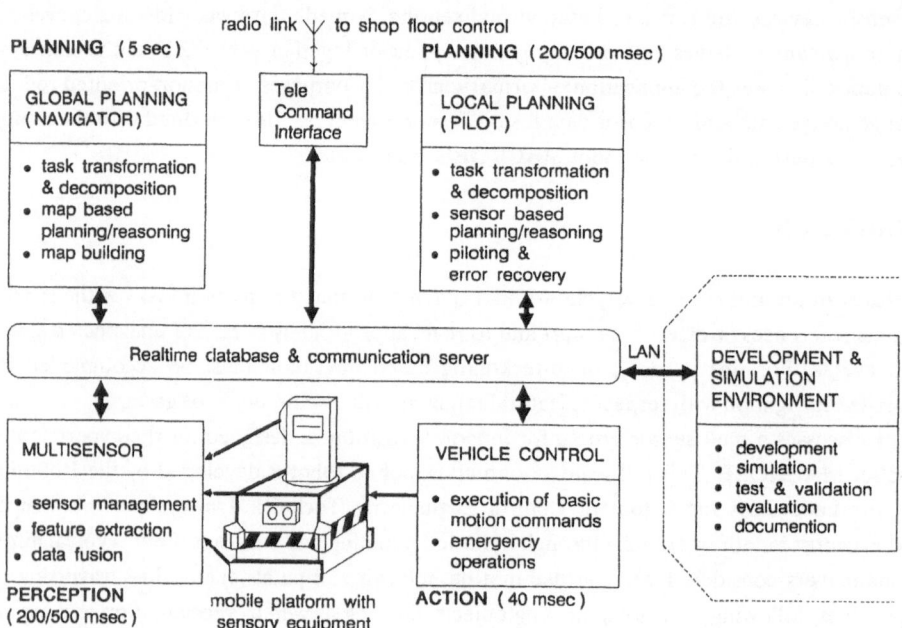

Fig. 1: Motion Planning and Control System of MACROBE

2. THE MULTISENSOR SYSTEM

Due to MACROBE's navigation speed of up to 1 m/s, a sufficiently large area in front of the mobile robot has to be monitored by physical sensor devices. The sector may contain objects made out of various material with various textures. Typical autonomous maneuvers like guarded motion, following of path's center line and traversing of narrow passages require a sensor field of view with a spatial angle of about 60° x 60°, a reasonable accuracy (2 cm) and repeatability of range measurements, a reasonable spatial resolution within a range up to 10 meters and a frame rate of several 3-D range images per second. In addition to these demands, exploration and reorientation tasks may require a broader field of view of about 180° x 60°, a higher accuracy with range measurement and a better spatial resolution within a range up to 50 meters.

Diversity of sensor assistance (Tab. 1) concerning updating period, accuracy and repeatability with measurements and monitoring area requires varying measurement techniques with specific modes of feature extraction. To provide sufficient robustness and reliability with the execution of autonomous maneuvers parallel in time, a multisensor system as a key subsystem of the full scale autonomous mobile robot MACROBE was developed.

maneuver	updating period	accuracy	monitoring range	sensor principle		
				sonar	laser	microwave radar
guarded move	< 0.5 s	< 3 cm	< 4 m	X	X	
following path's center line	< 1 s	< 2 cm	4 m		X	
			1 m	X		
traversing gates	< 2 s	< 2 cm	< 10 m			X
			< 5 m	X	X	
docking	< 2 s	< 1 cm	< 10 m			X
			< 4 m		X	
			< 1 m	X		
reorientation	< 10 s	< 1 cm	< 50 m			X
			< 4 m	X	X	
exploration	---------	< 1 cm	< 50 m			X
			< 4 m	X	X	

Tab. 1: Basic requirements with autonomous maneuvers in indoor environments

Main topics in the development of a multisensor system are the selection of suitable physical sensor devices, algorithms (virtual sensors) for real-time sensor data processing and the integration of both, physical sensor devices and virtual sensors, into a real-time system.

2.1 Physical sensor devices

The usual types of sensors providing 2-D or 3-D information of the environment and supporting maneuvers operated by autonomous vehicles are standard video, stereo vision and simple acoustical proximity sensors. While passive video sensors provide excellent resolution at very high data rates, their ability to detect objects under changing illumination and weak texture conditions is limited. Providing range data with a minimum computation time, active sensing techniques are more attractive for real-time navigation. For these reasons, physical sensor devices based on active range measurement techniques using infrared light or ultrasonic pulses were developed and integrated in the perception system of MACROBE (Fig. 1).

3-D laser range camera

In contrast to high power laser ranging systems developed for outdoor applications [13],[15],[16], the laser range camera (Fig. 2) developed by our group [2],[3],[4] is designed for eyesafe (1 mW) indoor operation. A collimated laser beam (λ = 810nm) is directed to the scene to be monitored, and back-scattered light is sensed. The distance between the camera and an object point is determined by measuring phase shift between the emitted and the received beams of modulated laser light (10/80 MHz). A two-frequency phase-shift method provides satisfactory resolution (up to 7 mm) over a range up to 15 meters. Full scene 3-D range data is obtained by scanning the transmitted laser beam, using two synchronized mirrors. Rotational speed of the mirrors are selected such that a measurement rate of 2.5 range images per second, consisting of 41 scan lines with 161 range pixels each, can be guaranteed (Tab. 2).

To adapt the laser range cameras limited field of view (60° x 60°) to motion relevant observation areas of the environment, the laser range camera is mounted on top of a platform, which can be servo-controlled over ±360°.

The resulting range image is represented in a 2-D indexed matrix, containing as its elements the distance from the laser camera to the nearest target point in a discrete spatial direction, and a value representing reflected signal quality.

Laserhead:		3-D Deflection System:	
IR-semiconductor laser (810nm)		Beam deflection through a set of two mirrors	
Hybrid APD photosensor		Field of view	: 60°x 60°
Emitted laser power	: 1mW	Spatial resolution	: 161 x 41 range pixels
Receiver Electronics:		Frame rate	: 2.5 3-D range
2-frequency phase shift method			images
Modulation frequencies	: 10/80MHz		
Resolution	: up to 7mm		
Working distance	: 0.1m - 15m		

Tab. 2: Characteristics of the 3-D laser range camera

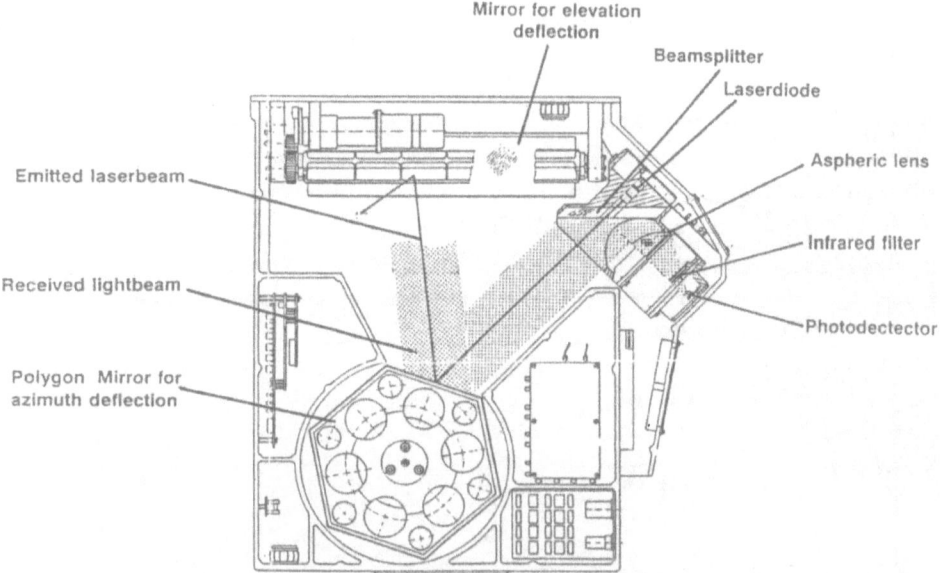

Fig. 2: Mechanical and optical design of the 3-D Laser range camera

Thin bars, reflecting surfaces or glass may cause problems in acquisition of the environment by the 3-D laser range camera.

2-D Multisonar System

In order to compensate some of these uncertainties a 2-D multisonar system was developed. Due to the required navigation speed of up to 1 m/s, detection, as well as a coarse localisation of any object within an observation area down to 1 meter in front of the vehicle is required.

The multisonar system consists of five autonomous, concurrently operating sensor units. Units S1 and S2 are identical. They are fixed at the left and right side of the vehicle's frontend. A 200 kHz pulse is transmitted with a 145° lobe by using a specially formed reflector. The echo, back-scattered from a target is evaluated by five independent receivers with 15° lobes. They are mounted at an 30° angle relative to each other. With this arrangement, a segmented detection area within a maximum range up to 1.2 m is achieved. The unit S3 fixed at the center of the vehicle's frontend uses an 80 kHz emitter and receiver. This frequency allows range measurements up to 8.5 m, so that any obstacle occuring on the vehicle's path can be detected in time.

Sensor units S4 and S5 are fixed on both sides of the vehicle. Each unit consists of three 200 kHz side looking ultrasonic range finders, measuring lateral distances up to a maximum range of 1.4 m. The cycle time of range measurement is 10 ms for units S1, S2, S4 and S5 and 60 ms for S3. Range data accuracy of the sensors is in the order of 0.5%. The multisonar system's total field of view is depicted in Fig. 3.

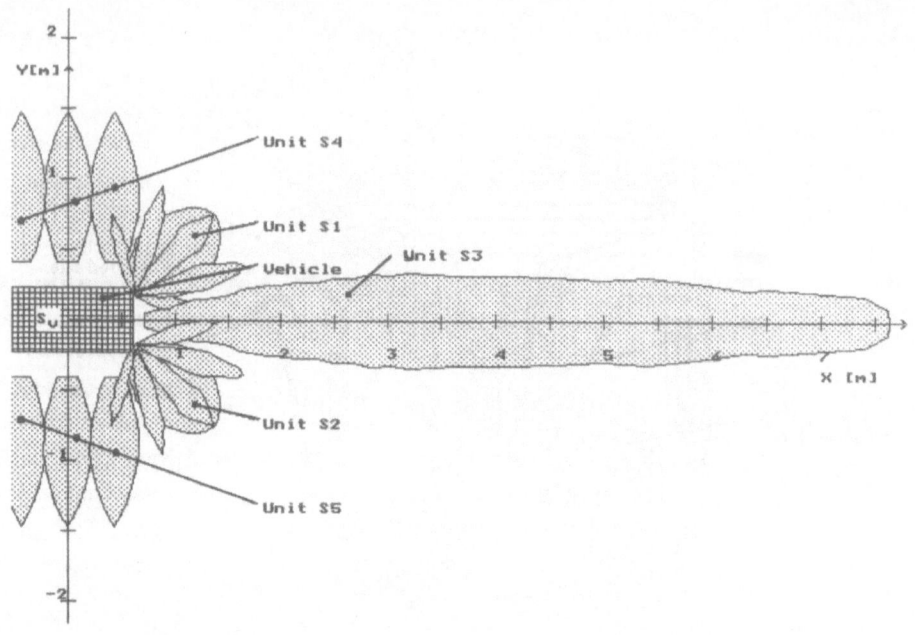

Fig. 3: Field of View of the Multisonar System

Additional physical sensor devices

Information from a newly developed microwave range imaging device as described in [5], and from interpreted CCD-camera images [6] will be incorporated in the multisensor system in future (Tab. 3).

Sensor	Range	Principle	Rate	Remark
3D - Laser Range Camera	0.1 - 15m	"optical, active" < 1 mW phase shift	0,4 sec / image	main sensor geometry of enviroment
Multi - Sonar front	0.1 - 1.2 m	"acoustic, active" time of flight	< 20 ms	collision avoidance
	0.1 - 8.5 m		< 60 ms	object localisation
side	0.1 - 1.4 m		< 30 ms	measuring lateral distances
3D-Imaging Radar*	0.5 - 50 m	"electromagnetic, active" 5 µW time of flight	10^4Volume cells/s	geometry of environment velocity of objects
CCD - Camera*	0.1 - 15 m	"optical, passive" greylevel	40 ms / image	fusion with range images
Shaft Encoder		"optical"		position
Optical Fibre Gyro		"optical"		heading
Tactile Sensors	0 m	"mechanical"		emergency stop

* not yet integrated

Tab 3.: Physical sensor devices of MACROBE's multisensor system

2.2 Sensor data processing through virtual sensors

Sensor data processing is performed by means of so-called virtual sensors [1]. Virtual sensors represent basic algorithmic structures which process one or more input data sets into a single output data set. Input data may result from physical sensor devices and/or from other virtual sensors. Each virtual sensor is autonomous and simply reacts to information it receives, according to its specified function. An important feature of virtual sensors is that they exhibit a defined interface, which allows them to communicate with other virtual sensors. The processing, necessary to perform a virtual sensor function, is therefore encapsulated within a computational entity that has a uniform but versatile interface specification.

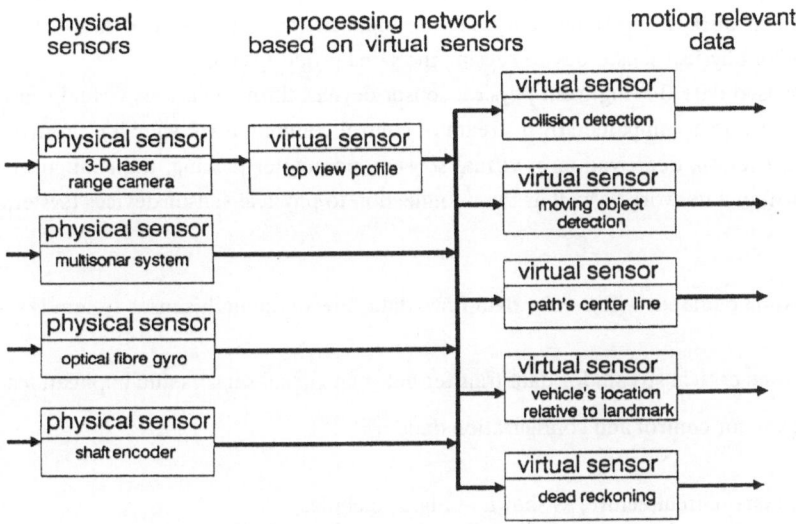

Fig. 4: Sensor data processing network based on virtual sensors

Some of the benefits a virtual sensor can offer, include its ability

- to transform sensor data from physical sensor devices into a task specific description ("abstraction"),

- to integrate information from two or more physical or virtual sensors and produce "new" relevant information,

- to respond to sequences of sensor events or conditions,

- to incorporate expectations and to respond whether or not expectations are met ("event triggered sensor data processing"),

- to define a uniform information interface for different sensors.

Information from a physical sensor device may, depending on the features to be extracted, undergo transformations by a number of virtual sensors. This results in a data-flow network (Fig. 4), where virtual sensors (nodes) may obtain their information from, and feed information to other nodes, so that integration, as well as transformation of sensor information can be achieved. To prevent deadlock at the data-flow level, cyclic paths in the network are avoided.

2.3 Architecture of the Multisensor System

The virtual sensors approach permits the dynamic reconfiguration of sensing resources for either some sort of adaptive sensing in response to a specific situation or fault tolerance. Depending on the actual environmental situation the physical sensor device is selected which is sensitive to the corresponding property to be sensed. In case of failure of some physical sensor device, the system may still be able to work, using redundancy, i.e. the input of the virtual sensor network can be switched to another physical sensor device sensing the same property (Fig. 4).

The stream of sensed data flowing from physical sensor devices through various virtual sensors to the application must be accompanied by a stream of control commands (or signals). Control commands include activation, deactivation of virtual sensors, parameter passing, the configuration of virtual sensors within a network as well as their connection to physical sensor devices (system configuration).

Due to the real-time demands with sensor data-flow, data-flow in the multisensor system is seperated into a

- horizontal, time critical stream for data transfer between virtual sensors and applications

- vertical stream, for control and configuration data.

The multisensor system architecture, as shown in Fig. 5, includes

- physical sensor devices,

- single virtual sensors and virtual sensor networks,

- a system configurator,

- a control unit,

- communication links with physical sensor devices and

- communication links with applications resulting from the motion planning and control system (Fig. 1).

Given the application specific command to be executed, the control unit of the multisensor system transforms the abstract (symbolic) command into sensor system specific commands. As a consequence, the connection manager within the control unit activates virtual sensors and connects them within the corresponding network. Depending on application specific parameters, the event manager passes parameters to the corresponding virtual sensors.

The system configurator configures the multisensor system by means of connecting data sources. Data sources may be physical sensor devices, e.g. during sensor-assisted navigation or units providing the multisensor system with offline stored or simulated sensor data, e.g. for development and evaluation purpose.

The functional modules of the multisensor system including virtual sensors form a real-time processing system. Due to the communication structure and the interfaces (links) between modules and virtual sensors, they may be implemented within a common real-time kernel system or on individual embedded processors (multi-processing).

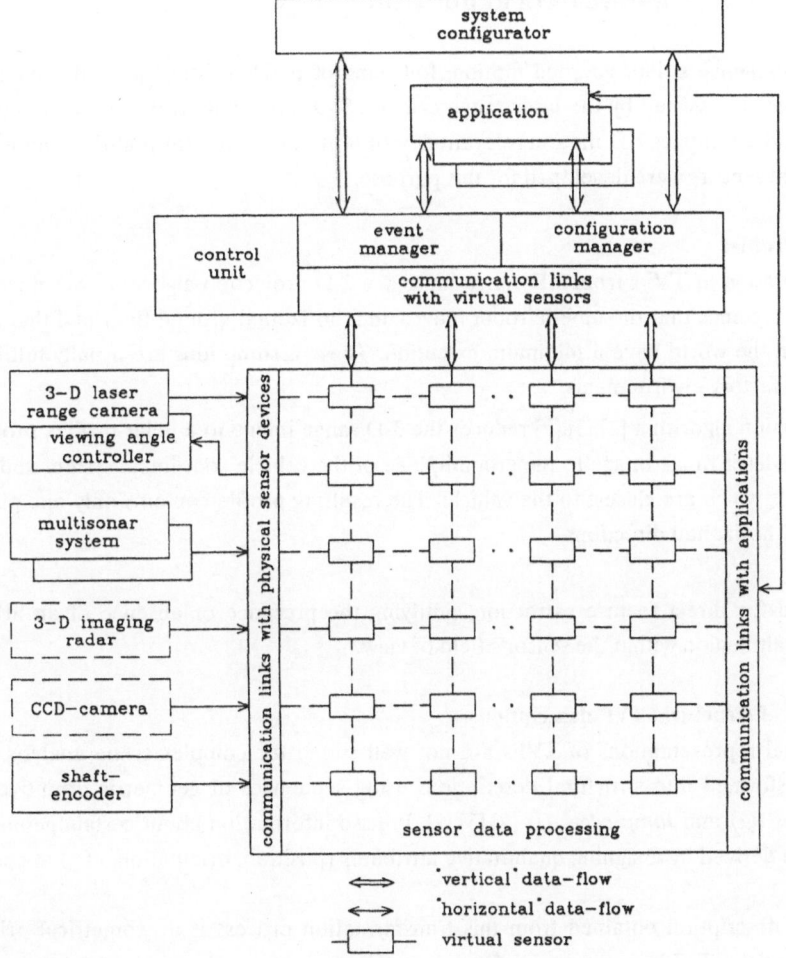

Fig. 5: Architecture of the multisensor system

With MACROBE's multisensor system integration of physical sensor devices as well as sensor specific range data preprocessing is achieved by means of frontend processors. Software modules for of the multisensor system are implemented on a real-time MicroVAX 3305 using VAXELN language. This approach allows a flexible hardware/software mix with the components of the multisensor system, which is capable of real-time sensor data processing. Furthermore, this approach permits a simple method of reconfiguration whenever virtual sensors or physical sensor devices are added to or removed from the system.

In the following sections, three typical applications of sensor data processing based on virtual sensors are reported.

3. MOTION-ORIENTED RANGE DATA REDUCTION

Basic vehicle maneuvers, like guarded motion, following of path's center line and traversing of narrow passages are assisted by the laser range camera. 3-D range data from the laser camera is transformed into a compact 2-D motion relevant description as required for real-time motion control. Two virtual sensors were developed for this purpose.

3.1 Top View Profiles
For the computation of TVPs from 3-D range images, a 2-D projection algorithm was developed. The algorithm assumes that the mobile robot moves on a horizontal ground-floor and that all isolated objects in the world have a minimum extention. These assumptions are usually fulfilled by laboratory and factory environments.
The 2-D projection algorithm [2],[3],[7] reduces the 3-D range image to a polar profile, projecting all pixels of the laser range image to the ground-plane of the vehicle coordinate system and selecting those pixels which are closest to the vehicle. The resulting profile contains only one pixel for each of the 161 horizontal directions.

TVPs are used for direct feature extraction, notifying the presence or absence of an arbitrary object in a certain region within the sensor's field of view.

3.2 Generation of structural TVP-descriptions
In general, pixel representations of TVPs are not well suited for complex scene analysis. Thus, TVPs are transformed into structural descriptions using sequences of geometric primitives, like *Lines (L)*, *Gaps (G)* and *Jump-edges (J)* [2],[3],[7]. Precise information about certain geometrical features can be derived by assigning quantitative attributes (position, orientation, etc.) to each primitive.
The structural description obtained from this transformation process is a geometrical primitive string, describing the TVP's shape from left to right. The primitive string derived from the scene presented in Fig. 6b is "Line, Gap, Line, Jump, Line", or in short notation "LGLJL".

TVPs as obtained from the 2-D projection algorithm and their structural description provide highly reliable information even with complex or semi-structured scenes. The described processing scheme may be applied to any 3-D range image independent of a physical sensor device.

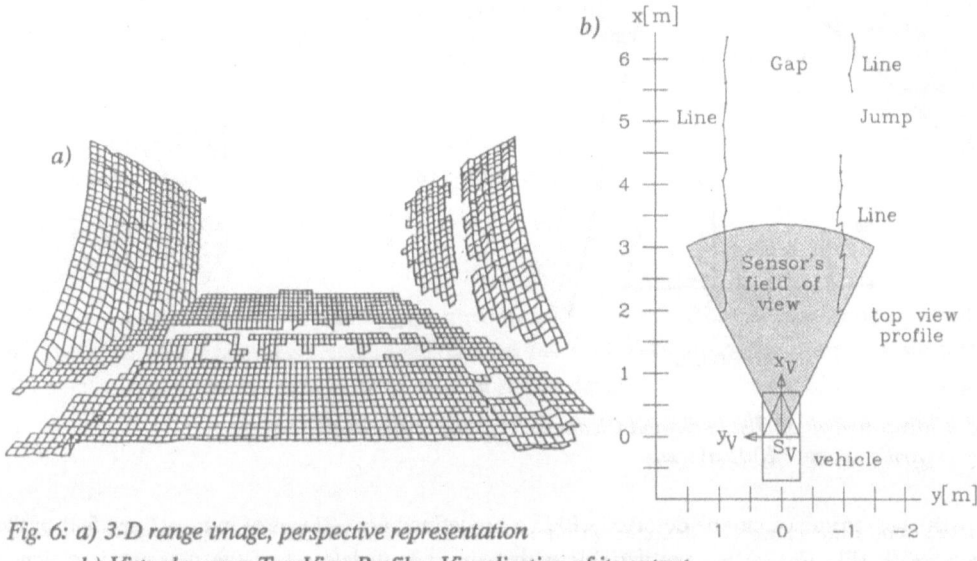

Fig. 6: a) 3-D range image, perspective representation
b) Virtual sensor: Top View Profile - Visualisation of its output

The experiment of following a path's center line as described in the next section demonstrates the application of both virtual sensors for sensor-assisted real-time navigation.

4. TOP VIEW PROFILES SUPPORTING SENSOR-ASSISTED NAVIGATION

A basic maneuver within indoor environments is to navigate MACROBE in the center of a path. Sensor-assistance, by means of calculating the actual deviation in heading and clearance from the vehicle's actual position to the path's center line (Fig. 7a), is realized by a repetitive processing scheme by means of the following five virtual sensors:

- Generation of actual TVP (virtual sensor 1)

- Transformation of actual TVP into a structural description (virtual sensor 2)

- Extracting left and right boundary (primitive *Line*) with respect to free motion space (primitive *Gap*) by means of least-square error line fitting and
  Computation of the actual deviation in clearance (C_l, C_r) and heading (φ_l, φ_r) relative to the left and right boundary (virtual sensor 3 and 4)

- Computation of the deviation in clearance and heading (C, φ) of the vehicle's actual position to the center line (virtual sensor 5)

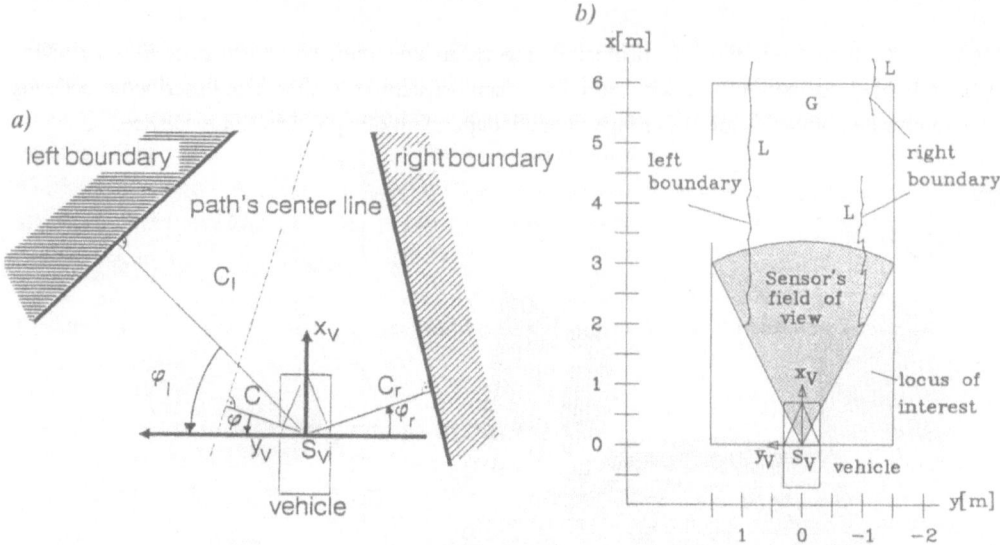

Fig. 7: Measurement of Heading φ and Clearance C to the center line
  a) Test scene   b) Real scene

If path's center line cannot be detected within a predefined area ("locus of interest", see 5.2) of the actual TVP (Fig. 7b), a boolean variable within the output data set of the network is assigned FALSE. Output data C and φ (Fig. 7a) are not updated until redetection of the center line. Then, the variable is set TRUE, indicating that output data is valid.

To achieve sufficient robustness with the extraction of the center line even within poorly structured environments, sequences of TVPs are evaluated by means of a median 5 filter.

These virtual sensors were implemented within the multisensor system based on the network shown in Fig. 8.

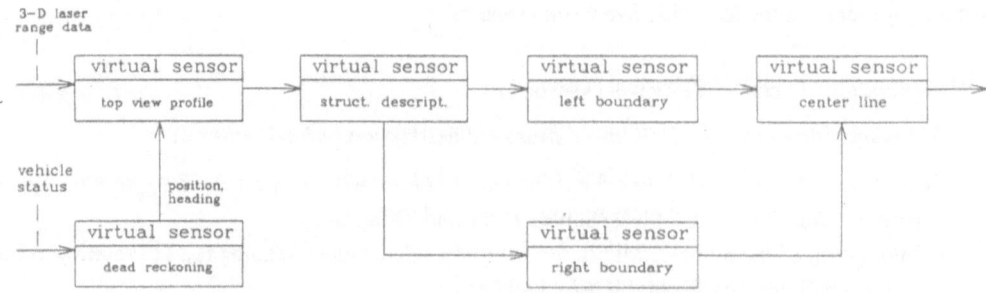

Fig. 8: Virtual sensor network within the multisensor system for the processing scheme

This algorithm was tested during several indoor experiments [7],[8],[9] in our laboratory (Fig. 6) and in a semi-structured factory environment [10]. With respect to the computation time of the sensor data processing network (200 ms) deviations from vehicle's actual position to center line of the path are limited to -2cm < C < 2cm for clearance and -2° < φ < 2° for heading (Fig. 7). Due to these results, real-time capability of the processing scheme is guaranteed with sensor-assisted navigation of MACROBE up to 1 m/s.

To increase accuracy and robustness with acquisition of information about the environment, sensor data of multiple physical sensor devices are fused.

5. FUSION OF SENSOR DATA

In order to compensate uncertainties with measurement of the 3-D laser range camera (e.g. thin bars, reflecting surfaces, glass) laser range data are fused with polar profiles of the multisonar system [11],[12],[13]. To fuse data of both physical sensor devices, fields of view, tilt, and pan angles of the sensors need to be modeled.

5.1 Modelling of sensor field of view
To achieve equivalent shapes for the field of view of the 3-D laser range camera and the 2-D multisonar system, laser range images are transformed into 2-D top view profiles which are comparable to the range profiles of the multisonar system.

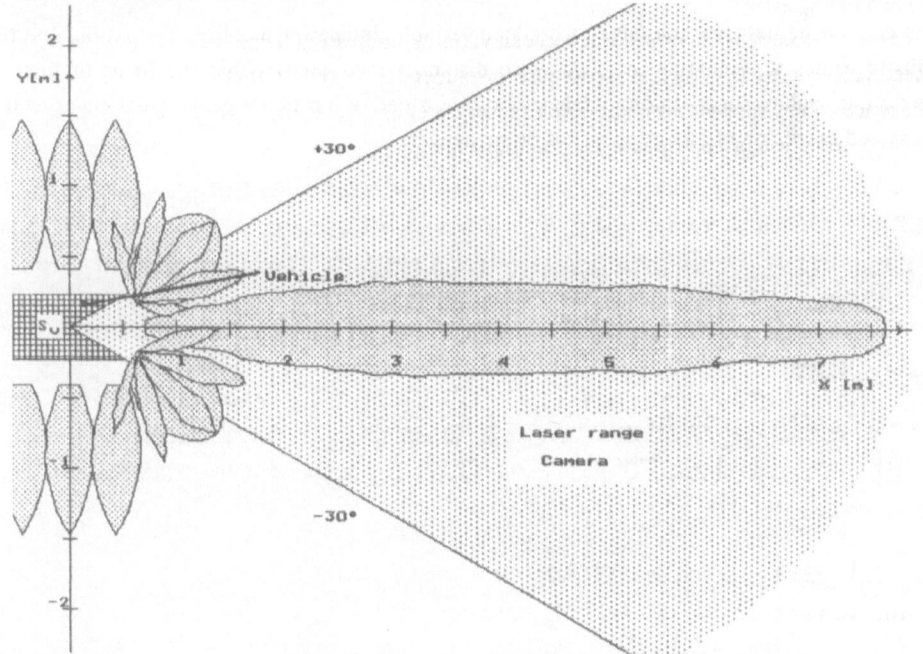

Fig. 9: Field of view: Top view profiles of laser range camera and multisonar system

Transformation of top view profiles and range profiles of the multisonar system into the vehicle coordinate system S_v is shown in Fig. 9.

As a result three different areas can be distinguished: overlapping areas, indicating competitive behaviour of laser range camera and multisonar system; excluding areas, indicating complementary behaviour and areas which cannot be sensed by any physical sensor device from a fixed sensor position.

5.2 Fusion of laser range and sonar data for guarded motion

Planning and execution of a guarded motion requires detection of potential collision situations within a certain region in front of a vehicle ("locus of interest"). The shape of the locus of interest (Fig. 7b) is computed offline and depends on the mobile robot's instantaneous speed and steering angle. Collision detection is performed by

- transforming both data sets into the vehicle coordinate system S_v,

- checking for collision situations by masking actual TVP and actual sonar range profile with the locus of interest and

- logical OR combination of detected objects within the locus of interest, based on a worst case assumption.

The processing scheme is implemented in the multisensor system by means of a virtual sensor network, shown in Fig. 10.

Output data of the network consists of a boolean variable, indicating a collision situation, and real coordinate values representing the worst-case distances to objects within the locus of interest. These distance values support nominal deceleration instead of hard emergency breaking as well as planning and execution of an obstacle evasion maneuver.

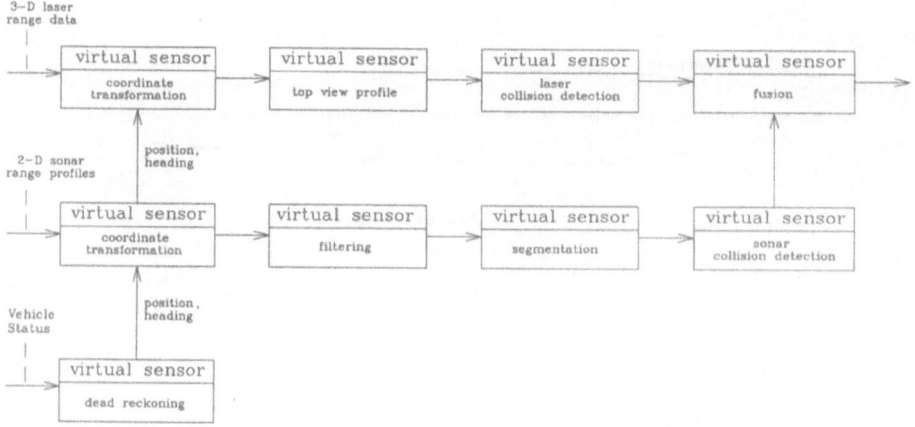

Fig. 10: Collision detection by means of TVPs and sonar range profiles

Experiments with MACROBE demonstrate that fusion of laser and sonar data lead to a highly reliable detection of potential collision situations, independent of the properties of typical indoor environments (e.g. reflectance characteristics of object surfaces, minimum object heights).

6. CONCLUSIONS

The multisensor system includes various physical sensor devices and a network of virtual sensors. Our approach for system implementation leads to a clear and expandable structure which is capable of multi-processing operations. Experiments have demonstrated the real-time capability of the multisensor system. For highly reliable collision detection, a simple form of sensor data fusion is based on a worst case assumption with distance measurements coming from a 3-D laser range camera and a multisonar system.

Accuracy and reliability with landmark recognition will be increased by means of model-based fusion. For this purpose, detailed models concerning physical sensor devices and the environment have to be developed. Additional information from a newly developed 3-D microwave imaging device [5] and from interpreted CCD-images [6] will also be incorporated in future. Another aspect of our research will be the detection of moving objects.

Validation of theoretical approaches through experiments with MACROBE will remain an essential element of our research work.

REFERENCES

[1]   Henderson, T.C.; Fai, W.S.; Hansen, C..: "MKS: A Multisensor Kernel System". IEEE Transactions on Systems, Man and Cybernetics, Vol. SMC-14, No. 5, (1984), pp. 784-791

[2]   Schmidt, G.; Karl, G.: "A 3-D Laser range camera for mobile robot motion control". Proceedings 1988 IEEE Internat. Workshop on Intellig. Robots and Systems (IROS'88), Toward the next generation of Robot and System. Tokyo, Science University of Tokyo, New York, IEEE 345, (1988), pp. 605-610

[3]   Karl, G.: "Eine 3-D Laserentfernungskamera zur Bewegungsführung mobiler Roboter". Dissertation am Lehrstuhl für Steuerungs- und Regelungstechnik, Technische Universität München, (1990)

[4]   Fröhlich Ch.; Freyberger, F.; Schmidt, G.: " A 3-D Laser Range Camera for Sensing the Environment of a Mobile Robot", Internat. Work. on EUROSENSORS IV, Sensoren, Technologie und Anwendung, Sensors and Actuators, Vol. 26, No. 1/3, (1991), pp. 453-458

[5]   Lange, M.; Detlefsen, J.: "94 GHz 3-D-Imaging Radar for Sensorbased Locomotion". Proc. Internat. Microwave Symp. IEEE MTT-S, Long Beach, (1989), pp. 1091-1094

[6]   Levi, P.; Munkelt, O.; Radig, B.; Sattler, R.: "Applications of Image Processing and Image Interpretation in manufacturing environments". Proc. of Internat. Workshop on Information Processing in Autonomous Mobile Robots, München, (1991), pp. 35 - 44

[7]   Fröhlich, Ch.; Karl, G.; Schmidt, G.: " Sensor Assisted Navigation of a Mobile Robot in Building Environments". Proc. of Internat. Workshop on SIFIR'89, Zaragoza, Spain, (1990), pp. 397-399

[8]   Fröhlich Ch.; Karl, G.; Schmidt, G.: "Wandfolgen als Grundfunktion autonom mobiler Roboter". 5. Fachgespräch "Autonome Mobile Systeme". Hrsg. P. Levi; U. Rembold. München, Bayerisches Forschungszentrum für wissensbasierte Systeme, (1989), pp. 1-10

[9]   Kampmann, P.; Schmidt, G.: "Mapping of Indoor Environments for Robot Navigation Based on 3D Range Images". Proc. IEEE Internat. Conf. on Rob. and Autom., Sacramento, (1991)

[10]   Schmidt, G.: "Towards Integration of Autonomous Subsystems for Assembly and Mobility into Flexible Manufacturing". Proc. of Internat. Workshop on Information Processing in Autonomous Mobile Robots, München, (1991), pp. 3 - 20

[11]   Cheng, V.; Sridhar, B.: "Integration of Active and Passive Sensors for Obstacle Avoidance". IEEE Control Systems Magazine, June (1990), pp.

[12]   Elfes, A.: "A stochastic spatial representation for Active Robot Perception". Proc. on the sixth Conference on Uncertainty and AI, AAAI, Cambridge, MA, July, (1990)

[13]   Hackett, J.K.; Shah, M.: "Multi-Sensor Fusion: A Perspective", IEEE Intern. Workshop on Intelligent Robots and Systems, Tsuchira, (1990), pp 1324-1330

[14]   Hinkel, R.; Weidmann, M.: "Real-time data processing in a laser radar". IARP, 1st Workshop on Multi-sensor Fusion and Environment Modelling, Toulouse, France, (1989)

[15]   Martin, W.N.; Aggarwal, j.K.: "Motion Understanding: Robot and Human Vision". Kluwer Internat. Series in Eng. and Comp. Science, Kluwer Academic Publ., Boston, MA, (1988)

[16]   Durrant-Whyte, H.F.: "Integration, Coordination and Control of Multi-Sensor Robot Systems". Kluwer Internat. Series in Eng. and Computer Science, Kluwer Academic Publ., Boston, MA, (1988)

[17]   Hebert, M.; Kanade, T.; Kweon, S.I.: "3-D Vision Techniques for Autonomous Vehicles". Analysis and Interpretation of Range Images, Springer Verlag, Berlin, (1990)

[18]   Thorpe, C; Kanade, T.: "First Year End Report for Perception for Outdoor Navigation". Techn. Report, Carnegie Mellon University, Pittsburgh, CMU-RI-TR-90-23, (1990)

[19]   Moravec, H.P.: "Sensor Fusion in Certainty Grids for Mobile Robots". AI-Magazine, Summer, (1988), pp. 61 - 74

Dynamic Control of Robot Perception Using Stochastic Spatial Models

A. Elfes

Abstract

Robot perception has traditionally been addressed as a passive and incidental activity, rather than an active and task-directed activity. Consequently, although sensor systems are essential to provide the information required by the decision-making and actuation components of a robot system, no explicit planning and control of the sensory activities of the robot is performed. This has lead to the development of sensor modules that are either excessively specialized, or inefficient and unfocused in their informational output. In this paper, we develop strategies for the dynamic control of robot perception, using stochastic sensor and spatial models to explicitly plan and control the sensory activities of an autonomous mobile robot, and to dynamically servo the robot and its sensors to acquire the information necessary for successful execution of robot tasks. We discuss the explicit characterization of robot task-specific information requirements, the use of information-theoretic measures to model the extent, accuracy and complexity of the robot's world model, and the representation of inferences about the robot's environment using the Inference Grid, a multi-property tesselated random field model. We describe the use of stochastic sensor models to determine the utility of sensory actions, and to compute the loci of observation of relevant information. These models allow the development of various perception control strategies, including attention control and focussing, perceptual responsiveness to varying spatial complexity, and control of multi-goal perceptual activities. We illustrate these methodologies using an autonomous multi-sensor mobile robot, and show the application of dynamic perception strategies to active exploration and multi-objective motion planning.

1 Introduction

Traditionally, most approaches to robot perception have cast the sensing and perceptual activities of a robot system in what is fundamentally a passive rôle: although the robot planning and action stages depend fundamentally on sensor-derived information, no explicit planning and control of the perceptual activities themselves is performed. Rather, the robot's sensing subsystems acquire data that incidentally happens to be available at the moment of sensory observation during the perception/planning/action cycle. No explicit connection is made between the sensor data acquired and the information required by the robot for the successful execution of a given task (Fig. 1). Robot systems that embody this *passive perception* approach tend to fall into two categories: those where the sensory subsystems are highly specialized and "hardwired" to the feedback control loop that handles the execution of a specific class of robot tasks; and those that embody "general-purpose" sensor understanding and world-modelling subsystems. Typically, the former category cannot reconfigure its sensory subsystems to adapt to different classes of tasks or to accomodate sensor degradation; while the latter category is populated by notoriously large and ineffective systems that are unfocused in their informational output, have slow reaction times, and are inefficient in the use of their sensory and computational resources. Because there is no explicit characterization of the

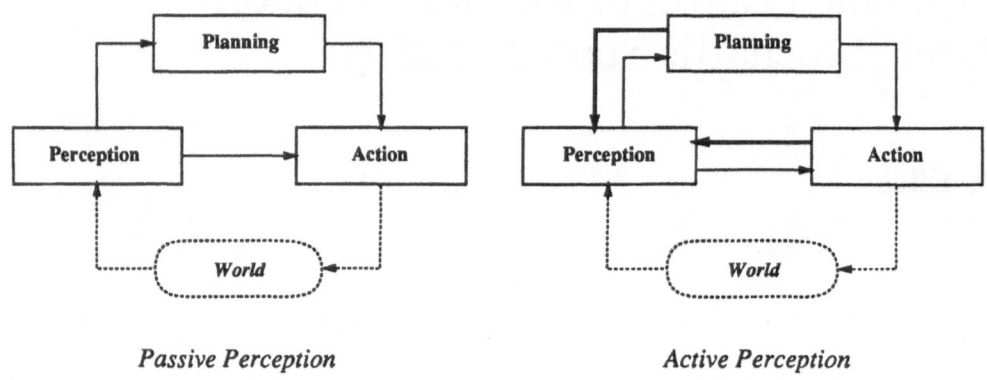

Passive Perception   *Active Perception*

Figure 1: Passive and Active Control of Robot Perception.

information really needed, over- or undersampling of the data may occur, and data irrelevant to the task at hand may be acquired.

More recently, researchers have identified the limitations intrinsic to the formulation of robot perception as a passive activity, and have argued for viewing it as an active process (Fig. 1). Bajcsy [2] discusses the need for *active sensing*, defined as the application of modelling and control strategies to the various layers of sensory processing required by a robot system. Elfes has developed an approach for *active mapping* using the Occupancy Grid framework [12, 14, 10]; this framework uses estimation-theoretic and Markov Random Field models to compose information from multiple sensors and multiple points of view, thereby addressing in a robust way the underconstrainedness of the sensor data. Aloimonos suggests that the composition of information from multiple views can be used to handle several ill-posed problems in Computer Vision [1]. Related research includes the development of recursive estimation procedures for robot perception [20, 15, 18]; methods for *active sensor control*, where specific parameters of the sensor system (such as camera aperture, focal distance, or sensor placement) can be changed under computer control [19, 25]; development of theoretical foundations for coordination, integration and control of sensor systems [8, 17, 16]; and generation of optimal and adaptive sensing strategies [7, 6, 25].

In this paper, we discuss *Dynamic Perception*, a framework for dynamic and adaptive planning and control of robot perception in response to the information needs of an autonomous robot system as it executes a given mission. The Dynamic Perception framework uses stochastic and information-theoretic sensor and world models to identify the information needs of the robot, plan the acquisition of task-specific data, dynamically servo the robot and its sensors to acquire the information needed for successful execution of the task, update the information acquisition goals as the robot progresses through sequences of tasks, and integrate the sensory tasks with the mission-specific tasks to be performed by the robot. We describe the use of a multi-property tesselated random field model, the Inference Grid, to encode inferences about the robot's environment. We illustrate the explicit characterization of robot task-specific information requirements, and the use of information-theoretic measures to model the extent, accuracy and complexity of the robot's world model. We also discuss the use of stochastic sensor models to evaluate the utility of sensory actions, and to compute the loci of observation of relevant information. We illustrate our approach using an autonomous multi-sensor mobile robot, and show the application of dynamic perception strategies to *active exploration* and *integrated motion planning*, combining perception and navigation goals. A more extensive discussion, with further experimental results, can be found in [13].

2 Dynamic Robot Perception

The work presented here is part of a long-term research effort that addresses the development of *agile* and *robust* autonomous robot systems, able to execute complex, multi-phase missions in unknown and unstructured real-world environments, while displaying real-time performance [10, 13]. For planning,

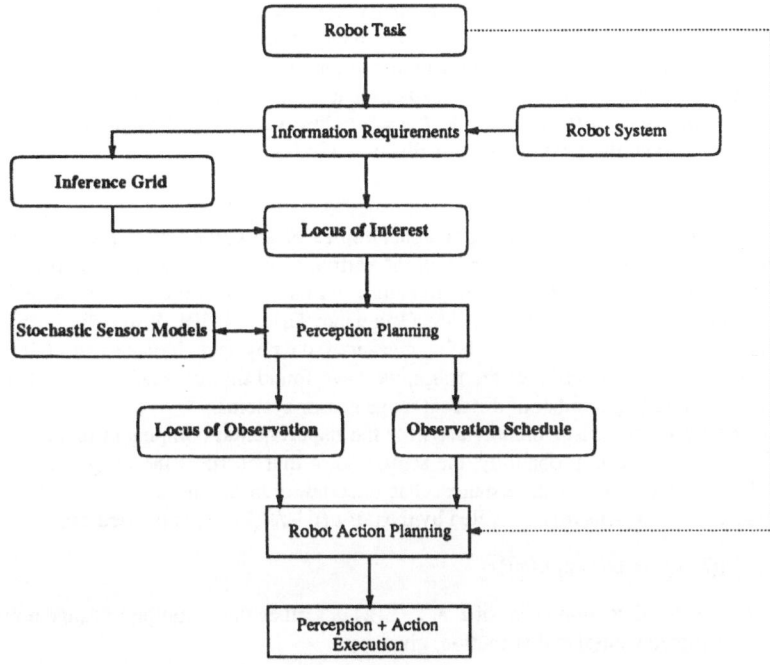

Figure 2: Components of the Dynamic Robot Perception Framework.

scheduling, execution and monitoring purposes, a specific robot *mission* is separated into *phases*, which are in turn decomposed into sets of *tasks*. These tasks can be scheduled in parallel and/or sequentially, depending on the nature of the component activities. As tasks are scheduled for detailed planning and execution by the robot, the information requirements posed by these tasks are used to dynamically update the perceptual goals of the robot. By maintaining explicit models of the task goals and the perceptual goals, the Dynamic Perception framework allows a closer integration between task planning and perception planning, and between task execution and perception execution. Using descriptions of the information required for successful execution of a specific task, appropriate sensors are selected, sensing strategies are formulated, and the robot's locomotion and sensing activities are planned so as to maximize the recovery of relevant information and accomplish the robot's task. Overall, the system's behaviour can be described as *servoing on required information*, as well as on the robotic task itself. We note that in control-theoretic terms this can be phrased as a problem in *dual control* [4, 13].

Some of the issues that have to be addressed in the context of the Dynamic Perception framework include *Sensor Modelling, Information Modelling, Perception Planning*, and the *Integration of Perception and Action*. This will allow us to enable an autonomous mobile robot with a number of important capabilities, including *attention control*, or the ability to efficiently acquire and process relevant data; *attention focussing* from larger sensing areas to smaller regions of specific interest; and development of *optimal information acquisition* and *exploration strategies*.

3 Components of the Dynamic Perception Framework

We have chosen to address the concerns outlined above through the development of estimation- and information-theoretic models. Contrary to the *ad hoc* AI-based methods still used in much of Computer Vision work, estimation-theoretic models have a long history of success and have been widely applied in signal processing and control tasks [4]. In this section, we describe some of the components of the Dynamic

Perception framework. In particular, we discuss the *stochastic sensor models* developed to describe the robot sensors; the use of a stochastic multi-property tesselated spatial representation, the *Inference Grid*, to encode inferences about the robot's world; the development of various information metrics to measure the robot's knowledge of the environment, and of complexity measures to determine the spatial variability of the environment; the use of *mutual information* to measure the utility of sensory actions; and the computation of the *Locus of Interest* and the *Locus of Observation* of relevant information [13]. The flow of computation between these components is illustrated in Fig. 2.

3.1 Stochastic Sensor Models

To enable us to reason about the perceptual capabilities of an autonomous robot system, mathematical models are needed to describe the behaviour of the various sensors of the robot. In previous work, we have discussed sensor models that are stochastic in nature and have shown their use in the recursive estimation of a tesselated spatial random field model, the Occupancy Grid [14, 10]. While the specific sensor models developed will depend on the properties being measured, the physical characteristics of the sensor systems and the environment of operation of the robot, we have found that the class of models briefly described below can be tailored to a number of different range sensor systems.

The stochastic range sensor models used for the experimental component of our research take into account the target detection probability, the sensor noise that corrupts the range measurements, and the variation of range uncertainty with distance. The uncertainty in the measurement r of a detected target T positioned at z from the sensor is modelled by the pdf $p(r \mid det(T, z))$ expressed as:

$$p(r \mid det(T, z)) = \tilde{G}(r, z, \Sigma(z)) \tag{1}$$

where $det(T, z)$ is the detection event of T at z, $\Sigma(z)$ describes the variation of range noise with distance, and $\tilde{G}()$ is a corrupted Gaussian distribution, given by

$$\tilde{G}() = (1 - \epsilon) G_1() + \epsilon G_2() \tag{2}$$

The distributions $G_1()$ and $G_2()$ are Gaussian, with $G_1()$ modelling the normal behaviour of the sensor, and $G_2()$ occasional gross errors. The parameter ϵ weighs the relative contributions of both terms.

To model the sensor detection behaviour, we use the target detection probability, $P[det(T, z) \mid \exists\, T \text{ at z}](z) = P_d(z)$, and the false alarm probability, $P[det(T, z) \mid \nexists\, T \text{ at z}](z) = P_f(z)$ (known as a Type I error [9]). The missed detection probability (Type II error) is of course given by $P[\neg det(T, z) \mid \exists\, T \text{ at z}](z) = 1 - P_d(z)$.

The target localization probability, $p(\exists\, T \text{ at z} \mid r)$, can be computed using Bayes estimation as:

$$p(\exists\, T \text{ at z} \mid r) = \frac{p(r \mid \exists\, T \text{ at z})\, P[\exists\, T \text{ at z}]}{p(r \mid \exists\, T \text{ at z})\, P[\exists\, T \text{ at z}] + p(r \mid \nexists\, T \text{ at z})\, (1 - P[\exists\, T \text{ at z}])} \tag{3}$$

where

$$p(r \mid \exists\, T \text{ at z}) = p(r \mid det(T, z))\, P_d(z) + p(r \mid \neg det(T, z))\, (1 - P_d(z)) \tag{4}$$

and

$$p(r \mid \nexists\, T \text{ at z}) = p(r \mid det(T, z))\, P_f(z) + p(r \mid \neg det(T, z))\, (1 - P_f(z)) \tag{5}$$

This class of stochastic sensor models can be applied to a large variety of range sensors, including infrared and laser scanners, sonar sensors, and stereo systems. The required parameters can be obtained through analysis of the physical characteristics of the sensor and through calibration (see, for example, [10, 20, 19, 21]). A typical set of curves showing the dependency of the detection probability and of the range variance on the distance to the target being imaged is given in Fig. 3, while Fig. 4 shows a plot of $p(r \mid \exists\, T \text{ at z})$. These stochastic sensor models are used in perception planning to determine the *Locus of Observation*, from which the robot can acquire spatial information of relevance to specific perceptual goals.

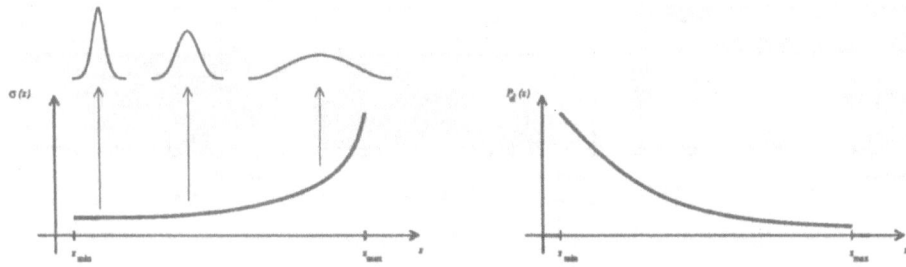

Figure 3: Sensor Range Variance and Detection Probability as a Function of Distance to the Target. These are typical curves for range estimation only.

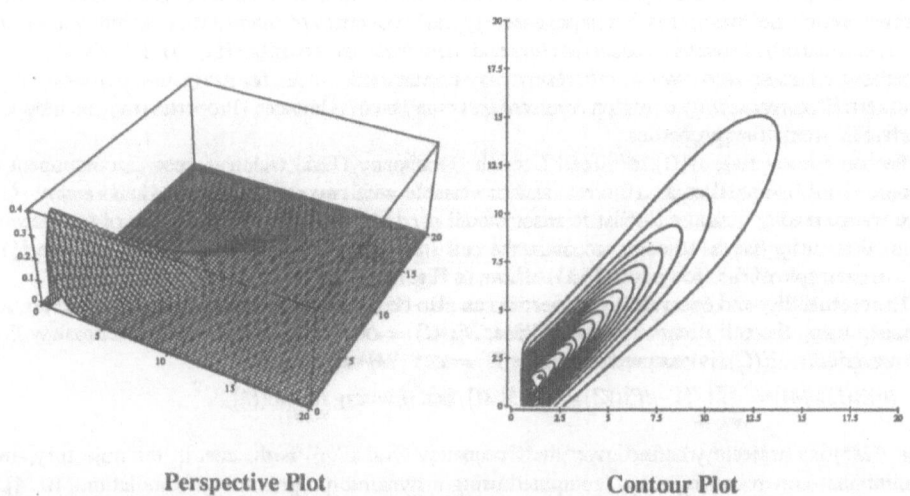

Perspective Plot   Contour Plot

Figure 4: Stochastic Range Sensor Model. The function $p(r \mid \exists T \text{ at } z)$ for range estimation only is shown, in both a perspective plot and a contour plot.

3.2 Inference Grids

In previous work, we have developed an approach to spatial robot perception and navigation called the *Occupancy Grid* framework [12, 14, 10]. The Occupancy Grid is a multi-dimensional discrete random field model that maintains probabilistic estimates of the occupancy state of each cell in a spatial lattice. Recursive Bayesian estimation mechanisms employing stochastic sensor models allow incremental updating of the Occupancy Grid using multi-view/multi-sensor data, as well as composition of multiple maps, decision-making, and incorporation of robot and sensor position uncertainty. The Occupancy Grid framework provides a unified approach to a variety of problems in the mobile robot domain, including autonomous mapping and navigation, sensor integration, path planning under uncertainty, motion estimation, handling of robot position uncertainty, and multi-level map generation. It has been successfully tested on several mobile robots, operating in real-time in real-world indoor and outdoor environments [10].

For the Dynamic Perception work we have generalized the Occupancy Grid representation, and have

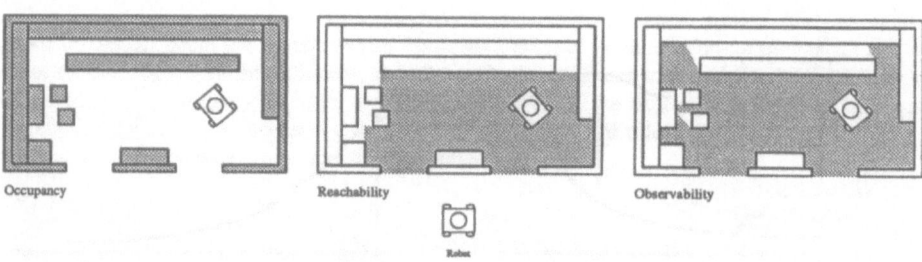

Occupancy   Reachability   Observability

Robot

Figure 5: Some Properties of the Inference Grid: Occupancy, Reachability, Observability. The shaded areas of each map indicate regions that are occupied, reachable by the robot, and observable by the robot.

developed a spatial model called the *Inference Grid*. The Inference Grid is a multi-property Markov Random Field defined over a discrete spatial lattice. By associating a random vector with each lattice cell, the Inference Grid allows the representation and estimation of multiple spatially distributed properties. For robot perception and spatial reasoning purposes, typical properties of interest may include the *occupancy* state of a lattice cell, as well as its *observability* and *reachability* by the robot (Fig. 5). For visual perception, properties such as surface *color* or *reflectance* may be estimated, while for navigation purposes properties such as terrain *traversability* or region *connectedness* may be of relevance. Properties may be independent, or derivable from other properties.

The occupancy state, $s(C)$, of a cell C of the Occupancy Grid (which is now a component of the Inference Grid) is modelled as a discrete random variable with two states, *occupied* and *empty*. Given a sensor range reading r, and a stochastic sensor model $p(r \mid z)$, the recursive estimation of the Occupancy Grid is done using Bayes' theorem to obtain the cell state probabilities $P[s(C) = \text{OCC} \mid r]$ (see [14, 10, 11]). An example of the Occupancy Grid is shown in Fig. 6.

The reachability and observability properties can also be treated as binary stochastic variables, and are estimated using the cell occupancy probabilities, $P[s(C) = \text{OCC} \mid M]$, stored in the Occupancy Grid M. Cell *reachability*, $\Xi(C)$, is computed using $P[s(C) = \text{OCC} \mid M]$ as:

$$P[\Xi(C) \mid M] = \prod_{\forall Z \in \eta} (1 - P[s(Z) = \text{OCC} \mid M]) \text{ s.t. } \eta = \arg \min_{\forall \xi(M)} \Gamma(\xi) \tag{6}$$

where $\xi(M)$ is a trajectory defined over the Occupancy Grid, $\Gamma(\xi)$ is the cost of the trajectory, and η is the minimum-cost robot trajectory, computed using a dynamic programming formulation [10, 4]. Cell *observability*, $\Psi(C)$, is estimated using $P[\Xi(C) \mid M]$ and the sensor model of Eq. 3 as:

$$P[\Psi(C) \mid M] = \max_{\forall Z \in M} P[\det(C) \mid s(C) = \text{OCC} \wedge \pi(R) = Z] P[\Xi(Z) \mid M] \tag{7}$$

where $\pi(R)$ denotes the position and orientation of the robot.

3.3 Information Measures

To guide the perceptual activities of a robot, we need metrics to evaluate the robot's world knowledge. The specific metric used depends on the robot task and the particular kind of information required for successful execution of the task. For tasks that involve spatial reasoning and navigation, we require measures of the extent and accuracy of the robot's sensory maps, while for target localization or shape recovery, precise surface position information is needed.

Map Uncertainty

For the Inference Grid model discussed above, an intuitive way of expressing the uncertainty in the spatial information encoded in cell occupancy estimates obtained from the sensor data is given by the cell uncertainty function [10]:

$$U(C) = 1 - 4 \left(P[s(C) = \text{OCC} \mid M] - \frac{1}{2} \right)^2 \tag{8}$$

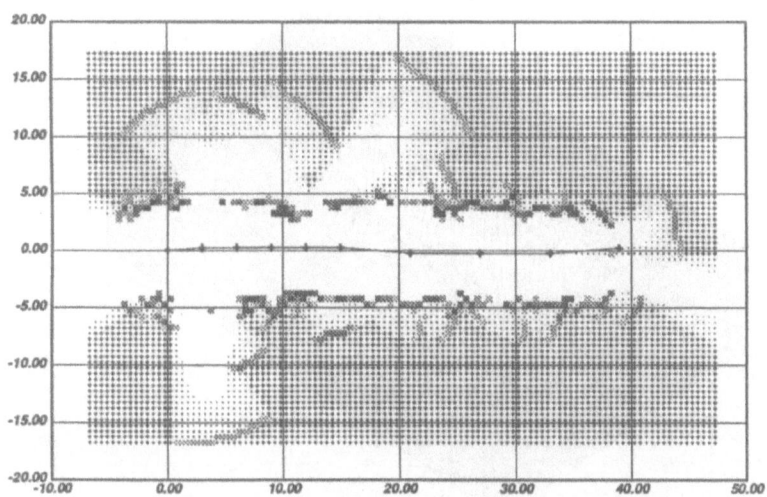

Figure 6: An Example of the Occupancy Grid. The map shows an Occupancy Grid built by a mobile robot using a sonar range sensor array. The robot is moving along a corridor, from left to the right. The gray cells correspond to high occupancy probability areas, while the areas marked with "·" correspond to cells with high emptyness probability. Areas not observed by the robot are identified using "+".

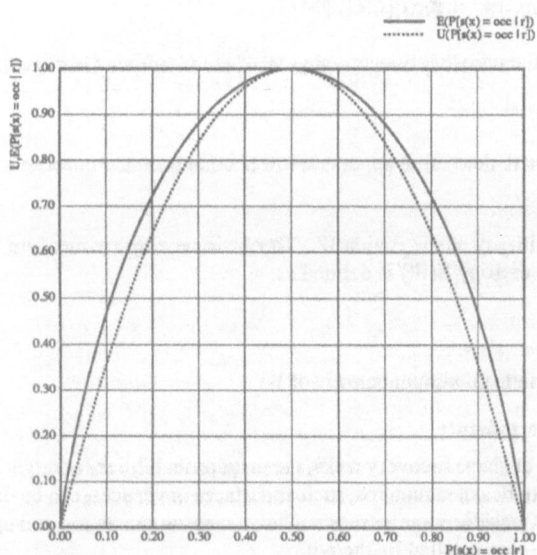

Figure 7: Occupancy Grid Cell Uncertainty Measures. The cell uncertainty function $U(C)$ and the cell entropy function $E(C)$ are shown.

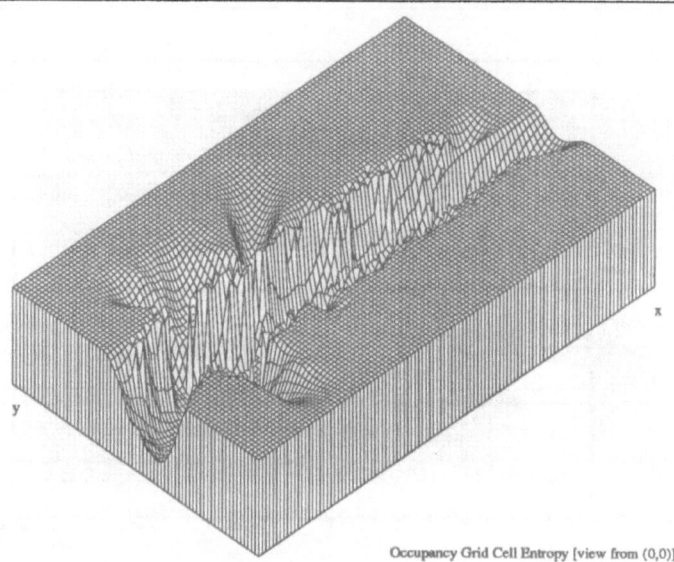

Occupancy Grid Cell Entropy [view from (0,0)]

Figure 8: Entropy of the Occupancy Grid. The cell entropies for the Occupancy Grid of Fig. 6 are shown.

A more generally used metric of uncertainty is the entropy of a random variable [5, 24]. The Inference Grid cell uncertainty can be measured using the entropy of the cell occupancy state estimate as (Figs. 7 and 8):

$$E(C) = -\sum_{s_i} P[s_i(C) \mid M] \log P[s_i(C) \mid M] \tag{9}$$

Using the cell entropy, the uncertainty over a region W of the Inference Grid can be computed as:

$$E(W) = \sum_{\forall\, C_i \in W} E(C_i) \tag{10}$$

This definition allows us to determine upper and lower bounds on the uncertainty of the region W:

$$0 \le E(W) \le \#(W) \tag{11}$$

where $\#(W)$ is the cardinality of the region W. To obtain an entropy measure that is independent of the region size, the average entropy $\overline{E}(W)$ is defined as:

$$\overline{E}(W) = \frac{E(W)}{E_0(W)} \tag{12}$$

where $E_0(W) = \#(W)$ is the maximum entropy of W.

Target Localization Uncertainty

For precise localization or shape recovery tasks, the error probabilities for target detection and the variance in the position estimate of detected features, such as surfaces or vertices, can be directly used as quantitative uncertainty measures. Consider a range sensor whose measurements are corrupted by Gaussian noise of zero mean and variance σ^2, modelled by the pdf:

$$p(r \mid z) = \frac{1}{\sqrt{2\pi}\sigma} \exp\left(\frac{-(r-z)^2}{2\sigma^2}\right) \tag{13}$$

The uncertainty in surface localization is given directly by σ, and the Type II error probability $1 - P_d(z)$ provides a direct measure of detection uncertainty.

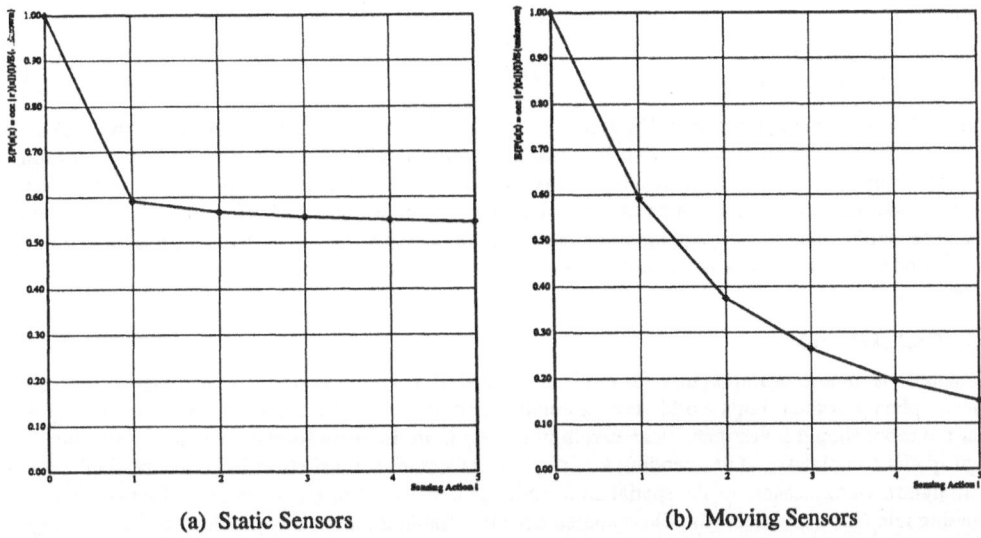

(a) Static Sensors (b) Moving Sensors

Figure 9: Occupancy Grid Entropy Change With Sequential Sensing Actions. Graphs (a) and (b) both show the change in OG uncertainty, measured as the average entropy $\overline{E}(M)$, as range sensor data is acquired sequentially and used to update the Occupancy Grid M. For graph (a), the robot and its sensors are static, and data is being acquired from a single view. In graph (b), the robot and its sensors are moving, and data is being acquired from multiple views. For this case, it can be seen that the uncertainty decay rate is faster, and that the asymptotic uncertainty limit is lower.

3.4 Information Provided by a Sensing Action

The information $\Delta I(\alpha_k)$ added to the Inference Grid by a sensing action α_k can be defined directly in terms of the decrease in uncertainty caused by that sensing action:

$$\Delta I(\alpha_k) = -(E_k(W) - E_{k-1}(W)) \tag{14}$$

where $E_{k-1}(W)$ is the entropy of a region W of the Occupancy Grid before the sensing action α_k, and E_k is the entropy of W after the sensing action α_k. This measure is known as *Mutual Information* [24].

Eq. 14 serves as an indicator of the efficiency of a specific sensor or sensing action, and can be used to implement "stopping rules" such as those used in statistical analysis [3]. Fig. 9(a) shows how the average entropy $\overline{E}(M)$ of an OG M decreases for the case of a static robot, where multiple scans are performed from a fixed location. In contrast, Fig. 9(b) shows how the average entropy decreases for a mobile robot that is taking sensor scans from multiple locations, as it moves around in the environment. The graphs make explicit and express in quantitative terms what would be intuitively expected, namely, that it is more useful in terms of world knowledge acquisition to integrate data obtained from multiple sensing locations than from a single view. Note that, for non-biased robot sensors, we have $\lim_{k\to\infty} I(\alpha_k) = 0$. In the case of a static robot, the average entropy $\overline{E}_k(W)$ will tend towards $\lim_{k\to\infty} \overline{E}_k(W) = 1 - \#(\omega)/\#(W)$, where $\omega \subset W$ is the region observable by the sensors from the robot's location, and W is the total extent of the region of interest. For the case of a moving robot, the average entropy $\overline{E}_k(W)$ will tend towards $\lim_{k\to\infty} \overline{E}_k(W) = 1 - \#(\Omega)/\#(W)$, where $\Omega \subset W$ is the region observable by the robot as its explores its environment, and W is again the total extent of the region of interest. This behaviour can be observed in Figs. 9(a) and (b).

As we already mentioned, in localization problems the quality of an estimate is usually measured by its variance. If n measurements are taken using a sensor described by the model of Eq. 13, the sample variance of the estimated parameter $\hat{r}$ will be $\sigma_{\hat{r}}^2 = \sigma^2/n$, and the information added by sensing action α_k is given by:

$$-\Delta\sigma_k^2 = \frac{\sigma^2}{k\,(k-1)} \tag{15}$$

It is straightforward to associate utility functions to the information measures mentioned above (Eqs. 14 and 15). Similarly, cost functions can be associated with the effort and risk involved in performing a sensing action [4]. For planning and decision-making purposes, expected values of ΔI_k and $-\Delta\sigma_k^2$, as well as of the sensing action costs, can be used to select optimal single-stage sensory actions, compute limited lookahead multi-stage sensing strategies, and determine stopping or termination conditions for sensor data acquisition [13]. An example of a single-stage optimal sensory action choice for attention control is shown in Fig. 13.

3.5 Spatial Complexity

An additional metric of importance for dynamic control of robot perception is a quantitative measure of the complexity of the robot's world. For exploration and navigation purposes, for example, it is intuitive that the robot should investigate more carefully and do more frequent measurements of areas that have high spatial complexity, while spending less sensory and computational effort in "uninteresting" regions. A straightforward measure of the spatial complexity of a region W of an OG is given by the mean zero-crossing rate (for a given threshold) computed over W. Another measure is provided by interpreting the sensing activity of the robot as a spatial sampling process performed over the robot's environment. Given the tesselated nature of the Inference Grid model and its representational similarity to images [10, 11], a position-dependent Fast Fourier Transform (FFT) $\mathcal{F}_W[M(x,y)]$ can be computed over a finite-size window W of the Occupancy Grid M to obtain the spatial frequency spectrum of a specific region. Appropriate window functions include the rectangular window and the Hamming window [22]. The spatial frequency spectrum gives us a metric of the spatial complexity of the regions being explored by the robot, and allows us to derive an optimal sensing strategy, by performing the sensing actions at spatial intervals determined using the Nyquist (or optimal sampling) rate [22]:

$$\Delta x = \frac{\pi}{\omega_x} \qquad \text{and} \qquad \Delta y = \frac{\pi}{\omega_y} \tag{16}$$

where ω_x and ω_y are the band limits of the spatial frequency spectrum. An example of the use of spatial complexity measures to plan the locomotion of a mobile robot is shown in Fig. 6. Using spatial variability estimates of its immediate surroundings as constraints on the distances between data acquisition stops, the robot is able to respond to the complexity of its environment: it stops more frequently in regions of high spatial variability, such as the two open doors on either side of the left portion of the corridor, and speeds up when the corridor becomes "dull".

4 Strategies for Dynamic Control of Robot Perception

We now turn to the application of the stochastic and information-theoretic models, discussed in the previous sections, in the development of strategies for dynamic control of robot perception. We will discuss the *Locus of Interest* and the *Locus of Observation*, outline methods for attention control and attention focussing, and illustrate the application of these strategies in autonomous robot exploration and in the integrated planning of robot navigation and robot perception.

4.1 Task-Directed Perception

Application scenarios for autonomous mobile robots require the execution of a variety of tasks related to spatial perception, reasoning and navigation, such as motion planning, detection and inspection of spatially distributed features, object recognition and pose determination, grasp planning, etc. In the work discussed here, the connection between robot task and robot perception is done by explicitly mapping the information needs of the task on the Inference Grid. Typical perceptual tasks may include observing a spatial feature with some minimum detection probability, localizing a spatial feature with some bound on the positional uncertainty, or selecting specific regions of interest, so that non-pertinent sensor data can be ignored.

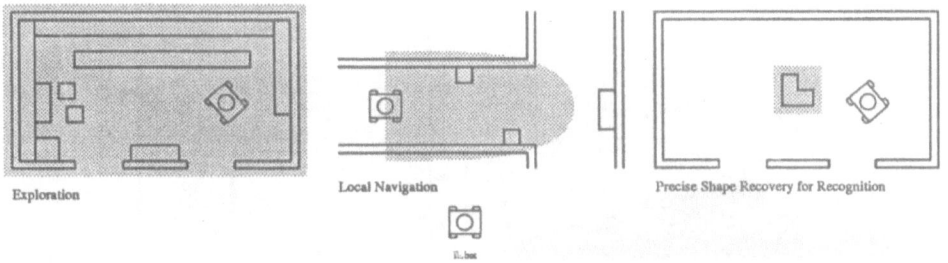

Figure 10: Locus of Interest for Several Robot Tasks. The regions of interest to the robot for *exploration*, *local navigation* and *precise shape recovery* tasks are shown as shaded areas.

Perception Constraints and the Locus of Interest

We define the *Locus of Interest* as a region specified on the Inference Grid that is fundamentally relevant for a specific robot task. It is determined by having the task define a utility function $R(W)$ over a region W of the Inference Grid, which measures the relevance of knowledge about W for successful task execution. Consequently, the Locus of Interest LI defined by task τ_i is computed as the region of the Inference Grid that exceeds a utility threshold u_t:

$$LI(M, \tau_i) = \{\forall C \in M \text{ s. t. } R(C) \geq u_t\} \tag{17}$$

Examples of the Locus of Interest for some specific robot tasks are shown in Fig. 10. In addition to region selection, tasks can impose additional constraints on the information to be acquired. For example, *Detection Constraints* and *Localization Constraints* of the form $det(T, z) \geq D_t$ and $\sigma(z) \leq \sigma_t$ can be defined, where D_t and σ_t are detection and range uncertainty thresholds (see example in Fig. 12). Note that if multiple goals are being pursued by the robot, the corresponding task-defined Loci of Interest and perceptual constraints can be merged and simultaneously represented on the same Inference Grid. It should also be mentioned that the robot system itself may impose LIs derived from robot-specific operational tasks.

The Locus of Observation

After determining the Locus of Interest, we can compute the *Locus of Observation*. The Locus of Observation is the configuration space region where the robot has to position a selected sensor or set of sensors to acquire the information needed by the robot's tasks and specified in the LI. The LO is computed as:

$$LO(M, \Theta[LI(M)]) = \{\forall C \in M \text{ s. t. } \pi(R) \in C \wedge \Theta[LI(M)] \geq \lambda\} \tag{18}$$

where $\Theta[LI(M)]$ is a predicate or utility function defined over the perceptual constraints, and which has to be above a threshold λ to be satisfactory. Fig. 11 shows the limits imposed on the region of operation of the stochastic sensor model of Eq. 4 by detection and range uncertainty constraints. An illustration of the Locus of Observation is given in Fig. 12. Industrially used mobile platforms, such as AGVs or sentry robots, frequently rely on the detection of specific landmarks, such as active beacons radiating on a known signature frequency, to determine the location of the vehicle or select the next segment of the path to be traversed. The shaded areas in Fig. 12 correspond to the Loci of Observation, and indicate the regions from which the beacons can be observed with $det(T, z) \geq D_t$.

4.2 Attention Control and Attention Focussing

In the Dynamic Perception framework, *Attention Control* is performed through the selection of regions in the Inference Grid that have both *high current relevance* as measured by the Locus of Interest and are *observable* as measured in the Inference Grid. An example of Attention Control is given in Fig. 13. The

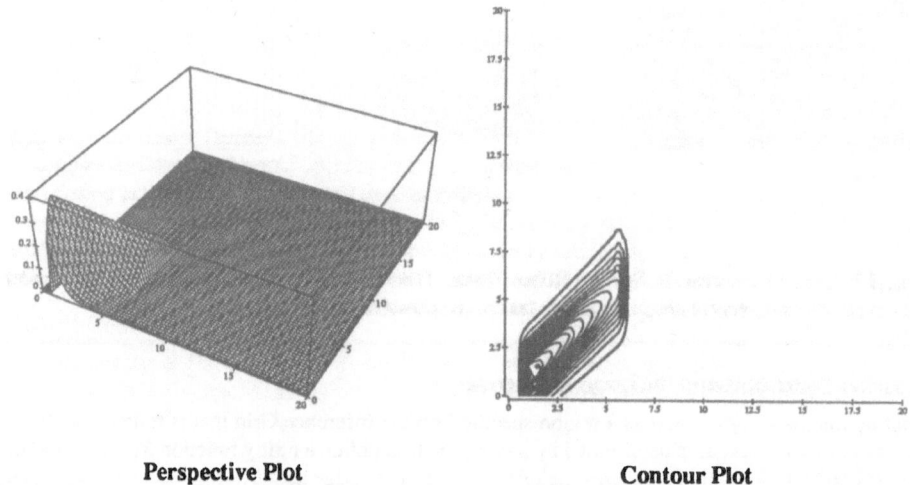

Perspective Plot   **Contour Plot**

Figure 11: Imposing Perceptual Constraints on the Stochastic Sensor Model. The stochastic sensor model of Eq. 4 is shown with the operational limits imposed by the perceptual constraints $det(T, z) \geq D_t$ and $\sigma(z) \leq \sigma_t$.

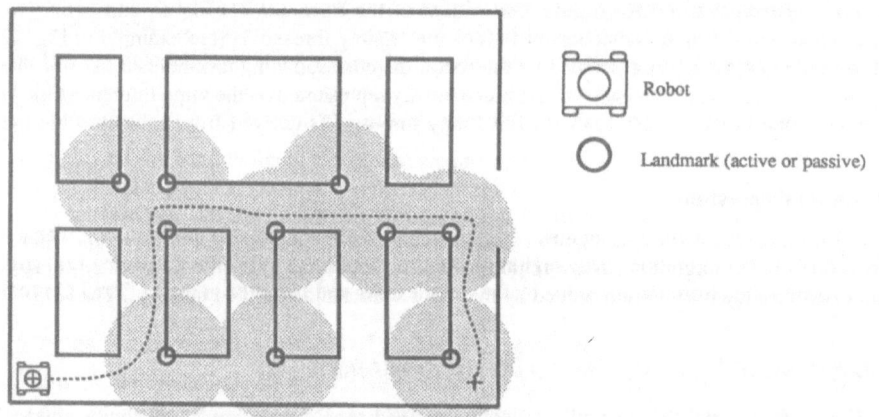

Figure 12: Locus of Observation for Landmark-Based Navigation. An environment instrumented with beacons (active landmarks) is shown. The shaded regions correspond to the Loci of Observation.

task is exploration, and the robot has already partially mapped the region of interest. After a perception planning cycle during which the observability of the cells of the Inference Grid was estimated, four regions were selected for further exploration. These regions correspond to windows that have both high average Occupancy Grid entropy and high average probability of being observable.

Given the uncertainty intrinsic to sensory observations and execution of robot actions, it is generally not possible to compute exact *a priori* measures of observability, sensor information, etc., and many

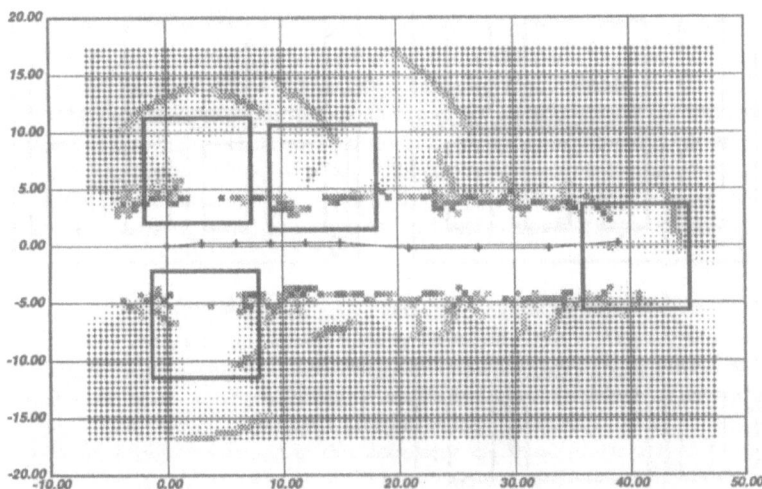

Figure 13: Controlling the Attention of the Robot. The same Occupancy Grid of Fig. 6 is shown. The robot's task is exploration, and the four areas to be investigated next are shown superimposed on the map.

of the measures used for perception planning are estimates or expected values of the actual parameters. Therefore, most scenarios require an iterative refinement approach, since at the beginning of the sensing and world modelling activity the robot will have only an incomplete map and partial information to work with. This situation leads naturally to the need for *Attention Focussing*, where the Locus of Interest covers a large area during the preliminary reconnaissance stage, but is narrowed down to more specific regions as more information becomes available. This is again illustrated in the exploration scenario presented in Fig. 13, where after a general reconnaissance phase the attention of the mapping system is now being narrowed to specific regions to be explored further.

5 Applications

Robot Exploration

We now outline two concrete scenarios. The first is the *exploration scenario*, whose components were already discussed in previous sections. As illustrated in Figs. 6 and 13, the Dynamic Perception framework allows the robot to react to the complexity of its environment, as well as to determine optimal exploration strategies, by reasoning explicitly about what the robot needs to know, what it does know, and what it doesn't know about its environment. A second exploration example is given in Fig. 14. It should be mentioned that the exploration problem we address can be seen as a stochastic generalization of the deterministic *Art Gallery Problem* [23].

Integration of Navigation and Perception

The second scenario involves *integrated perception and navigation planning*. As discussed in section 1, robot perception and robot task planning have generally been treated as separate stages of the robot's cycle of operation. This is clearly seen in the area of robot motion planning. Path planning methods have generally been limited to planning safe trajectories from a given starting point to a given goal, avoiding obstacles and taking into account the kinematic and dynamic characteristics of the robot. Other concerns, such as robot registration constraints, perceptual requirements, or environment complexity, are ignored.

The Dynamic Perception framework allows integrated navigation planning, where both perceptual and locomotion requirements can be taken into account. Simple obstacle-avoidance path planning is performed

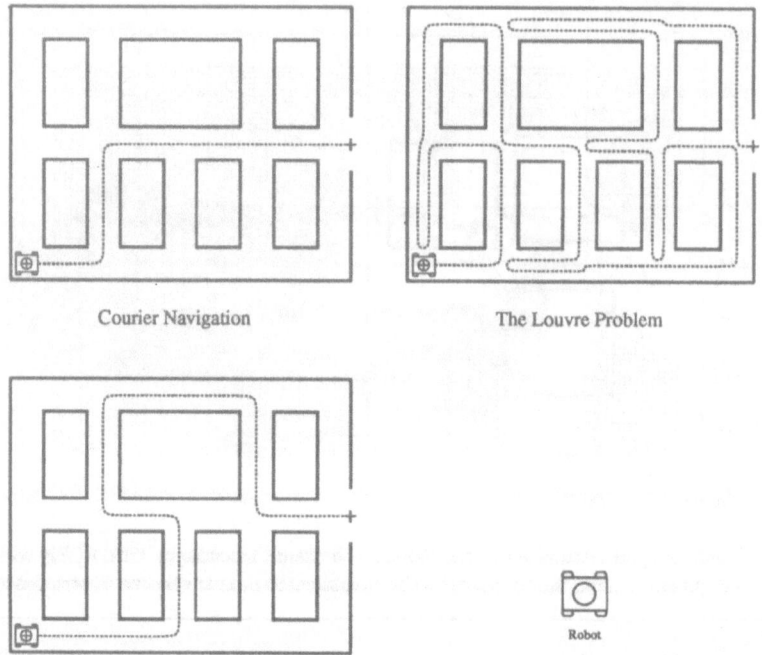

Courier Navigation

The Louvre Problem

Robot

The Tourist in the Louvre Problem

Figure 14: Integration of Perception and Navigation Tasks. The maps show the behaviour of a mobile robot in three cases: 1. Courier navigation tasks, when the goal is finding the fastest route. 2. Exploration tasks, when the primary goal is careful mapping of the robot's area of operation. 3. Integration of courier navigation and exploration, when both activities are given comparable importance, and the robot automatically adjusts its behaviour accordingly.

on the Inference Grid as the minimization of a dual-objective cost function [10]:

$$\min_{\mathbf{P}} f(\mathbf{P}) = w_d \, \text{length}(\mathbf{P}) + w_c \sum_{\forall C \in \mathbf{P}} \Gamma(C) \qquad (19)$$

where $\mathbf{P}$ is the robot path, w_d is the path length weight, w_c is the cell cost weight, and $\Gamma(C) = f_c(P[s(C) = \text{OCC}])$ is the cell traversal cost, defined directly in terms of the Occupancy Grid cell state estimates.

Integrated navigation planning for perceptual and locomotion tasks can be performed on the Inference Grid as the minimization of a multi-objective cost function:

$$\min_{\mathbf{P}} f(\mathbf{P}) = \sum_i w_i \, c_i(M) \qquad (20)$$

where the $c_i(M)$ are cost functions representing various perception and locomotion requirements, computed on the Inference Grid M, and w_i is the weight vector. An example of this approach is shown in Fig. 14, where different behaviours of a mobile robot are obtained by varying the relative importance of two tasks: *exploration*, where extent and accuracy of the resulting map are important, and *courier navigation*, where finding the shortest distance to the goal is essential.

6 Conclusions

The *Dynamic Robot Perception* framework outlined in this paper stresses the active and adaptive control of the perceptual activities of an autonomous robot. This is done by explicitly determining the evolving information requirements of the different tasks being addressed by the robot, and by planning appropriate sensing strategies to recover the information required for successful completion of these tasks. We discussed the development of strategies for dynamic control of robot perception that emphasize the use of stochastic and information-theoretic sensor interpretation and world modelling mechanisms, and explored the connection between specification of a task and its information requirements.

We have performed an initial experimental validation of the components of the Dynamic Perception framework discussed in this paper. Currently, our research group is finishing the software structure and sensor interfaces for a more powerful mobile robot. This vehicle is based on an omni-directional platform, and is equipped with a number of sensors, including infrared proximity sensors, a sonar sensor array, and an optical rotating range scanner. It will be used to conduct more extensive experimental work.

Acknowledgments

The author wishes to thank José Moura for useful insights into the information control literature, and Ingemar Cox for making me aware of the Art Gallery literature. The research discussed in this paper was performed at the Intelligent Robotics Laboratory, Computer Sciences Department, IBM T. J. Watson Research Center. It incorporates some results from research performed by the author during his association with the Mobile Robot Lab, Robotics Institute, Carnegie-Mellon University.

The views and conclusions contained in this document are those of the author and should not be interpreted as representing the official policies, either expressed or implied, of the sponsoring organizations.

References

[1] J. Aloimonos, I. Weiss, and A. Bandyophadyay. Active Vision. *International Journal of Computer Vision*, 1(4), January 1988.

[2] R. Bajcsy. Active Perception. *Proceedings of the IEEE*, 76(8), August 1988.

[3] J. O. Berger. *Statistical Decision Theory and Bayesian Analysis*. Springer-Verlag, Berlin, 1985. Second Edition.

[4] D. P. Bertsekas. *Dynamic Programming: Deterministic and Stochastic Models*. Prentice-Hall, Englewood Cliffs, NJ, 1987.

[5] R. E. Blahut. *Principles and Practice of Information Theory*. Addison-Wesley, Reading, MA, 1988.

[6] T. Dean, K. Basye, and M. Lejter. Planning and Active Perception. In *Proceedings of the DARPA Workshop on Innovative Approaches to Planning, Scheduling, and Control*, DARPA, 1990.

[7] T. Dean et al. Coping with Uncertainty in a Control System for Navigation and Exploration. In *Proceedings of the Eight National Conference on Artificial Intelligence*, AAAI, Boston, MA, July 1990.

[8] H. F. Durrant-Whyte. *Integration, Coordination, and Control of Multi-Sensor Robot Systems*. Kluwer International Series in Engineering and Computer Science, Kluwer Academic Publishers, Boston, MA, 1988.

[9] J. L. Eaves and E. K. Reedy. *Principles of Modern Radar*. Van Nostrand Reinhold, New York, 1987.

[10] A. Elfes. *Occupancy Grids: A Probabilistic Framework for Robot Perception and Navigation*. PhD thesis, Electrical and Computer Engineering Department/Robotics Institute, Carnegie-Mellon University, May 1989.

[11] A. Elfes. Occupancy Grids: A Stochastic Spatial Representation for Active Robot Perception. In *Proceedings of the Sixth Conference on Uncertainty and AI*, AAAI, Cambridge, MA, July 1990.

[12] A. Elfes. Sonar-Based Real-World Mapping and Navigation. *IEEE Journal of Robotics and Automation*, RA-3(3), June 1987.

[13] A. Elfes. *Strategies for Dynamic Robot Perception Using a Stochastic Spatial Model*. Research Report, IBM T. J. Watson Research Center, 1991. In preparation.

[14] A. Elfes. A Tesselated Probabilistic Representation for Spatial Robot Perception and Navigation. In *Proceedings of the 1989 NASA Conference on Space Telerobotics*, NASA/Jet Propulsion Laboratory, JPL, Pasadena, CA, January 1989.

[15] E. Grosso, G. Sandini, and M. Tistarelli. 3-D Object Reconstruction Using Stereo and Motion. *IEEE Transactions on Systems, Man, and Cybernetics*, 19(6), November/December 1989.

[16] G. Hager. *Information Maps for Active Sensor Control*. Technical Report MS-CIS-87-07, GRASP Lab, Department of Computer and Information Science, University of Pennsylvania, Philadelphia, PA, February 1987.

[17] G. Hager and M. Mintz. Estimation Procedures for Robust Sensor Control. In *Proceedings of the 1987 AAAI Workshop on Uncertainty in Artificial Intelligence*, AAAI, Seattle, WA, July 1987.

[18] M. Hebert, T. Kanade, and In So Kweon. 3-D Vision Techniques for Autonomous Vehicles. In *Analysis and Interpretation of Range Images*, Springer-Verlag, Berlin, 1990.

[19] E. P. Krotkov. *Active Computer Vision by Cooperative Focus and Stereo*. Springer-Verlag, Berlin, 1989.

[20] L. H. Matthies. *Dynamic Stereo Vision*. PhD thesis, Computer Science Department, Carnegie-Mellon University, 1989.

[21] G. L. Miller and E. R. Wagner. An Optical Rangefinder for Autonomous Robot Cart Navigation. In *Proceedings of the 1987 SPIE/IECON Conference*, SPIE, Boston, MA, November 1987.

[22] A. V. Oppenheim and R. W. Shafer. *Discrete-Time Signal Processing. Prentice Hall Signal Processing Series*, Prentice Hall, Englewood Cliffs, NJ, 1989.

[23] J. O'Rourke. *Art Gallery Theorems and Algorithms*. Volume 3 of *International Series of Monographs on Computer Science*, Oxford University Press, Oxford, 1987.

[24] A. Papoulis. *Probability, Random Variables, and Stochastic Processes*. McGraw-Hill, New York, 1984.

[25] H.-L. Wu and A. Cameron. A Bayesian Decision Theoretic Approach for Adaptive Goal-Directed Sensing. In *Proceedings of the Third International Conference on Computer Vision*, IEEE, Osaka, Japan, December 1990.

Contributions of a Microwave Radar Sensor to a Mutlisensor System Used for Autonomous Vehicles

J. Detlefsen, M. Rožmann and M. Lange

Abstract

Sensing 3-D geometrical properties of an environment is essential for a guideless autonomous vehicle operating in indoor situations like production plants. Collecting precise geometric information calls for the support by several complex active sensor systems [1-3], which have to cooperate under realtime conditions. The sensors which can be installed on the experimental mobile robot MACROBE are a 3-D range discriminating laser camera solving fast acquisition at high accuracy of geometrical environmental information within near and middle range and a 3-D multifunction radar sensor with range resolution capable of covering various operating modes e.g. search, tracking, moving target indication and Doppler evaluation for distances up to 50 m. The sensors which are used complement one another with respect to a part of their functions and overlap with respect to others giving redundance and a considerable increase of reliability. This is related to the physically different frequency regimes they use. In this paper system design and imaging results of a multitask 94 GHz pulse Doppler radar with 25 cm radial and 1.5° angular resolution, which has been described in greater detail elsewhere [4,5], are discussed under the aspects of their contribution to a multisensor system. To point out specific mm-wave scattering phenomena, radar images of typically structured indoor situations are presented. The results demonstrate that for guidance of autonomous vehicles, the extraction of geometrical properties of the environment by microwaves are a promising approach, especially in ranges exceeding a few meters. Although mm-wave frequencies at 94 GHz and very high modulation bandwidths are necessary for precise close range radar images, causing some tradeoffs in resolution and sensitivity, signal quality is sufficient for the given task. While 2-D scanning can be performed close to real time with a data flow of up to 10^4 volume cells per second, the acquisition of three-dimensional radar images at the present state is restricted to stationary scenes.

1. Introduction

Compared to current approaches with wire guidance, autonomous vehicles like mobile robots promise substantially improved flexibility for use in future production plants. Therefore, acquisition of relevant 3-D geometrical information through sensors is a key requirement for achieving genuine autonomy. The task of exploration usually requires a broad field of vision to generate maps and extract regions free of objects. For collision avoidance, object tracking and vehicle reorientation during navigation, however, a more limited field of view is sufficient. The environmental data obtained by the sensors are handled on different levels of abstractness, data reduction and representation, called layers. The primary data stream which corresponds to the physical layer must pass through one or more so called virtual sensors, which perform the tasks of signal processing and information reduction. Examples of such sophisticated tasks which are servicing the autonomous mobile robot are e.g. identifying the center of a lane, obstacle detection and recognitions, tracking of moving objects, preparation of maps and layouts of various kinds. The information coming from virtual sensors which are related to different physical sensors can be combined resulting in a multisensor frame, the information of which can be used for hierarchical higher functions like e.g. route planning and other navigational purposes. The current research activities which are described within this paper concern the development of tailored routines for radar image analysis and image processing, task related feature extraction, and the fusion of information received from different sensors.

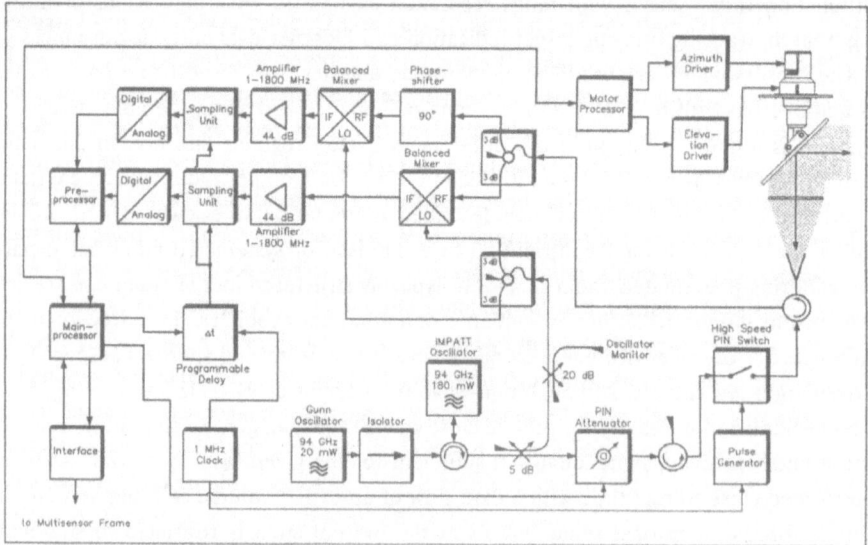

Fig.1   Block diagram of a 94 GHz homodyne pulse Doppler 3-D imaging radar for an autonomous vehicle, operating in indoor environments.

2. Design Considerations and System Implementation

Volume cells - measuring approximately 20 cm in each dimension - are considered a reasonable compromise between the data flow to be processed in real time and the accuracy necessary for collision avoidance or vehicle guidance. With an antenna diameter restricted to less than 20 cm, the required angular resolution can only be obtained at a high carrier frequency of 94 GHz. This also yields conveniently high Doppler frequencies of 625 Hz per 1 m/s, which enable accurate velocity measurements within a short observation time. Transmitter power is generated by a stable solid state 180 mW CW IMPATT source, as shown in Fig. 1, followed by an ultra fast PIN switch for coherent pulse generation [6], [7].

Table 1: System Parameters	
Principle	Active, coherent 94 GHz Pulse-Doppler-Radar
	Time of Flight Sensor
Deflecting System	1 scanning mirror(in 2 dimensions)
Primary information	Radial distance
	Elevation, azimuth angle
	Reflection amplitude
	Relative velocity
	(Shape related type of specular points)
Antenna beam diameter	15 cm
Measurement Domains	Distance: 0.1 ... 150 m
	Velocity: ±8 m/s (direct)
	Angle: Az.:-180° ... +180°
	El.: -25° ... +25°
Accuracy	Distance(absolute): <10mm
	Distance(incremental): <0.1mm,
	Angle: 0.125 ° (Az. u. El.)
	Amplitude: 0,7%
	Phase: 0.2°
	Velocity: <1 mm/s
Measurement Rates	Acquisition: 10.000 Voxels/s
	Offsetcorrected:: 3.000 Voxels/s
	Mirrorrotation: 3 rounds/s
Dataflow	Acquisition: 30 kByte / s
	Transmission > 300 kByte / s
	Radarmaps: 3*1000 Pixels /s
Effective Power	10 uW effektiv
Pulse Power	10 mW
Power Consumption	150 W
Dimensions	(H x B x T)
Sensor Front End	80 cm x 40 cm x 28 cm
Sensor Electronic Rack	19" x 27cm
Weight	10 kg

A fixed radar front end in combination with a scanning reflector was chosen to maintain low inertia. The far field beamwidth of 1.5^o in azimuth and elevation is generated by a lens of 168 mm diameter, illuminated by a corrugated horn for polarization decoupling. Due to the high antenna gain, scatterers within a range of approximately 10 m are observed under near field conditions. The reflector was designed to enable fast scanning modes as well as focussed looks at objects of interest via servo motors. Scattering data from up to 10^4 volume resolution cells - measured at 3 full azimuth turns or 3 looks to arbitrary directions - can be measured per second.

Three processors are in charge of radar signal processing as well as timing and data management of the system itself, so that low level functions like Doppler analysis and beam control can be performed simultaneously with high-level functions like recognition of paths free of obstacles, generation of radar maps, or navigation. Working in parallel, about 3000 measurements per second can be processed, when high-level functions are performed. For guidance of a platform, manoeuvreing with velocities up to 0.1 m/s, this figure turns out to be sufficient. The system parameters of the radar sensor can be characterized by the information given in Tab. 1.

3. Imaging Performance

Due to the requirement for real time data acquisition, the mm-wave image is generated by interpreting the received signal as a measure of the density of point scatterers within an observed volume cell. Based on this linear scattering model, where interactions between individual point scatterers are at this time not taken into account, effects like interference, multiple reflections, shadowing or absorption require a considerable amount of interpretation to derive the actual geometry from the measured mm-wave radar image.

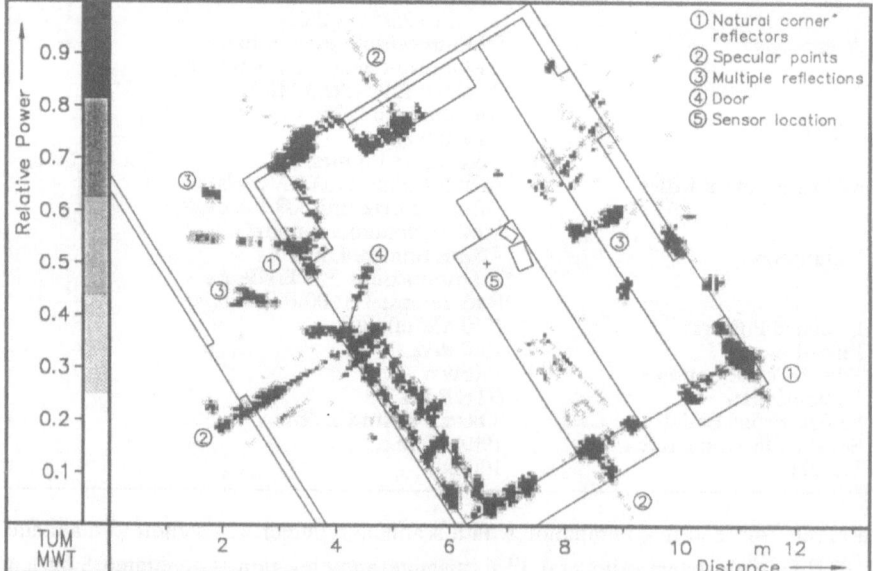

Fig. 2   2-D horizontal distribution of backscattered power at 94 GHz. For better orientation the actual laboratory contours are outlined.

Imaging experiments were carried out with the 94 GHz sensor for a laboratory scene, which was scanned with the sensor located in the middle of the room, covering 360° in azimuth. The obtained radar image in Fig. 2 shows a satisfying resolution and indicates that although diffuse scattering can be observed, specular reflections appear to be dominant, sometimes exceeding the receiver's dynamic range of nearly 70 dB. These spots appear either on flat surfaces, or frequently on dihedral or trihedral configurations where the transmitted beam is directly reflected to the antenna. Fortunately, artifacts caused by multiple reflections or receiver saturation always appear at greater distances. Therefore, obstacle detection algorithms, usually applied only up to the nearest reflection, are not effected. Consequently the evaluation of motion free space should only be based upon the nearest reflection detected.

To provide an example of the sensor's 3-D imaging abilities, the left corner of the laboratory scene, displayed in Fig. 2, was scanned from the same sensor location in range, azimuth and elevation. Fig. 3 displays a simplified CAD model of the structured scene, composed of walls, furniture and a doorway. Object extraction is realized first by transforming the polar data to a rectangular coordinate system aligned parallel with the actual geometry. Then thresholding is used to mark those volume cells in which the backscattered power exceeds the receiver noise level by typically 7 dB. All marked volume cells, each measuring 20 cm in each dimension, are displayed in Fig 3. For better visualization, grey scales are assigned manually to different layers on the z-axis. Although the detail of object reconstruction suffers, and several volume cells are missing due to lack of backscattered signals, the basic topology of the scene appears. A more solid representation of the scene can be achieved, when images are superimposed that are measured from different aspects.

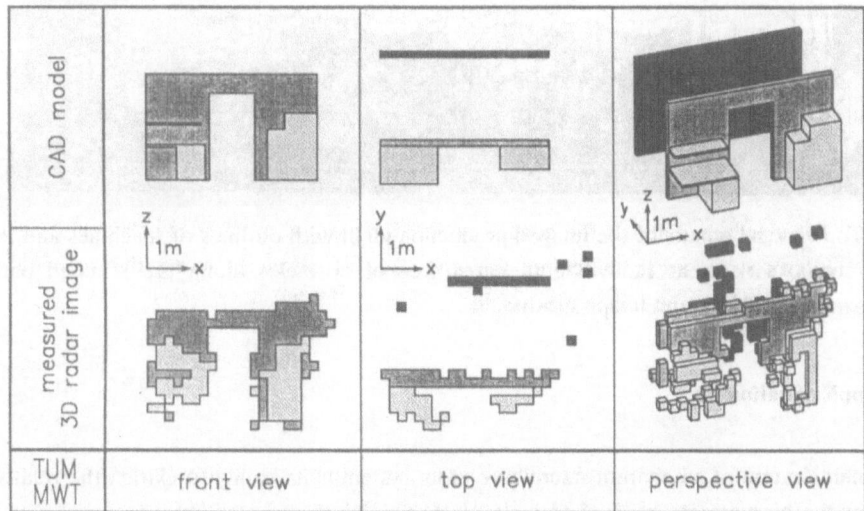

Fig. 3   Reconstructed 3-D geometry of an indoor scene (walls, door, table, locker) based on 20 x 20 x 20 cm³ volume cells in comparison with the actual geometry.

The next step towards vehicle guidance by mm-wave radar sensors is to take multiple looks of a stationary scene from different locations. A realistic environment for these trials was a production plant with various machines, and stationary robots with lanes for wire guided transportation vehicles in between them. The plant was scanned horizontally along a straight path from 15 locations, spaced by 2 m, as marked in Fig. 4. By incoherently superimposing all 15 full scans, a 2-D radar map can be generated for navigation and route planning. For collision avoidance, paths free of obstacles can be determined by converting the analog radar image to a binary radar image and convolving it with the outline of the vehicle. Adjacent pixels then form an area free of obstacles, as demonstrated in Fig. 4, where a circular vehicle outline is assumed. The inner boundaryline defines the maximum position to which the vehicle's center may be moved.

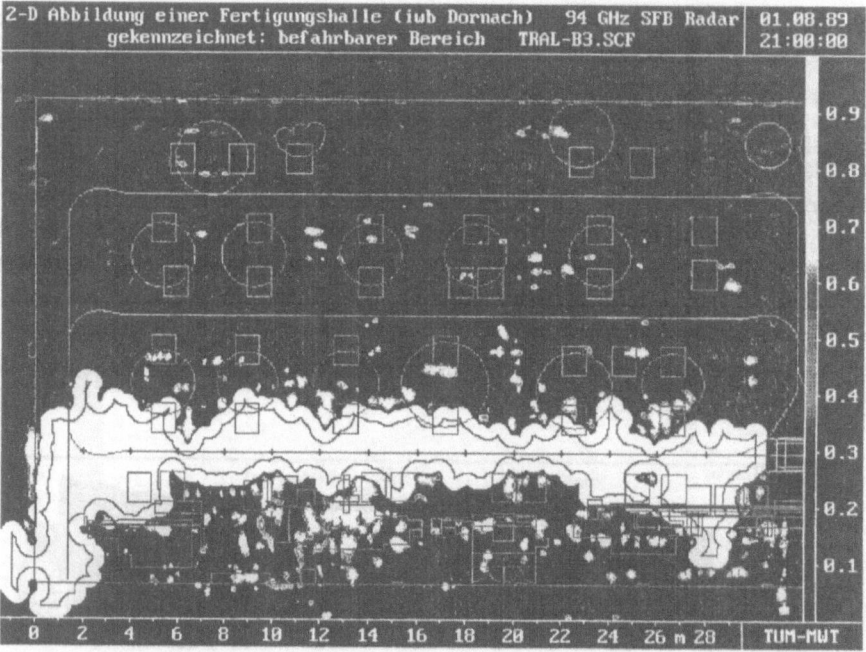

Fig. 4   Top view schematic of the imaged production plant with outlines of machines and the robot's work areas indicating the area free of obstacles along the scanned path, extracted by radar image processing.

3. Doppler Evaluation

A prominent feature of a coherent microwave radar system is its ability to extract the relative velocity at the hierarchical level of a single pixel. This gives the additional feature velocity which can not be obtained with comparable simplicity and accuracy by other noncoherent sensors. This information is obtained from the phase of the complex reflection amplitude, delivered only by a coherent system. For this evaluation the phase difference of two samples measured at two precisely spaced instants of time, which can be chosen according to the desired unambiguous range of velocities. As indicated in Fig. 5, a fixed reflection results in no phase difference, moving reflections show increasing or decreasing phase changes according to the direction of movement. Using this type of velocity evaluation which has been examined for the purposes of true ground speed measurement within another research project [8], a discrimination of fixed and moving objects within defined given velocity ranges is possible.

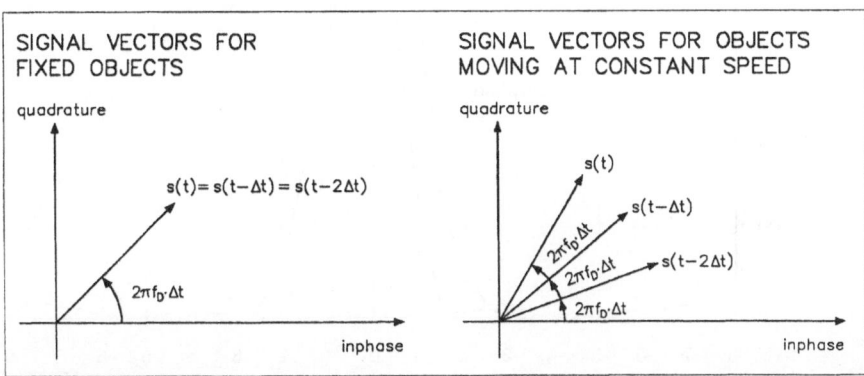

Fig. 5 Principle of velocity evaluation, s(t) : complex signal amplitude, Δt : time interval, f_D : Doppler frequency.

4. Extraction of shape-related features of specular reflection points

As the general features of radar maps are mainly characterized by the backscattering of specular points and the reflection of dihedral or trihedral structures, it is of importance to extract shape and curvature of the surface, the contributing scatterer is located on. This information can be obtained by evaluating the polarisation of the reflected signal or by a local assessment of the reflection characteristic in the vicinity of the scattering center. The latter is demonstrated in Fig. 6, showing the reflection characteristic of a dihedral reflector depending on azimuth respectively elevation angle. Taking into account the width of the radar beam, the difference in the widths of the reflection characteristics can be attributed to the different radii of curvature and the extents of the reflector, the edge of which was vertically oriented.

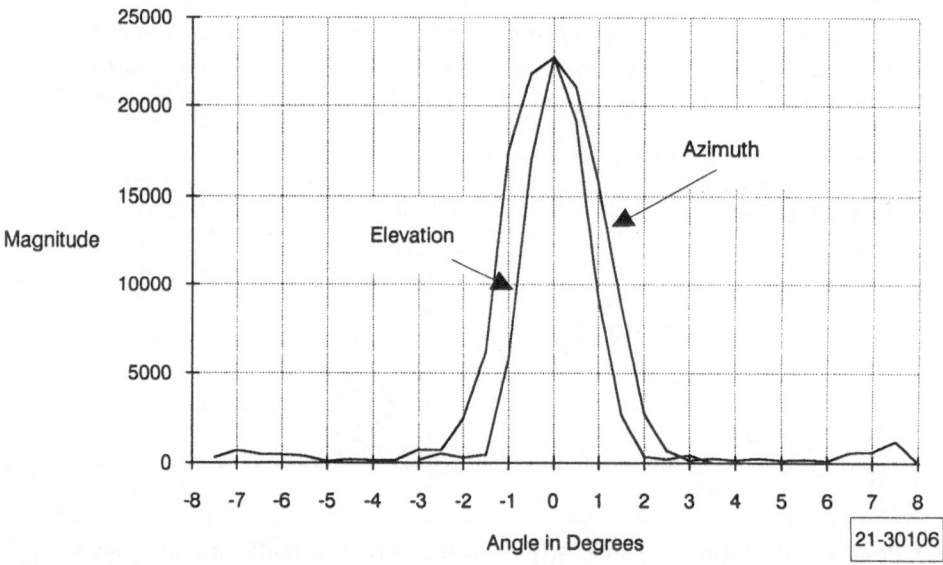

Fig. 6: Reflection characteristic of a measured dihedral reflector with vertically aligned edge. Response in azimuth and elevation.

5. Conclusion

The discussed results demonstrate that for guidance of autonomous vehicles, the extraction of geometrical properties of the environment by microwave sensing is a promising approach, especially also for ranges exceeding a few meters. Although mm-wave frequencies at 94 GHz and very high modulation bandwidths are necessary for precise close range radar images, causing sensitivity tradeoffs, the observable features seem to be well suited for the collection of geometric sensory information. The features are partially redundant to other sensors giving the possibility to compare and aggregate observations, but submits also unique features related to velocity and shape. While 2-D scanning can be performed close to real time with a data flow of up to 10^4 volume cells per second, at the present state limitations still exist for extended areas of three dimensions. The next step will be the demonstration of the cooperative use of the radar sensor information within the multisensor frame of the Sonderforschungsbereich 331.

References

[1]   Durrant-Whyte H.: Integration, Coordination and Control of Multi-Sensor Robot Systems, Kluwer Academic Publ. Boston, 1988.

[2]   Ch. Fröhlich, G. Karl and G.Schmidt, "Sensor AssistedNavigation of a Mobile Robot in Building Environments", IEEE Int. Workshop on Sensorial Integration for Industrial-Robots Architectures + Applications (Zaragoza), Nov. 1989.

[3]   G. Karl, F. Freyberger and G. Schmidt "Data Processingfor a 3-D Range Imaging Laser Camera applied to Real-time Operations of a Mobile Robot" in Proc. of 2nd Int.Workshop on Manipulators, Sensors and Steps towardsMobility (Manchester), 1988, pp. 18.1 - 18.18.

[4]   M. Lange and J. Detlefsen, "94 GHz 3-D-Imaging Radar for Sensorbased Locomotion" Proc. Int. Microwave Symp. IEEE MTT-S (Long Beach), June 1989, pp. 1091-1094.

[5]   J. Detlefsen, M. Lange and M. Bockmair, "Evaluation of Near Range 94 GHz Radar Images for Autonomous Vehicles" in Proc. Int. Conf. on Radar (Paris), April 1989, pp. 203-208.

[6]   M. Lange and J. Detlefsen, "A 35 GHz homodyne pulse-Doppler radar with very high resolution for short range application" in Proc. Int. Conf. on Radar (Paris), Mai 1984, pp. 210-214.

[7]   M. Lange, J. Detlefsen and M. Bockmair, " Resonant Pulse Amplification for Radar Imaging Applications" in Proc.15th Europ. Microwave Conf. (Paris), Sept. 1985,pp. 1005 - 1009.

[8]   Detlefsen, J.; Weinberger, M.: Eigengeschwindigkeitsmessung von Fahrzeugen über natürlichen Oberflächen mit Mikrowellen, Abschlußbericht über Forschungsvorhaben DE 270/4-1, gefördert durch die DFG, Nov. 1990, in German.

Navigation and Control

Hierarchical Control of Free-Navigation AGVs

H. Van Brussel, J. De Schutter and K.T. Song

ABSTRACT : The main drawback of traditional automatic guided vehicles (AGVs) is their confinement to a pre-laid out wire track in the floor. This paper describes an alternative solution, developed in view of the growing need for autonomy and freedom of trajectory. The experimental system, called EMIR-1, contains a hierarchically structured motion control system. It is based on odometry for position measurement. A centralised path following controller eliminates position errors of the vehicle. For planning purposes, an ultrasonic sensor is used to map the environment; it also detects unexpected obstacles and guides the vehicle around them. A few solutions for enhancing the position accuracy, by frequent recalibration using absolute position sensors are also outlined.

1. INTRODUCTION

AGVs (Automatic Guided Vehicles) have become very important components in flexible manufacturing facilities and, more broadly, in automated transportation systems in general. The vehicles used in present day systems are mostly guided by inductive wires buried in the ground. Such guiding systems are costly and the trajectories are difficult and costly to be modified. Modern plants and transportation facilities require more flexibility in the motion control of AGVs, due to the ever larger varieties of products to be handled in ever smaller batches. Easily reprogrammable free ranging vehicles are needed to satisfy these flexibility requirements.

Mobile robots are able to navigate autonomously from a start position to a destination without colliding with any object in the environment with the help of sensory information. As such they are good candidates to serve as free ranging AGVs. Functionally, a mobile robot acts through a combination of high level decision making strategies with a lower motion control level. While the high level AI-based planning algorithms are applied at the task specification stage, an accurate and robust motion control system is responsible for carrying out the planned actions.

Because of the obvious advantages of unmanned free ranging vehicles over current AGV systems it is not surprising that various approaches towards free ranging have been suggested and - in some cases - tested. The technologically more adventurous ideas come from universities, whereas the AGV manufacturers seem to be more conservative. Proposed ideas vary from systems using active and passive beacons based on infra-red light or lasers, on-board ultra-sonic range finders, on-board cameras with image processors and inertial navigation to much simpler solutions that combine conventional AGV technology with odometry to create pseudo free-ranging capabilities.

Presently, none of those proposed solutions are actually operational. In all cases the actual 'measurements' necessary for finding the vehicle's position cause problems: beacons give problems with obstacles between beacon and vehicle, image processing is not yet feasible in real-time, ultrasonic range-finding systems are said to be not accurate enough and error prone.

In this paper a free ranging AGV is discussed which contains a hierarchically structured motion control system. The focus will be on the path following controller which eliminates position errors of the vehicle, and which forms the heart of the motion control system. The system also contains an ultrasonic sensor which is used to make a map of the environment. This map is used to plan a collision free path for the AGV. In addition the sensor detects unexpected obstacles.

2. THE EMIR-1 CONCEPT

A mobile robot named EMIR-1 (Experimental Mobile Intelligent Robot 1) has been designed and constructed to serve as a test-bed for sensors and navigation strategies.

EMIR-1 is powered by two 12-volt batteries. It has two independent wheels in the front driven by DC-motors and a free castor in the rear for balance. This driving mechanism was selected as a result of a detailed comparative study, [2], because of its mechanical simplicity and manoeuvrability. A hierarchical planning and control architecture has been implemented on EMIR-1 enabling it to navigate in the real

world.   The higher level tasks such as world modelling and path
planning are implemented in a resident host computer, a Microvax 2000.
An Amiga 500 microcomputer is put on-board the robot to handle the
lower level tasks such as motion control and sensor system control.   A
communication system is made between the two computers based on a RS-
232C serial link.   Fig. 1 illustrates this planning and control
architecture.

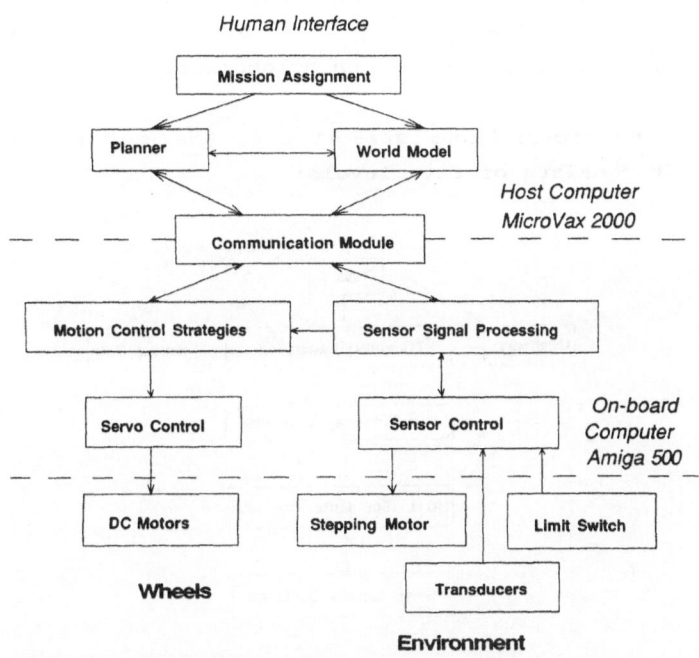

Fig. 1:   The planning and control of architecture of EMIR-1.

3. THE STRUCTURE OF THE MOTION CONTROL SYSTEM

The motion control system of a mobile robot must exhibit following
features:
- accurate path following capability.
    A small orientational error will introduce accumulated position
    errors of the vehicle.   Therefore accurate control of the
    orientation is necessary in order to follow the planned path
    accurately.
- possibility to cope with sensor data.
    The motion of the mobile robot is to be influenced by external

sensor data.   A mobile robot is constantly interacting with the
environment and the motion control system must be able to respond
to the information from the sensor system in real-time.
- motion in three degrees of freedom.
The mobile robot must be able to travel to any position with any
specified orientation in the working area.   There are three
degrees of freedom for the motion in a plane, i.e. two
translations in the plane and one rotation around an axis
perpendicular to the plane.   As there are only two control
inputs, the EMIR-1 system is non-holonomous.

Fig. 2 shows the hierarchical structure of the motion control system
of EMIR-1.   It consists of five levels:

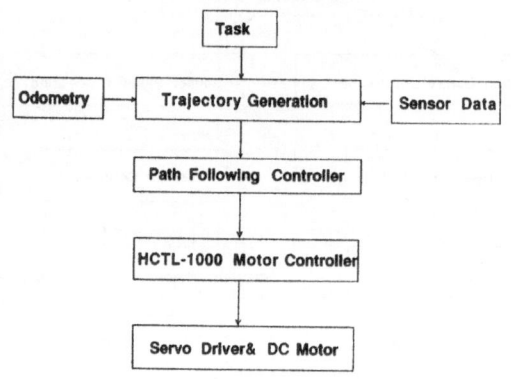

Fig. 2: The structure of the motion control system.

Level 1: *task specification*.
At task specification level, the optimal overall robot trajectory is
planned to execute a particular job.   This path is passed to the on-
board motion control system to be executed.   A task for the motion
control system usually consists of moving from one point to another in
the plane of motion.

Level 2: *trajectory generation*.
A trajectory is then generated to carry out the specified task.   This
trajectory consists of a combination of motion primitives such as
linear motions and blending motions connecting two path segments
(circles, clothoïds).

For each of these motions the desired linear velocity $V_d(t)$ and
angular velocity $\omega_d(t)$ are determined in function of time, and passed
on to the path following controller.

Level 3 : *the path-following controller.*
A path-following controller is working at the centre of the control
hierarchy. This controller is developed from a kinematic state space
model of the vehicle. The desired linear velocity V_d and angular
velocity ω_d are fed forward to this controller. In addition the
controller contains feedback which eliminates tracking errors (i.e.
the orientational error, the lateral position error and the position
error along the travelling direction).

Level 4 : *servo control of the wheel motors.*
In order to reduce the computational burden of the on-board control
computer, a pair of HCTL-1000 chips are implemented for servo control
of the two dc motors, which drive the two main wheels.

Level 5 : *servo drivers and actuators.*
At the lowest level of the hierarchy are the servo drives and the
actuators for driving the wheels. Two PWM (Pulse Width Modulation)
amplifiers are used to drive the dc motors.

Sections 4 to 7 describe the operation of the path following
controller in more detail.

4. FEEDFORWARD CONTROL

As explained in the previous section the command velocities for the
left and right wheels, V_{dl} and V_{dr}, consist of a feedforward term and
a state feedback term:

$$V_{dl} = V_{ffl} + V_{sl} \tag{1}$$
$$V_{dr} = V_{ffr} + V_{sr} \tag{2}$$

The feedback terms, V_{sl} and V_{sr}, are discussed in sections 5 and 6.
The feedforward terms are easily calculated based on the desired
linear and angular velocities of the vehicle V_d and ω_d (figure 3):

$$\omega_d = \frac{V_{ffr} - V_{ffl}}{E} \tag{3}$$

$$v_d = \frac{v_{ffr} + v_{ffl}}{E} \qquad (4)$$

with E the distance between the driving wheels. Hence, the feedforward velocities are:

$$v_{ffl} = v_d - \frac{\omega_d E}{2} \qquad (5)$$

$$v_{ffr} = v_d + \frac{\omega_d E}{2} \qquad (6)$$

This can be written as:

$$\begin{bmatrix} v_{ffl} \\ v_{ffr} \end{bmatrix} = F . \begin{bmatrix} v_d \\ \omega_d \end{bmatrix} , \qquad (7)$$

$$\text{with } F = \begin{bmatrix} 1 & -\dfrac{E}{2} \\ 1 & \dfrac{E}{2} \end{bmatrix} \qquad (8)$$

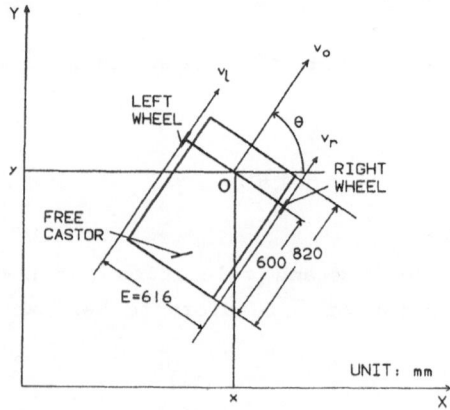

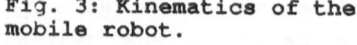

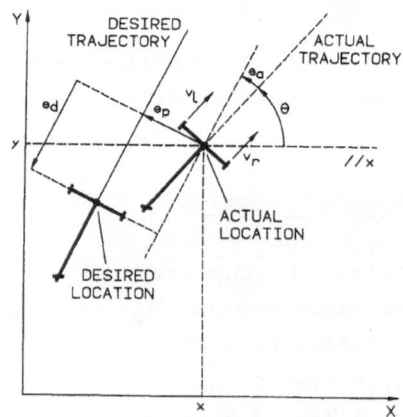

Fig. 3: Kinematics of the mobile robot.

Fig. 4: A geometric model of the errors of motion in a plane.

5. STATE SPACE MODEL

The feedback strategy is based on a state space model of the vehicle. In this model the dynamics of motors and amplifiers are neglected, because their mechanical time constant is about 10 ms, whereas the sampling period for the digital feedback controller is selected at 20 ms. Hence, the actual wheel velocities are assumed to equal the commanded velocities,

$$v_l = V_{dl} \; ; \; v_r = V_{dr} \; ,$$

and only the kinematics of the vehicle are involved in the model. The kinematic model is derived from figure 4.

$$\dot{x} = \frac{v_l + v_r}{2} \cos \theta \tag{9}$$

$$\dot{y} = \frac{v_l + v_r}{2} \sin \theta \tag{10}$$

$$\dot{\theta} = \frac{v_r - v_l}{E} \tag{11}$$

These are nonlinear equations which can be linearized about the nominal trajectory by defining three errors of motion (figure 4): the angular error e_a, the lateral position error e_p, and the error along the direction of motion or tangential error e_d.

The following equations are derived for the errors of motion:

$$\frac{de_a}{dt} = \frac{v_l - v_r}{E} + \omega_d \tag{12}$$

$$\frac{de_p}{dt} = \frac{v_l + v_r}{2} \sin(e_a) \tag{13}$$

$$\frac{de_d}{dt} = \frac{v_l + v_r}{2} \cos(e_a) - V_d \tag{14}$$

For normal operation, the orientational error e_a is very small. I.e. $\sin(e_a) \simeq e_a$ and $\cos(e_a) \simeq 1$. Hence:

$$\frac{de_a}{dt} = \frac{v_l + v_r}{E} + \omega_d \tag{15}$$

$$\frac{de_p}{dt} = \frac{v_l + v_r}{2} e_a = v_o \cdot e_a \tag{16}$$

$$\frac{de_d}{dt} = \frac{v_l + v_r}{2} - V_d \tag{17}$$

where V_o represents the nominal tangential vehicle velocity.

In state space representation, the above equations become:

$$
\begin{bmatrix} \dot{e}_a \\ \dot{e}_p \\ \dot{e}_d \end{bmatrix} = \begin{bmatrix} 0 & 0 & 0 \\ V_o & 0 & 0 \\ 0 & 0 & 0 \end{bmatrix} \begin{bmatrix} e_a \\ e_p \\ e_d \end{bmatrix} + \begin{bmatrix} \frac{1}{E} & \frac{-1}{E} \\ 0 & 0 \\ \frac{1}{2} & \frac{1}{2} \end{bmatrix} \begin{bmatrix} v_l \\ v_r \end{bmatrix} + \begin{bmatrix} 0 & 1 \\ 0 & 0 \\ -1 & 0 \end{bmatrix} \begin{bmatrix} V_d \\ \omega_d \end{bmatrix} \quad (18)
$$

or

$$
\begin{bmatrix} \dot{e}_a & \dot{e}_p & \dot{e}_d \end{bmatrix}^T = A \begin{bmatrix} e_a & e_p & e_d \end{bmatrix}^T + B \begin{bmatrix} v_l v_r \end{bmatrix}^T + G \begin{bmatrix} V_d & \omega_d \end{bmatrix}^T \quad (19)
$$

A state space control model is constructed as shown in fig. 5. In the figure, A, B, and G are given by the above equation. K is the state feedback matrix, which is determined in the next section, and F is the feedforward matrix as defined in section 4. Notice that F has been chosen as:

$$BF = -G, \quad (20)$$

such that the feedforward terms cancel the third term of equation (18).

6. STATE FEEDBACK CONTROL

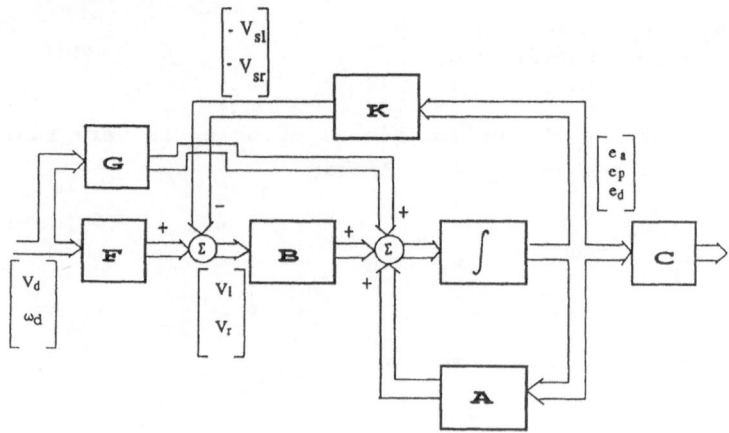

Fig. 5: A state space model for the path following control.

A linear combination of the states is fed back to each of the driving wheels:

$$
\begin{bmatrix} v_{sl} \\ v_{sr} \end{bmatrix} = - \begin{bmatrix} k_{1a} & k_{1p} & k_{1d} \\ k_{2a} & k_{2p} & k_{2d} \end{bmatrix} \begin{bmatrix} e_a \\ e_p \\ e_d \end{bmatrix} \tag{21}
$$

or $\quad \begin{bmatrix} v_{sl} & v_{sr} \end{bmatrix}^T = K \begin{bmatrix} e_a & e_p & e_d \end{bmatrix}^T \tag{22}$

The elements of the feedback matrix are determined using pole placement. However, only three poles can be assigned which is not sufficient to determine uniquely the six feedback constants. In order to solve this, three additional constraints are specified, which are quite obvious in view of the symmetry of the vehicle:
- left and right wheel have an opposite effect on the orientational error e_a and hence also on the lateral e_p (cf. equations 15 and 16). Therefore we put:

$$k_{1a} = - k_{2a} \tag{23}$$
$$k_{1p} = - k_{2p} \tag{24}$$

- left and right wheel have an equal effect on the tangential error e_d (cf. equation 17). Hence we choose:

$$k_{1d} = - k_{2d} \tag{25}$$

Notice that the feedback constants are dependent on the nominal tangential velocity V_o which can be approximated by the desired velocity V_d.

Because none of the state variables is directly measurable, a state estimator is needed to estimate the state variables. An open-loop estimator, which uses the information from the measured velocities is constructed by directly integrating the error equations of motion. That is :

$$
e_a = \int \left(\frac{v_l - v_r}{E} + \omega_d \right) dt \tag{26}
$$

$$
e_p = \int \left(\frac{v_l - v_r}{2} \cdot e_a \right) dt \tag{27}
$$

$$
e_d = \int \left(\frac{v_l - v_r}{2} - v_d \right) dt \tag{28}
$$

The velocities are measured by counting the encoder pulses in each sampling period.

By this state feedback the motion errors due to the load disturbances on the driving wheels are eliminated and the mobile robot follows the planned path accurately. However errors due to wheel slippage remain. To this end, the estimated errors are reset each time the actual errors e_a, e_p and e_d can be calculated based on an external reference (see section 8).

Combination of feedforward and feedback control yields the final control law:

$$\begin{bmatrix} v_{dl} \\ v_{dr} \end{bmatrix} = F \begin{bmatrix} v_d \\ \omega_d \end{bmatrix} + \begin{bmatrix} v_{sl} \\ v_{sr} \end{bmatrix} \qquad (29)$$

7. PRACTICAL IMPLEMENTATION AND THE EXPERIMENTAL RESULTS

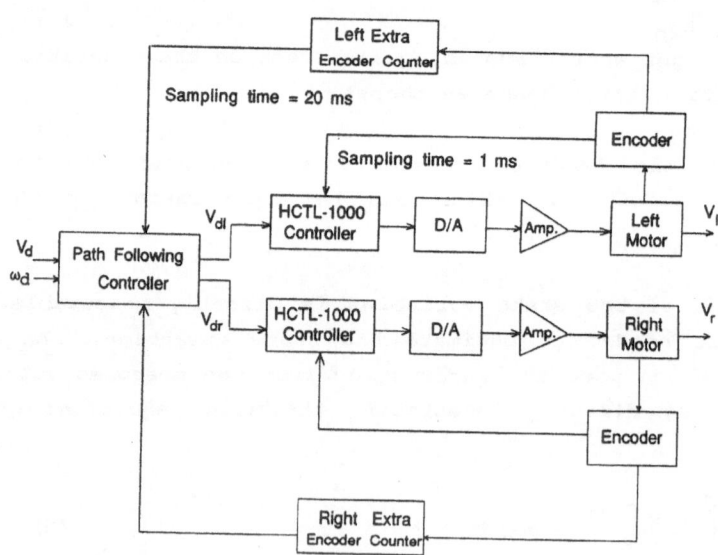

Fig. 6: The two control loops of the motion control system.

The path-following controller is implemented in the on-board control computer Amiga. The outputs of the path-following controller are sent to the servo controller for commanding the wheel motors. Two HCTL-1000 chips from Hewlett-Packard are used. The advantage of using these chips is that they reduce the computational burden of the on-board computer. Fig. 6 shows the block diagram of the two control loops. While the sampling time of the servo control of the chips is 1 ms, the path-following controller is implemented with sampling time 20 ms.

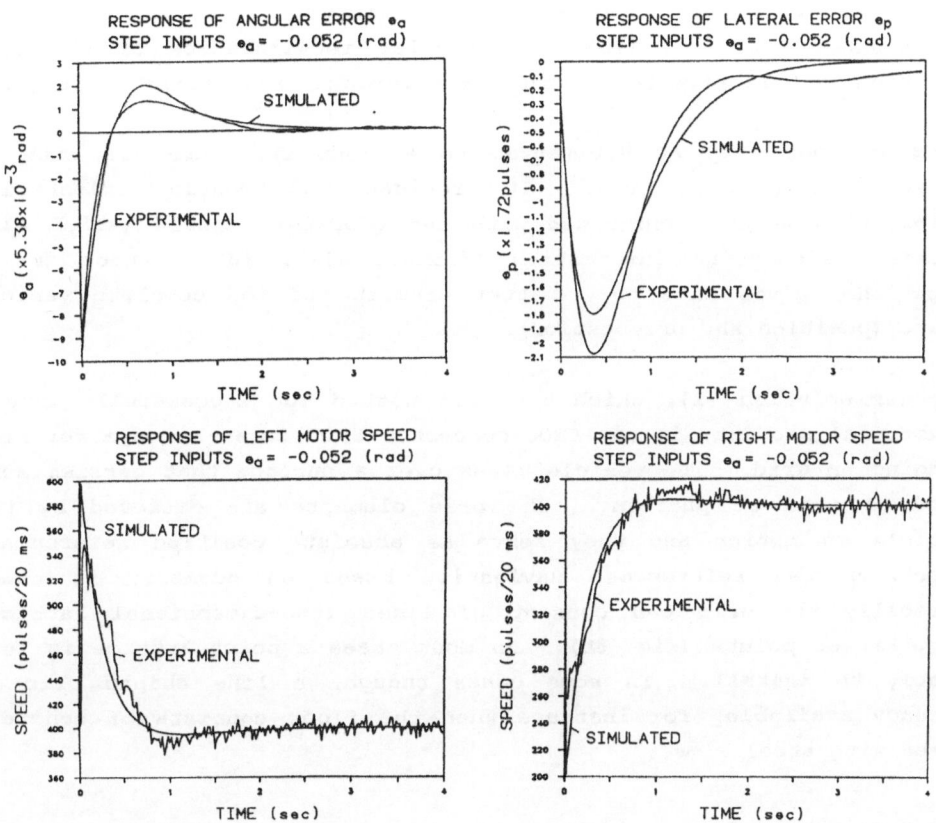

Fig. 7: Experimental and simulation results of the responses for step input: e_a = 0.052 rad., v_d = 16.64 cm/sec (400 pulses/20 ms), ω_b= 0.0

Fig. 7 shows the experimental results of this controller when a step angular error e_a = -0.052 *radians* (3˚) is introduced. The desired linear velocity V_d is 16.64 cm/sec (about 400 encoder pulses/20 ms). The desired angular velocity ω_d is zero. The step input angular error induces the lateral position error e_p. All encoders decrease to zero as expected. After eliminating the errors, the velocity of the left and right wheels get back to their nominal values.

8. ABSOLUTE SENSORS

The above path following controller makes optimal use of the position information coming from the wheel encoders. Nevertheless, errors due to wheel slippage, changing wheel diameter, etc.... cannot be compensated for. To this end, regular recalibration of the real position with respect to an absolute reference is a desirable feature.

Work is under way at K.U.Leuven to augment the odometers with an optical fibre laser gyro. It provides high-accuracy orientation information which, confronted with the odometer signals (which also provide orientation information (formula 11)), in a sensor fusion algorithm, gives rise to a better estimate of the complete vehicle state (position and orientation).

Another solution [1], which has been worked out successfully into a commercial product, is the FROG-concept. FROG is an acronym for Free Ranging on Grid. The vehicle moves over a surface that carries some kind of grid or pattern. The grid elements are detected by the vehicle in motion and they serve as absolute position references. Inbetween two references, navigation based on odometry is used. Basically the grid can consist of lines (two-dimensional lattice, fig. 8a) or points (fig. 8b). In most cases a point lattice is less costly to install. In some cases though, a line shaped grid is already available, for instance when the floor consists of concrete slabs with steel rims.

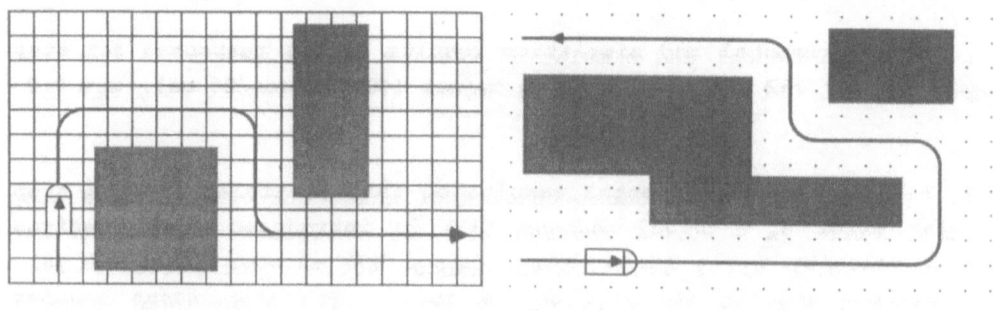

a.   b.

Fig. 8: Navigation over a line grid (a), or a point grid (b).

In [1], electronic tags (transponders) are used, buried in the ground, to form a point lattice. These transponders are passive devices, containing an absolute code. They are energised from the vehicle for identification. A special detector circuit on-board the vehicle allows accurate absolute position determination of the vehicle with respect to the tag. Accuracies of one cm have been obtained. The distance between transponders in the grid depends on many parameters, like vehicle speed, floor condition, etc. The system is now being implemented in the harbour of Rotterdam, for container transport.

9. NAVIGATION

The task of a navigation system is to plan a collision-free path from the start position to the destination. In addition, if an unexpected obstacle is met during the execution of the planned path, the navigation system must be able to modify the plan and avoid the obstacle.
The method used by the path planner is derived from the concept of configuration space. The objects in the real world are represented as extended polygons. The robot is therefore shrunk to a point and the path is defined for this reference point of the mobile robot. Visibility graphs and a modified A* graph search algorithm are used to search for the shortest collision free path from the start point to the goal point [2]. Owing to its reasoning nature, this planner is implemented Prolog.

The planner contains both global and local strategies. While the global path planning is based on the existing world model, the local strategies handle local motion planning in case an unexpected obstacle is detected by the sensor system.

The ultrasonic sensor of EMIR-1 is modified from the Polaroid ultrasonic ranging system. The range of this sensor is from about 30 cm to 9 meters. The time elapsed between the starting of transmission and the first detected echo is recorded and the distance between the sensor and the target can be calculated. A stepping motor is used to rotate the sensor. A scanning system is implemented to point the sensor in an arbitrary direction with a resolution of 0.3 degrees. The advantages of this type of sensors are that range data are easily obtained at low cost. Fig. 9 shows a typical world model obtained by the ultrasonic sensor.

In order to cope with a dynamic environment, the sensor system is designed to work in two modes:

- *command mode*

  In this mode the sensor executes the command from the planner, e.g. in order to check the presence of an existing obstacle or avoid an unexpected obstacle.

- *watch mode*

  During the motion of the mobile robot, the sensor is always watching the immediate environment in front of it. The purpose is to detect unexpected obstacles and to work as a *soft bumper*.

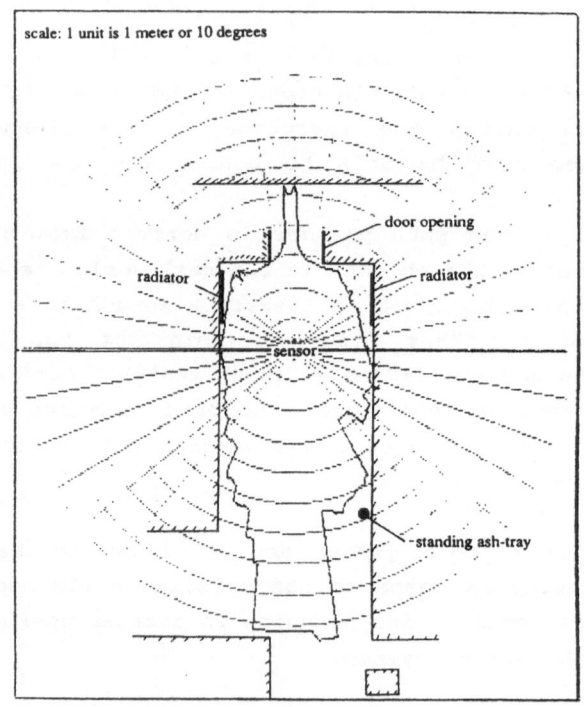

Fig. 9: A world model made by the ultrasonic sensor.

10. CONCLUSION

A hierarchical control system has been developed for a free ranging AGV.
A path following controller forms the heart of this motion control system. The path following controller is based on a kinematic state

space model of the vehicle. Experimental results confirm the theoretically expected behavior.

A high level navigation system plans a collision free path from a start to a destination position. In addition it is able to modify the planned path in case an unexpected obstacle is detected by the ultrasonic sensor system.

REFERENCES
(1) VAN BRUSSEL, H., C.C. VAN HELSDINGEN, K. MACHIELS, "FROG: Free Ranging on Grid. New Perspectives in Automated Transport", Proc. AGV6, IFS/Springer, 1988.
(2) SONG, K.T., "Planning and Control of a Mobile Robot based on an Ultrasonic Sensor", Ph.D.Thesis, K.U.Leuven, 1989.

Performance Data of Dead Reckoning Procedures for Non Guided Vehicles

A. Merklinger

Introduction

The search for more flexibility in all parts of industrial manufacturing increasingly affects material handling and distribution. For the field of Automatically Guided Vehicles (AGVs) this especially means:

- replacement of fixed guidance lines by flexible driving-paths,
- renounciation of costly installations,
- extension of driving paths also in out door areas,
- acceleration of driving speed,
- improvement of safety installations and
- flexible programming.

Efforts of research institutes and manufacturers aspire to develop vehicles without any guidance line, so-called non-guided vehicles. Thereby a special look out is to be drawn to the localization aspect. In most caes this unit is realised by using dead-reckoning systems in combination with various other methods of position detection.

Different methods of dead-reckoning localization have been examined at the free ranging AGV IPAMAR of the Fraunhofer Institute IPA. The results are presented hereafter.

Methods for Localization

Fundamentals

The expression "Navigation" originates in nautics and comprises all activities necessary to pilot a vehicle from starting position to destination, i.e.:

- path planning (time, area, consumption,...),
- localization,
- position update measurements,
- path control and
- collision avoidance.

There are different ways to classify localization procedures:

a) Use of External Installations

    Some localization procedures require additional external installations such as transmitters, marks, reflectors etc.
    Other methods only use natural environmental references, e.g. dynamic pressure of the air or the velocity relative to the ground.

b) Model-Based

    A localization procedure using a model of the environment, such as the position of reference marks or transmitters, is classified as "model-based".

c) Position Measurement - Dead Reckoning Localization

    The determination of vehicle position from geometrical data (angles and distances) is called "position measurement".
    Dead reckoning localization computes the vehicle position from its state of movement (velocities and accelerations).
    The extrapolation of consecutive position measurements can also be used for dead reckoning localization.

d) Continous Procedures - Spot Measurements

    Continous procedures need a non-intermittent data flow to compute the actual position.
    Spot measurements only use information taken at a certain fix.

e) Environment Related - Inertial

Environment related procedures obtain necessary data out of observations relative to the environment.

Inertial procedures compute the vehicle's location by using values measured with respect to inertial space (velocities and accelerations).

Classification Characteristic (Necessary Condition)	use of external installations		model-based		position measurement	dead-reckoning localization	continous	spot measurement	environment related	inertial
	yes	no	yes	no						
Procedure										
marks, barrettes	X		X		X			X	X	
grid	X			X		X	X		X	
guidance lines -segments	X		X		X		X		X	
transmitter	X		X		X			X	X	
beacon	X		X		X			X	X	
satellites	X		X		X			X	X	
environment related measurements optical,...)		X	X		X			X	X	
environmental comparison		X	X		X		X		X	
relative measurement of the state of movement (odometers, correlation)		X		X		X	X		X	
inertial measurement of the state of movement (accelerometers, gyros)		X		X		X	X			X

Fig. 1: Classification possibilities of different localization procedures

Very often different procedures for localization are used alternately. The advantages of environmental spot measurements are taken to compensate the accumulating errors of dead reckoning localization.

The test vehicles IPAMAR I and II developed at the Fraunhofer Institute IPA are provided with odometers and steering angle sensors respectively partly inertial with odometers and gyros for localization purposes. The location is computed using dead reckoning methods and is updated by ultrasonic measurements relative to the environment. Two dimensional procedures are used and discussed hereafter, thereby diminishing the degrees of freedom from six to three.

The moving gear's kinematic structure (three-wheel with steered and driven front-wheel) again reduces the independant degrees of freedom to two. Hereafter only this mostly applied configuration is discussed.

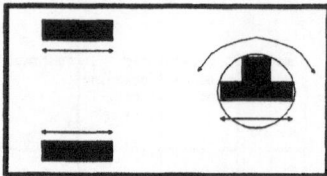

Fig. 2: Three-wheel kinematic

The movement data necessary for a well defined description of the vehicle's path are:

- translational velocity and
- rotational velocity around the gear axis.

Different types of sensors provide these velocities purely odometric or partly inertial:

- measurement of steering angle and path length of the front wheel,
- measurement of both rear wheel's path lengths or
- measurement of rear axle's path length and rotational velocity.

The accuracy of odometric dead reckoning localization is mainly limited by effects occuring at the boundary between wheel and floor such as:

- forces between wheel and ground causing slip effects in directions parallel and perpendicular to the wheel.

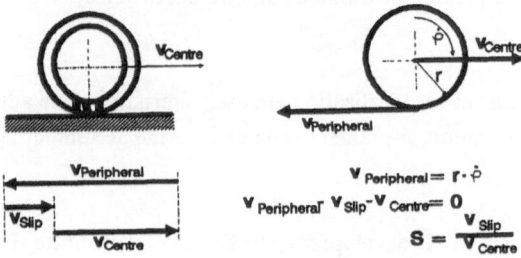

Fig.3: Wheel slip definition

- the width of the wheel's contact surface on the ground in connection with the ground's nature and quality cause conditions not exactly defined and also variable in time and space. So the geometrical parameters of the vehicle's moving gear are uncertain, thus is the dead reckoning localization result.

- manufacturing tolerances and abration affect the parameter uncertainties of dead reckoning methods in a similar manner.

However the limited resolution of real sensors almost doesn't affect the accuracy of dead reckoning localization because of its integrating behaviour.

Discretionary errors of numerical computation are too small for a relevant influence on practical applications (see /3/).

Front Wheel Localization

The steered and driven front wheel is supported with an odometer and an absolute encoder at the live ring of the driving motor.

Centrifugal forces in turns and the driving force caused by the driving torque of the motor result in elastic deformations parallel and perpendicular to the wheel direction of both the wheel and the ground and thus to slip effects.
In state of straight path and constant velocity the slip-caused error can easily be compensated. In case of accelerations or turns this is much more complex and the real situation has to be taken into account.

In turns there is a movement of the centre of turn and a variation of its radius due to slip effects. This results in position errors longitudinal and lateral to the vehicle as well as in misorientation, causing position errors increasing proportional to the subsequent straight path length.

The contact point of the front wheel with the ground being not exactly known results in the same way: a given wheel-width of 50 mm allows the contact point to vary at a maximum of 25 mm. Thus there is a resulting radius error of the front wheel of 2.5%. This means an orientation error after a 90° turn of 2.2°.

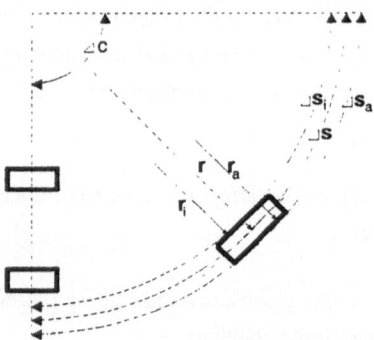

Fig. 4: Contact point variation of the front wheel

An unprecisely measured or manufactured moving device effects the orientation similarly: taken the dates of the above example, a long wheel base (LWB) of 0.5 m and a LWB error of 5 mm there is a radius error of 1%, that means an orientation error of 0.9°.

An orientation error of 3° leads to a lateral position error of 1 m after a straight ride of 20 m.

Rear Wheel Localization

When locating with the non driven rear wheels there is almost no effect of slip when driving straight ahead. Also tangential forces in the boundary between wheel and ground are negligible and thus are slip effects in wheel direction. Only centrifugal forces in turns may cause wheel slip perpendicular to the driving direction and so cause orientation errors.

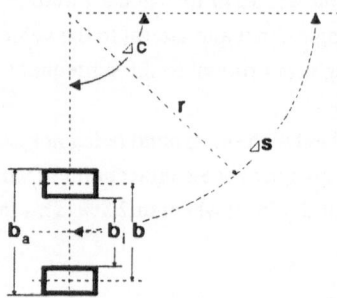

Fig. 5: Contact point variation of the rear wheels

The main cause of orientation errors when locating with the rear wheels is the uncertainty of the points of surface contact of these wheels. Taken the data of the before mentioned example, a front wheel turn radius of 1 m, a LBW of 0.5 m, a wheel width of 50 mm and an axle width of 0.5 m, there is a radius of the middle of the axle of 0.87 m. Based on this data the contact points of the rear wheels may vary ± one axle width thus creating a maximum orientation error of 10°. LBW errors affect in the same way.

Given a misorientation of 10° a straight ride of 20 m would cause a lateral position error of 3.5 m.

Gyro Supported Localization

Here odometers at both non driven rear wheels measure the path length of the rear axle. So there is no influence of slip effects (except lateral slip). An averaging of both odometer values is also diminishing possible errors. The remaining orientation error is dependant on the sensor used to detect the angular rate.

In case of IPAMAR the sensor quality is of about 5°/h, that means:
When riding straight with a velocity of 0.5 m/s it will take 40 s for a distance of 20 m. The possible orientation error occuring therby will be about 0.06°, that means a lateral position error of 0.01 m.

The main advantage of a gyro supported localization is, without any doubt, the complete independence of any effect influencing the contact between wheel and ground. The gyro senses orientation changes out of these reasons in the same way as it senses control initiated changes of orientation.

Conclusions

The above worst-case calculations of some localizational influences evidently show the importance of all process data influencing the contact between wheel and ground when locating purely by odometry. Any small obstruction such as stones, cables, etc. is reducing the accuracy of the dead reckoning localization.
When using gyros these influences are sensed in the same way as regular turns and thus don't impact the locational accuracy. The path length measurement is done similarly in all cases. An increasing localization precision is achieved when using non-driven wheels for odometric measurements and when averaging path length values.

All procedures discussed so far contain as an additional source of errors the deviation from two dimensional case. A lateral or longitudinal inclination will produce a change in all three angles of orientation when turning around only one of the vehicle's axis. Usually in manufacturing environment this influence is negligable because of the small inclinations of the ground.

Test Results

Wheel Slip

Wheel slip is dependant on forces in the contact surface between wheel and ground and therefore on weight and acceleration of the vehicle. The vehicle's mass and centre of gravity are known, its accelerations are calculated from the odometrical measurements. The developed error model allows to compute the wheel slip and so to compensate the arising localization error.

For test purposes the measured odometric values of front and rear wheels have been compared. Fig. 6 and 7 show the deviation in path length and the respective calculated slip between front and rear wheel. All measurements have been taken on carpeted floor at straight ride. The complete movement profile consists of acceleration, constant velocity and deceleration to zero.

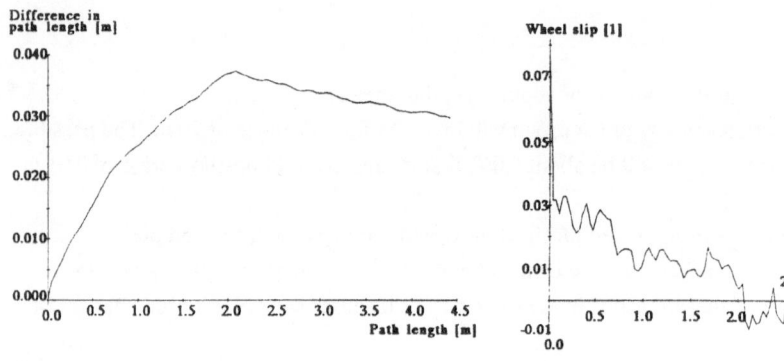

Fig. 6: Path length difference   Fig. 7: Wheel slip

Fig. 8 shows the calculated result compared to the measured differences between front and rear wheels.

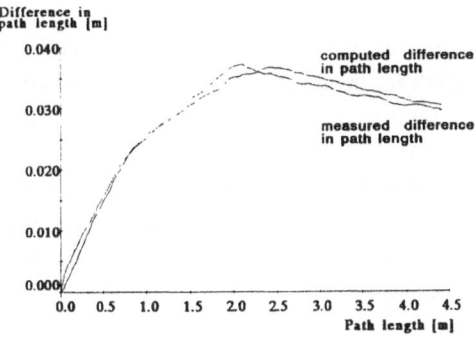

Fig. 8: Comparison of differences in path length

The path length differences between front and rear wheel of **IPAMAR** I ist at a maximum of 40 mm in the shown example. The calculation provides a very good approximation of the measured data. The remaining difference between computed and measured path length differences is about 4 mm. That means a reduction of the path length error of 90%.

Such a on-line calculation of wheel slip depends on the kind of ground, the vehicle's load and additional outer forces due to turns, ground inclinations, trailers, etc. The assumption or approximation of these data is necessary to allow an effective compensation. Remaining lateral slip of non driven wheels depends on centrifugal forces and so on turn velocity. It results in a variation of the vehicle's centre of turn with the effect of of position errors, lateral as well as longitudinal, after the end of the turn.

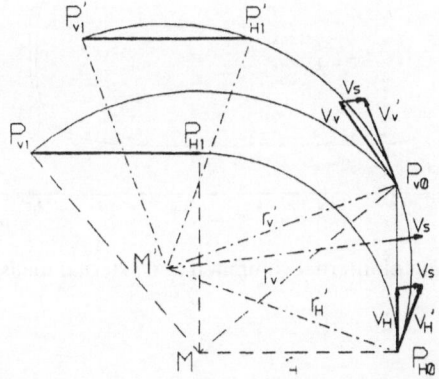

Fig. 9: Slip caused variation of the centre of turn

Localization

The whole localization system of the vehicles IPAMAR I and II has been examined with a test equipment developed especially for this purpose. This reference measuring system "Wire-Track" allows to detect the vehicle's position during its ride without intermediate stop. It was realized by using thin wires and measuring the wire length to different vehicle points (see Fig. 10). These distances provide the vehicle's coordinates in a user chosen coordinate system based on the geometrical relations. These data compared to the on board calculated position data lead to position differences in three coordinates. The on-board calculated path and the external referred path of a closed course is shown in Fig. 11.

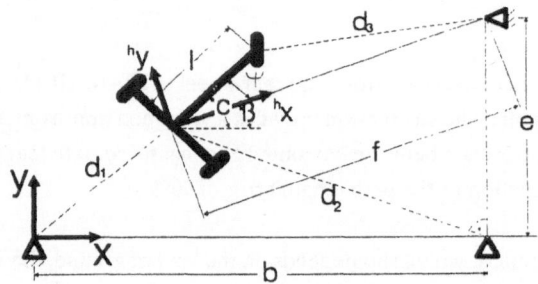

Fig. 10: Reference system "Wire-Track"

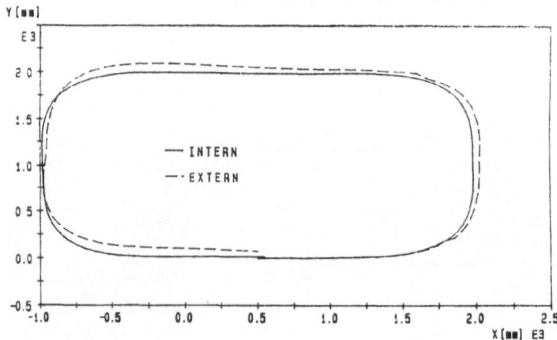

Fig. 11: Comparison of internal computed and external measured driving paths

Front Wheel Localization

There are lateral deviations of about 20 cm to 30 cm after a straight path length of about 15 m ride. Longitudinal deviations stay below 20 mm. Misorientations occur up to 2°.

In turns there is also a considerable misorientation up to 2°.

The performance data of this dead reckoning method ist limited by these orientation errors.

The above data are average test results on rigid PVC or on wash floor. On other kinds of floor the performance data vary up to factor 3 or 4.

Rear Wheel Localization

Here the performance is a little better when driving straight. The lateral deviation after a path length of 15 m is 10 cm - 15 cm. Lateral deviations are only about 5 mm. Misorientation doesn't exceed 1°.

In turns however the performance becomes much worse compared to front wheel localization. After a 90°-turn there is a misorientation up to 5°.

The performances are once again very much floor dependant.

Gyro Supported Localization

In this case the above equipment "Wire-Track" could not be used: the performance data of the on-board dead reckoning system could not be detected because of the reference system's limited accuracy (about 5 mm). Therefore the path length of the vehicle's course was enlarged up to about 80 m. The position error at the end of this course, longitudinal as well as lateral, did not exceed 2 cm.

Even more remarkable: the misorientation angle after an elapsed time of about 5 minutes is 0.1° - 0.3°.

Similar results have been achieved on carpeted floors and on concrete and asphalt surface in out-door-areas.

So the above worst-case error calculations are confined in quality by the test results. The real data differ a lot and are of about half the value of the estimated differences.

Summary and Outlook

The limited performance of pure odometric procedures for localization is shown evidently by the test results. Position updates within distances below 20 m of path length are necessary when navigating exclusively with odometric sensors to ensure tracking within a bandwidth of about ± 20 cm. This especially requires either a suitable environment or special installations to refer on.
For driving long distances (> 100 m) without position update it is essential to locate by gyro support. Also rides from starting position to destination without intermediate reference are possible provided the distances are not too long or the admissible position tolerance is respectively wide.

Assessment-Characteristic	path length [m]	position errors [mm]		misorientation [degrees]	
		longitudinal	lateral	straight	90°-turn
Procedure					
front-wheel localization	15	15	<300	2	2
rear-wheel localization	15	5	<150	1	5
gyro supported localization	80	15	15	0.3	0.1

Fig. 12: Comparison of typical performance data of dead reckoning procedures for AGVs on industrial wash floor

Suitable gyros, especially mechanical rate gyros, can be found not only in military but also in civil, especially civil nautic applications.
Advanced technologies, e.g. fibre optic rate gyros, give hope for the future once to have cheap, rugged and precise mass products. Nowadays however they are available as prototypes for test purposes only.

Further tests with different rate gyros, also fibre optic gyros, are beeing prepared at IPA.

New applications for AGVs are possible with the technology of gyro supported dead reckoning localization which are a domain for traced vehicles up to now. Especially for long path lengths without position update, e.g. for out door AGVs, high performance, flexible and cheap solutions come into view.

Bibliography

/1/ G. Drunk: Sensor- und Steuerungssystem für die leitlinienlose Führung automatischer
   Flurförderzeuge, IPA-IAO Forschung und Praxis, Volume 147, Springer Verlag 1990

/2/ R.D. Schraft, G. Drunk, A. Merklinger, S. Forster: Mobile Autonomous Robot IPAMAR
   Performs Free Ranging AGV Operation, proceedings of AGVS-conference in Berlin, june
   1989

/3/ M. Jantzer: Bahnverhalten und Regelung fahrerloser Transportsysteme ohne Spurbindung,
   ISW Forschung und Praxis, Volume 82, Springer Verlag 1990

Guide Dog Robot Harunobu-5
- Stereotyped Motion and Navigation -

H. Mori

Abstract

Paradigm of two vision system visuo-motor system and vision analysis system is introduced in this paper. To apply this paradigm to guide dog robot, visuo-motor system is realized by a set of sign pattern-based stereotyped motions, Moving-Along, Moving-Toward, Following-a-person and Moving-for-Sighting. Instead of vision analysis system guide dog robot has an interactive navigation system which starts and stops each stereotyped motion exclusively, allocates vision resources to points of attention, and matches the current location with a digitized map. The robot has a guide map expert system and voice interface to communicate interactively with the visually impaired person.

1. Introduction

Harunobu project for mobile robot development has been started since 1982 in Yamanashi University. This project has two aims.

The first aim of the project is the investigation of human visual processing. From 1960s to 1970s according to the experiments of animal brain's ablations, two visual systems, *visuo-motor system* and *vision analysis system*, were proposed to explain human behavior; The former is concerned with visual orientation and the latter is with visual discrimination. In a series of studies of the golden hamster Schneider drew a distinction between visual orientation and visual discrimination [1,2]. He showed that lesions to the superior colliculus selectively destroy the hamster's ability for visual orientation, and lesions of the visual cortex selectively disturb pattern discrimination. Trevarthen concluded that vision system is decomposed into two subsystems; focal vision and ambient vision system. Table 1 shows a list of characteristics of two vision system which is originally written by Trevarthen[3,4] with a little modification in this paper.

Table 1 Two visions

Vision analysis	Visuo-motor	Vision analysis	Visuo-motor
focal vision	ambient vision	what	where
day eye	night eye	analysis	localize
central field	peripheral field	goal object	cue object
cone system	rod system	object holding	field holding

The vision analysis system which lies on retina-geniculo-striate cortex pathway processes stimuli projected on central field (fovea), and the visuo-motor system which lies on retino-superior colliculus pathway processes stimuli projected on peripheral field. Rolls of vision analyzing system are the identification of object and the 3D perception of environment. On the other hand rolls of visuo-motor system are guidances of behavior such as head motion, eye movement, handling and walking.

Tinbergen called the specific visual behavior behaviors of animals which are inherent by species "fixed action pattern"[5]. The fixed action pattern is defined by a temporal and spatial contraction patterns of a group of muscles which are released when a specific stimulus is given or physiological level of the nervous system caused by hormone exceeds a certain level. Tinbergen proposed a model of nervous system of fixed action pattern.

It is assumed that even in higher animal voluntary and non voluntary visuo-motor actions are decomposed into fixed action patterns which are more flexible and rich in variety.

However a conceptual model and function of two visual systems have been proposed, computational model and algorithms haven't been well-known yet.

It is assumed that mobile robots are classified into two groups; vision analysis oriented robot and visuo-motor oriented robot. Each group prefer paradigms and terms listed in Table 2. NAVlab developed by CMU[6] and ALVin developed by Martchin Marietta[7] may be examples of vision analysis robot, and Seymor developed by MIT[8] and Harunobu[9,10,11,12] may be examples of visuo-motor oriented robot.

Table 2 Paradigms and terms preferred by two group

Vision analysis oriented robot (NAVlab,ALVin)	Visiuo-motor oriented robot (Seymour, Harunobu)
Black board model	Autonomous distributed control
Path planning	Stereotyped motion, Reflective behavior
Environment representation	Sign pattern
Sensor fusion	Active sensing
Act after thinking	Act while thinking

In Harunobu project, functions and rolls of visuo-motor system in road following have been studied and implemented on a mobile robot and tested in real world. Little parts of vision analyzing system is realized in Harunobu-4.

The second aim of Harunobu project is the development of small outdoor robots with the following advantages:
(1) Intelligent robot applicable to common road environment: The robot can move in variety kind of road environments, not only asphalt paved road but also color-tile paved sidewalk. It can detect and avoid obstacles and distinguish them from shadows of buildings and trees.
(2) Town robot: The robot moves roads or sidewalks of town at almost the same speed as man walks. It performs simple tasks such as carrying goods from home to home, scavenging road, painting lane mark and guiding visually impaired person.
(3) Economical robot: The cost of robot is less than three years's salary of a worker.
(4) Small size and light weight robot: The robot is small enough to move sidewalks and also light enough not to injure man and environment accidentally.
(5) Low electricity consumption robot: The robot consumes less than 400 watts in electric power, 200 watts for motors and 200 watts for computer system including vision system.

In Japan among 250,000 visually impaired peoples who are more than 18 years, about 130,000 are almost blind and require a guide dog. However guide dog training requires much effort and skill, guide dog in Japan does not increase remarkably. Now the number of guide dogs are about 600 , 0.5% of total demands.

The seven years project for guide dog robot was done in Mechanical Engineering Laboratory in Japan from 1977 [13]. Guide dog robot MELDOG followed white bars painted on road by its CCD sensor. Because the computer power and ability of 10 years ago was not enough strong, the robot didn't reached to a practical point.

But now the rapid progress of 32 bits microcomputer. and TV camera have caused the advanced study of autonomous mobile robots. Since 1989, as one of the applications of mobile robots in Yamanashi university *guide dog robot Harunobu-5* has been developing.

2. Guide dog robot outline

A combination of visually impaired person and guide dog performs an excellent locomotive behavior. Whenever the master(visually impaired person) wants to go to a place, he makes a planning of route from his house to the destination on a cognitive map in his head, then he give the dog such a command as go right or left. The kind of commands which the master give to the dog are surprisingly small as shown in Table.3.

Table 3 Guide dog command list

Start stop command	GO, STOP
Orientation command	BACK, RIGHT, LEFT, GO-CLOSE
Estimation command	GOOD

The guide dog stops at every crossing or corner, he matches current location with a cognitive map which he made before the beginning of locomotion, and verify the environment of the location with his remaining senses. He judged whether the dog stops at the crossing or corner where he predicted, then he give the dog the command which shows the relevant orientation to the destination. When there is an obstacle on their way, the dog makes an avoiding path which is enough in space for master to walk. When a car coming the dog waits against master's command until the car passes by.

It seems that the visually impaired person and guide dog share the role of knowledge-based analyzing behavior and visuo-motor reflective behavior.

These sharing principle is followed in guide dog robot designing as listed below.

1) Assisting master to keep balance: By grasping a handle attached to the rear of the robot, the master can get information about abrupt changes in the ground to keep his balance against them. Fig.1 and Table.4 show the outline of the robot and its components.

2) Matching speed of robot with masters: A handle is attached on the rear of the robot. By touching the handle, the master can adjust robot's movement to his walk.

3) Sign pattern based stereotyped motions: As primitive visuo-motor actions, four kinds of stereotyped motions are implemented; Moving-Along, Moving-Toward, Moving-for-Sighting and Following-a-Person.

4) Avoiding stationary obstacles: Usually permanent obstacles such as litter, tree and pole are fixed in their locations. But the prediction of temporary obstacles such as parking cars and bicycles is not possible. The robot has to detect not only permanent obstacles but also temporary ones mainly by image processing and avoid them.

5) Avoiding coming car: When a driver catches the robot in his visual field and has enough time, he can steer to avoid the robot. But when the robot is going to run behind a parking car to the center of the road, or going to enter from a small passage to a large road, the driver of coming car will not have enough time to find the robot. Therefore the robot has to look at a distance and finds whether there a car is coming or not. When the coming car has passed by, the robot can enter the road or avoid the parking car.

6) Moving across zebra mark: In Japan a zebra mark is painted in front of the crossing. When the robot is far from a crossing, the zebra mark becomes a sign pattern for moving toward the crossing. When the robot moves across the crossing, the zebra mark informs the width and direction of the passage to move across.

7) Voice interface: Instead of keyboard and display the robot has synthesized voice maker and word recognizer to communicate with the master.

8) Interactive navigation: The robot tells the next motion to be done through voice

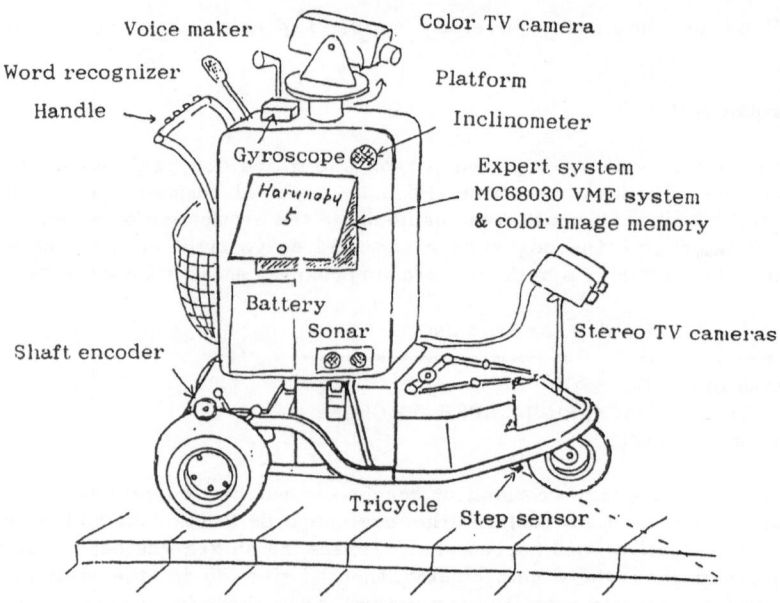

Fig.1 Guide dog robot " Harunobu-4

Table 4 System components

Undercarriage system	
Tricycle	Suzuki Co. SENIOR CAR ET-10
	(Front wheel steering, rear wheels driving)
Locomotion control	NEC Co. Lap top computer PC9801N(8086 10MHz)
Sonar	Ultrasonic sensers for lateral obstacle detection
Step sensor	LED sensers
Inclinometers	Lucas sensing system Inc. ACCUSTAR II
Gyroscop Scope	Hitachi Cable Ltd. OPTICAL GYROSCOPE OFG-3
Expert system and Voice interface	
Expert system	Epson Co. Lap top computer PC286-L(8086 10MHz)
Voice maker	ASCII CO. VOICE MAKER
Word recognizer	Sanyo Co. KIKIWAKEKUN
Image processor	
CPU	MC 68030 (25MHz) VMEsystem 4MB
OS	OS-9
Color image memory	AVAL DATA Co. TVME-338 (1024(H)*512(V)*8(bits)*3(RGB))
Color TV camera	Sony Co. 8mm Handycam
Stereo TV cameras	Victor Ltd. TK-S300
Platform	Maid-to-order by Bit Engineering Co. (2 worm geared DC servo
	motor with shaft encorder, gyroscope for yaw, inclinometers
	for pitch and roll)

maker. The master judges the relevance of the command by knowledge-based integration of remaining senses and answers "OK" or "NO".

9) <u>Guide map expert system</u>: Robot has a digitized road map around the neighborhood of the master's living. In the map crossings, public facilities and shops are located with their names. When the master asks the robot "Where ?", the expert system of the robot answers the current location. Fig.2 illustrates functional shares of locomotion between the master and the robot.

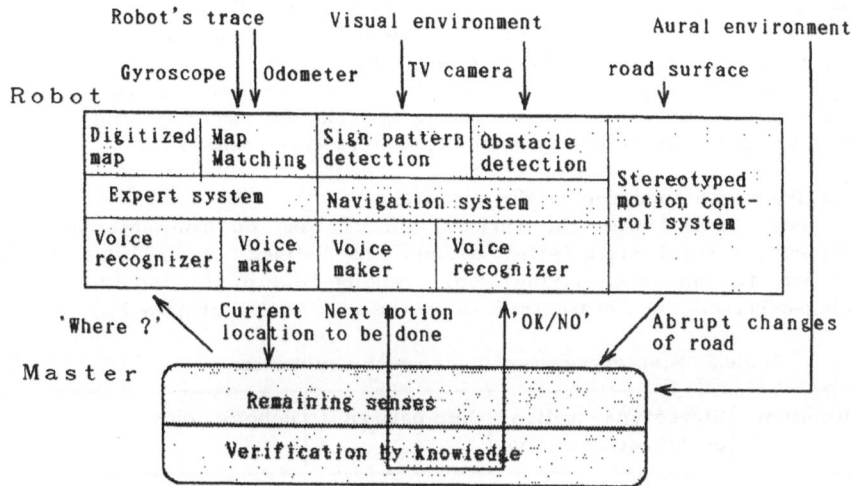

Fig.2 Functional architectures of guide dog robot " Harunobu-5

2. Visuo-Motor System in Guide Dog Robot

Swedish psychologist Jannson[14] defined vision-motor behavior with four primitives; Keeping-balance, Moving-toward, Moving-along and Moving-around in order to explain locomotion of visually impaired person.

Infuluenced by the study of ethologists especially Tinbergen, a paradigm for autonomous mobile robot called "Sign pattern based stereotyped motion" was proposed and applied to a motor tricycle size autonomous robot "Harunobu-4"[9,10,11].

Stereotyped motion is defined by a fixed motion which the robot performs when it encounters a certain environment and situation. A visual pattern which is utilized as a guide for stereotyped motion is called *sign pattern.* In other word stereotyped motion is performed by a fixed feed forward control pattern which utilizes the sign pattern as a target. Assume that a driver is going to steer his car into the garage. He will steers always in the same fixed steering pattern looking at not the whole of the garage but a part of the garage such as wall or pole. This steering is called stereotyped motion and the part is called sign pattern.

The stereotyped motion of mobile robot is defined not only the motion of undercarriage but also the motion of TV camera which coordinates with that of undercarriage. To use a sign pattern as a guide, TV camera should trace the sign against robot's pitching, rolling and yawing.

Stereotyped motion is different from reactive behavior[15] or reflexive behavior[16]

Merits of stereotyped motion against path planning are as follows.

(1) <u>Flexible to various environment</u>: Stereotyped motions identify only sign patterns and ignore all the other visual pattern, therefore they are flexible to various environment.

(2) <u>Easy to implement</u>: A sign pattern is specified by a long edge lying on the road, a long edge standing on the road or a moving object on the road. The robot detects only the road and sign pattern and ignore all the other patterns such as buildings,
 trees and sky. Detection algorithms of road and sign pattern are very simple and easy to implement.

(3) <u>Quick in processing cycle</u>: Very simple algorithms result in quick response to visual input. In dynamic vision an image of road scene is not so different from the last image, locations and attributes of the road and sign pattern are predictable [13].

(4) <u>Low cost</u>: To implement these simple algorithms, a distributed microcomputer system, TV camera system, ultrasonic sensors and touch sensors are required. Any kind of expensive computer or additional devices are not required.

(5) <u>Useful in practical application</u>: In practical usage such as construction application, the robot moves the same course repeatedly. The movement of this course can be specified by a sequence of sign pattern-based stereotyped motions.

2.1 Primitive stereotyped motions for locomotion

In order to deal with the various kind of road environments and degree of crowdedness, several stereotyped motions are defined as in Table 5. Primitive stereotyped motions are Moving-Along, Moving-Toward, Moving-for-Sighting and Following-a-Person. The fist two motions are the same as that proposed by Jansson[14].

Table.5 Stereotyped motions for each environment

Environment	Stereotyped motion or Reflective motion	Sign pattern	Robot's voice
Low crowded straight road	Moving-Along	Lane mark, Road boundary	'Find an obstacle' 'Near a crossing'
Shaded or partially wet straight road	Moving-Straight by dead reckoning	None	
Moderately crowded road	Following-a-Person	Pedestrian	'Start following' 'Change following' 'Stop following'
Near crossing	Moving-Toward	Zebra mark	'Near a crossing' 'Stop at crossing'
At crossing (Go straight)	Moving-Straight	Zebra mark	'Let's go'
At crossing (Turn)	Wait and Moving-for-sighting	Lane mark	'Wait' 'Lets turn right'
In front of obstacle	Avoiding-Obstacle (A chain of stereo-typed motions)	Lane mark, Obstacle	'Near obstacle' 'Wait' 'Let's go aside' 'Free from obstacle'

Moving-Along. Moving-Along is a locomotion strategy guided by a sign pattern lying on road in longitudinal direction. The robot will move more than 90 % of the course by this stereotyped motion.

Moving-Toward: Moving-Toward is a primitive stereotyped motion that is directed to a sign pattern which may be such a visual pattern standing on the ground as an entrance of building or stairs or a visual pattern lying on the road as traffic mark. In moving toward a crossing, the zebra mark in front of the crossing can a good sign be pattern.

Following a person: When the road is crowded with pedestrians, the robot will not be able to see the visual guide or target. In such situation following a person who is walking to the same direction that the robot is going to is a good locomotion strategy, because the path that the person has moved guarantees safety and free from obstacles. The robot can neglects complicated processings of obstacle and hole detections.

Moving-for-Sighting: To search a new sign pattern and locate robot itself to the starting point for the new stereotyped motion based on the new sign pattern, a stereotyped motion named Moving-for-Sighting is defined. It is applied when the robot is going to turn a corner as shown in Fig.3 or avoiding an obstacle as shown in Fig.3. The robot keeps its motion along or toward an old sign pattern while TV camera is directed to a new sign pattern, until the new sign pattern is caught in the center of TV camera.

In addition to the above sign pattern-based stereotyped motion several reflexive motion are devised.

Avoiding-Unevend-Ground: A sidewalk often has a small abrupt change on its surface. To make a smooth locomotion on the sidewalk, the robot should detect the change by sensor and move around it reflexively. To detect abrapt changes of road surface, a passive stereo method[19] is implemented.

Watching-Step: To protect the robot against false step from sidewalk to lane, reflexive motion Watching-Step which is activated by step sensors is devised.

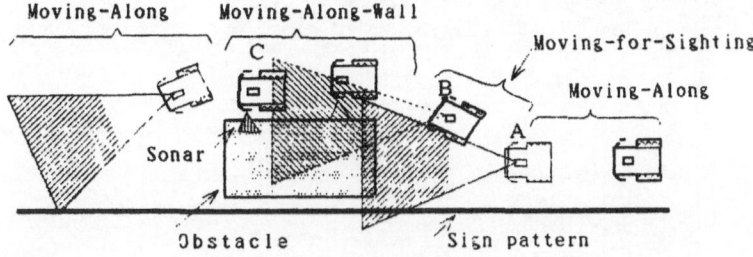

Fig.3 Avoiding-Obstacle

2.2 stereotyped motion chain

Complex motions such as turning corner and avoiding obstacles are performed by a stereotyped motion chain, in other word, a fixed sequence of stereotyped motions. For example, Avoiding-Obstacle is defined as shown in Fig.3. At first the robot is in Moving-Along and finds an obstacle at point A where a distant part of the road is hidden by the obstacle. The first tasks to be done are to localize the obstacle and to look the hidden part of road. So the first stereotype motion is Moving-for-Sighting in which the right side of obstacle is specified as a temporal sign pattern. It guides the robot to point B where the temporal sign pattern is seen in the center of TV image Frame. At point B the distant part of the road and the right side is in view. Moving-Along-Wall will be done until an ultrasonic sensor attached to the left side of robot detects nothing. Then robot resumes the interrupted Moving-Along and returns to the old course.

2,3 Dynamic Window

Dynamic window setting in Moving-for-Sighting are shown in Fig.4. At first a window is set up on the right side of an image frame waiting for a new sign to appear in the frame as shown in Fig.4(a). Sign pattern detection method is applied on the image in windows[Mori,1988]. A sign pattern is represented as shown in Fig.5(a) by

Ucos θ +Vsin θ =D

In Fig.5(b) values of θ and D are plotted along robot movement. In Fig.4(a) and (c) edges of wall and pole are detected as sign patterns at first but identified non-sign patterns by their angle θ and distance D. In Fig.4(d),(e) and (f) θ and D of the new sign pattern lying on road gradually decrease to zero when the robot approaches to the corner.

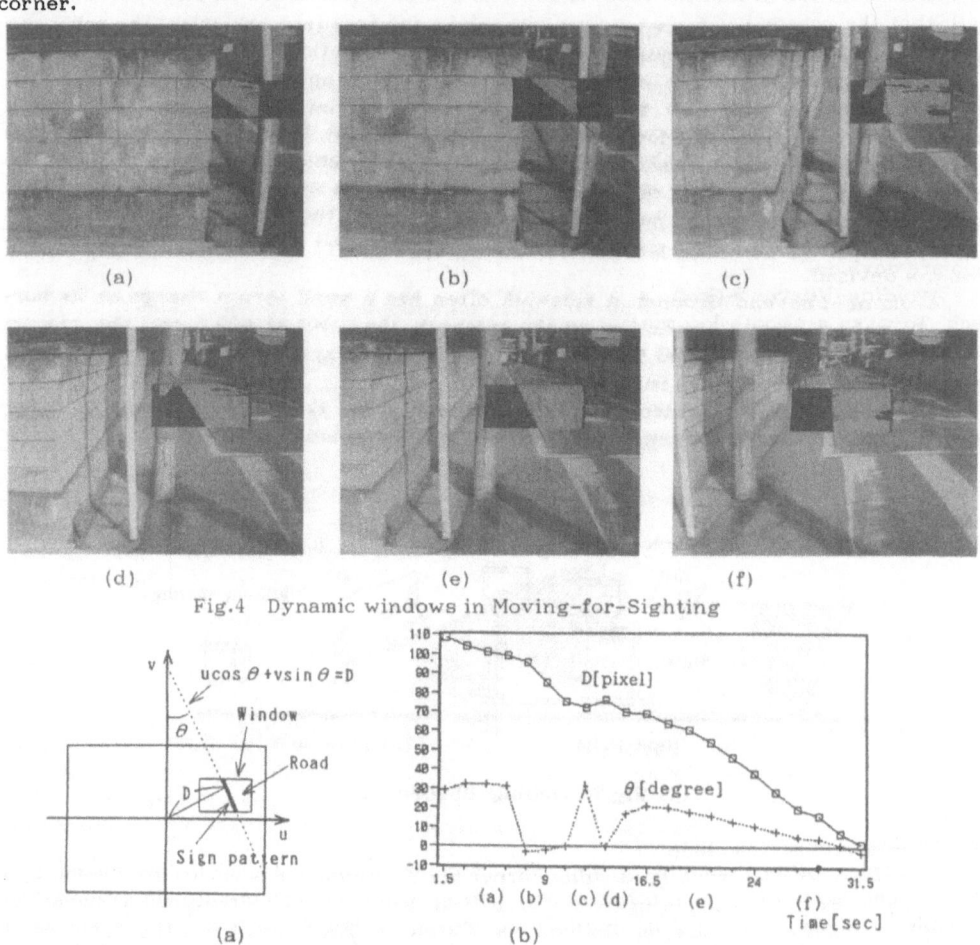

Fig.4 Dynamic windows in Moving-for-Sighting

Fig.5 D and θ curves in Moving-for-Sighting

2.4 Active sensing in stereotyped motion

In this paper active sensing is defined by the sensor and motion coordinative contol which guides the robot to the place where a sign pattern may be seen, and which directs TV camera to the direction in which the object may be in view.

The active sensing in stereotyped motion is different from animate vision[17] in planning TV camera control. In animate vision the pattern of TV camera control which is called gazing control is not fixed and has to be planned to fit each case, but in stereotyped motion the control patterns of locomotion and TV camera movement are fixed. For instance in Moving-for-Sighting, the locomotion control system guides the robot to the point where a new sign pattern will be in view, while the TV camera control system turns the TV camera to the predicted direction of the new sign pattern. In this case only the new sign pattern's parameters are fed to the two systems.

Advantages of fixed active sensing are as follows:

(1) Decrease of searching failures: The Robot often misses a sign pattern or confuses other pattern as the sign pattern by the cause of narrow visual angle and poor recognition function. Searching failures and mean searching time will be decreased, by fixing camera control patterns which are fitted to the constraints of robot size, height of TV camera, visual angle and image processing speed.

For outdoor mobile robots, heuristic path searching methods for minimum set of elementary paths in distance or time have been investigated by many researchers. However in real environment a robots usually cannot identify the path ahead when an obstacle stands in front of the robot. In this case it will be very difficult to make path planning based on ambiguous information about the path ahead.

(2) Shortening of image processing time: By filling camera control model with the knowledge of sign pattern's location and orientation, camera control pattern is obtained. The control pattern is able to limit searching area of sign pattern in TV image and decrease its processing time.

(3) Step by step improvement of localization: By adding observation motions to objective stereotyped motion, the robot can improve localization errors of sign pattern and obstacle.

2.5 Stabilization of locomotion and TV camera system

To implement active sensing, Locomotion and TV camera systems must be stabilized in posture. Common roads have irregular surfaces and lateral slopes for drainage as well as longitudinal slope. The mobile robot will swing in pitch, roll and yaw when it moves in these roads. These swings cause larger localization errors of objects than lateral motion and up-and-down motion. Those kinds of swings even cause sign pattern missing when the swings are large in magnitude. Stabilization of locomotion and TV camera movement are necessary for active sensing.

2.5.1 TV camera stabilizing

When the robot is moving around an obstacle in Avoiding-Obstacle state, its heading will be gradually changed and fixed TV camera will miss the sign pattern or the obstacle. To keep looking for the sign pattern or the obstacle, the TV camera should be directed parallel to its direction.

The robot swings usually about 2 or 3 degrees in rolling and pitching on the asphalt road. This swing causes much errors in object localization by vision. In harunobu-4 a TV camera with wide angle lens of 40 degrees in horizontal visual angle is fixed at the head of the robot about 140cm in height and connected to an image frame memory of 512×512 pixels. When TV camera turns 10 degrees in tilt and localizes an obstacle 2.6m ahead, location errors by rolling and pitching are considered as follows.

$\Delta x \alpha$ =0.22cm/degree, $\Delta y \alpha$ =2.1cm/degree

$\Delta x \beta$ =2.1cm/degree, $\Delta y \beta$ =20.7cm/degree

Where Δx and Δy denote location error of lateral and depth, α and β denote rolling and pitching. Quantizing errors caused by image memory pixel size are as follows.

Δ xh=0.4cm/pixel, Δ yh=0.0cm/pixel, Δ xv=0.2cm/pixel, Δ yv=2.1cm/pixel

Where h and v denote horizontal and vertical. Localization errors caused by rolling and pitching are considerably more than those by quantizing errors. The stabilizing of TV camera is very important for object localization.

These small swings need much time to correct mechanically, therefore localization errors caused by a small swinging are corrected by computational calculation

3. Shadow elimination

In road following of a vision based mobile robot, the problems of shadow are very important and difficult. The first problem is that it is difficult to detect sign pattern accurately on shaded road. The second problem is that robot may misidentify a shadow as an obstacle and avoid it, or not be able to find a sign pattern.

Several research groups have reported color vision-based approaches for shadow problems[6,7]. In NAVlab a road following algorithm deals with the first problem [6]. This algorithm classifies each pixel of image into road and nonroad by its color models and uses the result of classification to select the bestfit road position, and then updates color models based on the detected road and nonroad regions. However in u

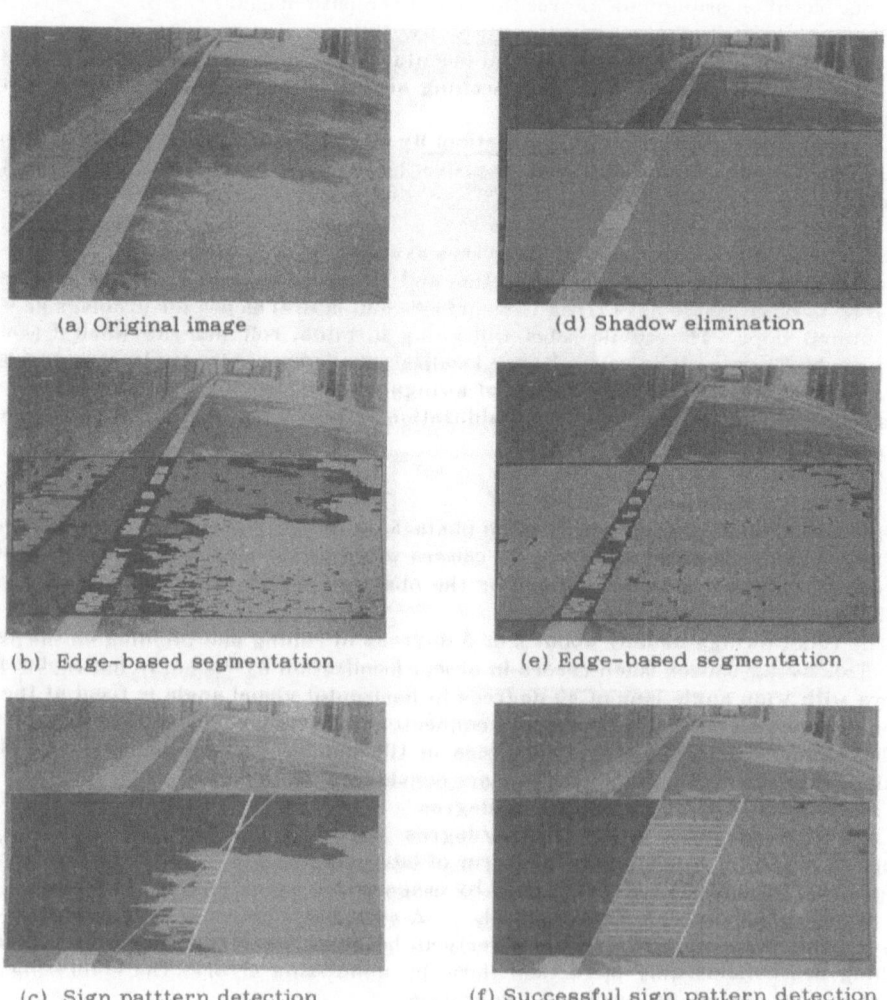

(a) Original image   (d) Shadow elimination

(b) Edge-based segmentation   (e) Edge-based segmentation

(c) Sign patttern detection   (f) Successful sign pattern detection

Fig.6 Shadow elimination

sual road scenes it is very difficult to classifies road and nonroad pixel by their colors. In order to solve the second problem, NAVlab and ALVin utilize laser range sensors.

In Harunobu-4 color vision system is utilized to process shadow image[9]. On sunny day, shadows on a road causes edges between sunny and shaded regions as shown in Fig.6(a). An edge detecting filter is operated on the image. After edge-based segmentation, the road will be separated into several different regions as shown in Fig.6(b). A largest regions in area is selected as a road candidate regions. Holes in the candidate region are filled for noise elimination. Sign pattern is detected by line scanning in road regions as shown in Fig.6(c). Where the detected sign pattern in the last scene is used to predict the next sign pattern, The road candidate region is scanned horizontally from the predicted sign pattern to the boundary of the region. Boundary points obtained by horizontal scanning are approximated by a line (expressed by $U\cos\theta +V\sin\theta =D$) which becomes a new sign pattern candidate. In this example sign pattern candidate is judged to be a false sign pattern, because θ and D of the sign pattern are different from the predicted one.

To detect the real sign pattern, a shadow elimination method was developed[9]. In this method, each pixel is classified into shadow or non shadow groups by its brightness ($Y = R+G+B$) and normalized red ($r=R/Y$) and blue ($b=B/Y$) components. Those pixels that are classified as shadow group are changed in RGB components from the original one to the average RGB components of sunny road region. Thus shadows on the road can be eliminated as shown in Fig.6(d). After this an edge-based segmentation is shadow elimination is done as shown in Fig.6(e), and a sign pattern will be detected successfully as shown in Fig.6(f).

3.1 Shadow elimination algorithm

Sunny roads are illuminated mainly by sun light of about 5,000 ° K. On the other hand shaded roads are illuminated by blue sky of about 14,000 ° K. Hue of shaded road slightly shifts to blue. In other word average b component $\mu b1$ of shaded region is greater in its value than $\mu b0$ of sunny region, whereas average r component $\mu r1$ of shaded region is smaller in value than $\mu r0$ of sunny region as shown in Fig.7. Caused by poor S/N ratio of video signal, standard deviations $\sigma r1$ and $\sigma b1$ of shaded regions are much larger in values than $\sigma r0$ and $\sigma b0$ of sunny region. If a pixel falls into the box of which center is fixed at ($\mu r1, \mu b1$) the pixel is judged to be a shaded pixel.

Fig.8 shows dr(= $\mu r0- \mu r1$) and db(= $\mu b1- \mu b0$), $\sigma r1$ and $\sigma b1$ for the scene of a road shaded by trees. $\sigma r1$ and $\sigma b1$ show almost constant, while dr and db show slight changes. In general sunny region of a new image is detected successfully, therefore ($\mu r1, \mu b1$) of shaded regions are predicted by ($\mu r0, \mu b0$) and (dr, db).

(14

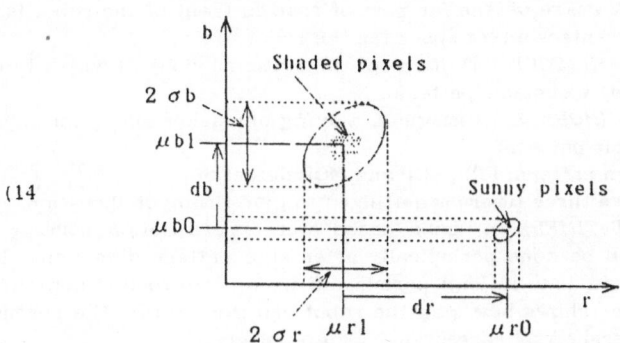

Fig.7 r-b scattering diagram

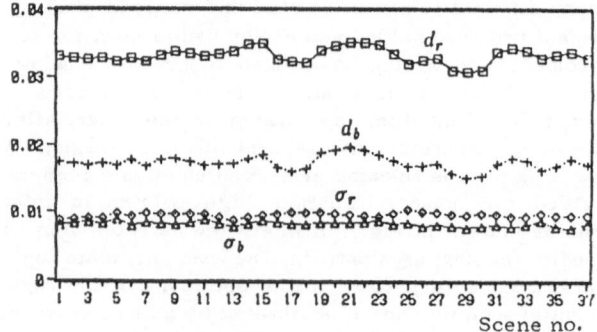

Fig.8 dr and db, σ rl and σ bl changes in a Moving-Along shaded road4.

Interactive Navigation System in Guide Dog Robot

Roles of interactive navigation in guide dog robot are as follows
(1) Switching of stereotyped motions: Navigation system(NAV) switches stereotyped motion module(STM) at the end of each sign pattern by referring to the digitized map given before start.
(2) Allocation of vision resources to points of attention: NAV switches windows for sign pattern detection, free space detection and obstacle detection by the timeout signals of timers .
(3) Map matching: NAV localizes the current location which is expressed by (x,y, θ) on the map, where (x,y) denotes abscissa and ordinate values of digitized map coordinate system, and θ denotes robot's heading. The current location is corrected by map matching at every crossing or corner.
(4) Voice interaction: Before sending a command to the stereotyped motion control in the undercarriage, navigation system informs the master the command through voice maker and asks him to verify the command to be relevant or not with his remaining senses. After receiving "OK" through voice recognizer, the navigation system makes the motion control performs the command. When the system receives "NO", the second candidate of relevant command is issued.
(5) Location correction: The current location (x,y, θ) is obtained by dead reckoning and corrected periodically by sign pattern-based localization and corrected at every crossing by map matching-based localization.
To detect dangerous objects which have the possibility to damage the robot, three kinds of patterns are defined.
Free pattern (FRP): A TV image of the far part of road in front of the robot is called free pattern. Because it guarantees a free space for the robot to move.
Stationary obstacle pattern (SOP): A TV image of the stationary object on road in front of the robot is called stationary obstacle pattern.
Moving obstacle pattern (MOP): A TV image of moving obstacle such as car or pedestrian is called moving obstacle pattern:
Fig.9 shows windows for sign pattern, FRP, SOP and MOP detection.
To allocate vision resource three timers are utilized to give timing of allocation.
Dead reckoning timer (D-TIMER): In vision-based stereotyped motion, corrections of lateral offset of location will be done periodically after sign pattern detections. In flat smooth road the correction is done in long period and in uneven road is done in short period. Dead reckoning timer shows how long the robot can move within the permissible lateral offset by after a lateral offset correction by sign pattern.
The navigation system(NAV) gets a certainty factor γ $(0.0=< \gamma <=1.0)$ from the sign

Sign pattern   Free pattern   Stationary or moving
                                obstacle pattern

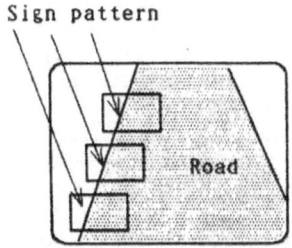 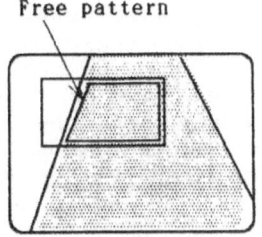 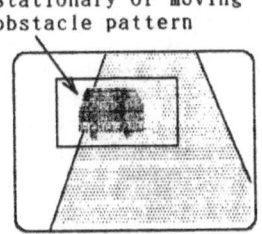

(a) Windows for sign pattern (b) Window for FRP   (c) Window for MOP
Fig.9 Windows for sign pattern, FRP, SOP and MOP detection

pattern detection module (SPD). The factor shows the reliability of the detected sign
pattern. NAV assigns γ Td to the timer. Where Td is given as follows.

Assume that the robot moves y[m] along a straight line by dead reckoning method,
and makes x[m] lateral offset. Lateral offset ratio is given by k=x/y. Mean of k equal
to zero. σ k,the standard deviation of k, is depend on unevenness of road surface.
Suppose that the maximum of permissible lateral offset Xlimit[m], the speed of robot V
[m/sec] and processing time of sign pattern detection τ s have been given. For 5% of
error of the first kind the lateral offset is within 2 σ k*v*Td which must be less than
Xlimit.

2 σ k*V*Td < Xlimit

Td < Xlimit/(2 σ k*V)

By considering τ s, Td is given by

Td=Xlimit/(2 σ k*V)- τ s [sec]

Collision timer (C-TIMER): Collision timer shows how long dose it take for the
detected obstacle to collide with the robot. The setting time is different by the moving
speed of the obstacle.

Assume that by processing of a TV image taken at t=0 [sec], the distance of an obstacle D $\pm \Delta$ D [m] and the speed of the obstacle Vo $\pm \Delta$ Vo [m/sec] are obtained, and the
speed of the robot itself Vr[m/sec] is known, the time of collision Tc'[sec] is given by the
following equation.

Tc' =(D $\pm \Delta$ D)/(Vo $\pm \Delta$ Vo +Vr)

By considering image processing time for obstacle detection τ o, NAV assigns collision
timer Tc by the following equation.

Tc = (D- Δ D)/(Vo+ Δ Vo+Vr) - τ o

The allocation of vision resources is done as follows.

(1) NAV send a sign pattern detection signal to the stereotyped motion control system
(STM) and receives 'SP_FOUND' and γ from STM, calurate Td and set D-TIMER at γ Td.

(2) NAV sends a free pattern detection signal to STM and receives 'FRP_FOUND', this
process is repeated, until D-TIMER is zero in value. Then the process goes back to (1).

(3) When NAV receives 'SOP_FOUND' or 'MOP_FOUND' in free pattern detection cycle,
NAV calculate Tc and set C-TIMER at Tc. NAV sends a signal of stationary or moving
obstacle detection and reset C-TIMER by the newly calculated Tc. When Tc is less than
τ 0, iteration stops. Where τ 0   is the minimum time to avoid the obstacle in safe.

5. Concluding remarks

All the necessary data to design the guide dog robot is obtained through outdoor tests
of Harunobu-4.

(1) <u>Rolling and Pitching</u>: Rolling and pitching of Harunobu-5 in general road is less than
five degrees.

(2) <u>Lateral offset ratio</u>: Lateral offset of dead reckoning-based locomotion in general road can be decreased by a direction sensor such as gyroscope or electro-magnetic compass. Experimental results of Harunobu-4 which has an electro-magnetic compass showed that moving by 20cm/sec on the campus road in which the robot swings less than 3 degrees in pitching and less than 4 degrees in rolling, the standard deviation of lateral offset ratio σ k will be less than 0.01, which means that the robot can move without lateral correction by vision about Td=Xlimit/(2 σ k*V)- τ s [sec]. (For Xlimit=20cm and τ s=1sec, Td=49sec)

(3) <u>Image processing speed</u>: It takes 1.5sect to detect a sign pattern by MC68020(16.7MHz) VME system. Processing speed will be improved by substituting MC68040(25MHz, 13.5MIPS) for MC68020(16.7MHz, 2.7MIPS). To make better use of MC68040, VME bus has to be changed by future bus.

Many difficult problems of guide dog robot remain untouched such as:

(a) <u>Maintenance</u>: How to maintain the guide dog robot by visually impaired person himself. To keep the robot ready to use, the person has to check all the sensor and motors and has to supply electric power by his remaining sensors. A maintenance expert system will be required.

(b) <u>Power consumption</u>: In Harunobu-4 the undercarriage system consumes about 90 watts, however image processing system including TV camera and monitor consumes 265 Watts. To supply electric power by batteries, power consumption of image processing system must be reduced remarkably.

(c) <u>Portability</u>: Guide dog robot described in this paper is too large to go into a shop or restaurant, and too powerless to go up and down stairs. The usage of the robot will be limited to guiding of neighborhood.

Acknowledgement

I would like to thank to ten students of master course who developed and have been developing since 1985. Moving-for-Sighting is developed by S.Nakayama and shadow e-limination by H.Chen. I would also like to thank members of Harunobu support company group for financial and technical support; especially Sokkisya Co. Ltd., Tokyo Electric Co. Ltd., Fuji Zerox Co. Ltd., Touyou-douro-shisetu Ltd., Bit Engineering Co. Ltd., Suzuki Co. Ltd.

References

1 Schneider G.E.: " Contrasting visuomotor functions of tectum and cortex in the golden hamster " Psychol. Forsch., 31, 1967, pp.52-62.
2 Schneider G.E.: " Two visual systems " , Science, 163, 1969, pp.895-902.
3 Trevarthen C.B.: " Two mechanism of vision in primates " , Psychol.Forsch., 31a, 1968, pp.299-337
4 Trevarthen C.B. & Sperry R.W.: " Perceptual unity of the ambient visual field in human commissurotomy patients " , Brain, 96, 1973, pp.547-570
5 Tinbergen N.: " The Study of Instinct " , 1951, Oxford University press.
6 Thorpe C., Herbert M.H., Kanade T. & Shafer S.A.: " Vision and Navigation for the Carnegie-Mellon Navlab " , IEEE Trans. on PAMI, Vol.10, No.3, 1988, pp.362-373.
7 Turk M.A., Morgenthaler D.G., Gremban K.D. & Marr M.: " VITS - A Vision System for Autonormous Land Vehicle Navigation " , ibid., pp.342-361.
8 Brooks R.A & Flynn A.M.: " Robot Beings " , Proc. of IEEE Int'l Workshop on Intelligent Robots and Systems IROS'89, 1989, pp.2-10
9 Chen H. & Mori H.: " Sign Pattern Detection on Shaded Road " , ibid., pp.350-357
10 Mori H.: " A Mobile Robot Strategy-Stereotyped Motion by Sign Pattern " ,in 5 Robotics Research, ed. by H.Miura & S.Arimoto, (MIT Press, 1990), pp.161-172
11 Mori H.: " Active Sensing in Stereotyped Motion " , Proc. of the IEEE International Workshop on Intelligent Motion Control, 1990, IP-11-19.
12 Mori H.: " Active sensing in Vision-based Stereotyped Motion " , IEEE Int'l workshop on Intelligent Robots ans Systems IROS'90, 1990, pp.167-174

13 Tachi S & Komoriya K.: " Guide Dog Robot " , in Second International Symposium of Robotics Research, ed. by Hanafusa H. & Inoue H., (MIT Press, 1985), pp.333-340

14 Jansson G.: " Perceptual Information for Orientation and Mobility " , in Contemporary Psychology-Biological Processes and Theoretical Issues ed. by J.L.McGaugh, (North-Holland, 1985), 225-2413

15 Arkin R.C. & Lawton D.T.: " Reactive Behavioral Support for Qualitive Visual Naviga tion " ,Proc. of the IEEE International Workshop on Intelligent Motion Control, 1990, IP-21-27

16 Anderson T.L. & Donath M. : " Autonomous Robots and Emergency Behavior: A set of Primitive Behaviors for Mobile Robot Control " , Proc. of the IEEE International Workshop on Intelligent Robots and Systems IROS'90, 1990, pp.723-730.

17 Ballard D.H.: " Reference Frames for Animate Vision " , Proc. 11th Int'l Conf. Artificial Intelligence, 1989, pp.1635-1641

18 Graefe V.: " Dynamic Vision Systems for Autonomous Mobile Robots " , Proc. of IEEE Int'l Workshop on Intelligent Robots and Systems IROS'89, pp.12-23

19 Ferrai F., Grsso E., Sandini G. & Magrassi M.: " A stereo vision system for real time obstacle avoidance in unknown environment " IROS'90, 1990, pp.703-708

Indoor Navigation of Mobile Robots by Use of Learned Maps

P. Kampmann and G. Schmidt

Abstract: Mobile robots can perform various kinds of transportation and inspection tasks in industrial factory plants, e.g. flexible manufactoring systems (FMS) as well as in similar environments such as laboratories, hospitals or office buildings. Although a typical robot mission is tightly scheduled by a central supervisory system, it is impossible to preplan detailed movements within a real-world environment. Therefore, the robot must be able to navigate autonomously around it's environment and to cope with typical problem situations, such as blocked roads or partially unknown or even changing environments.

Navigation of a mobile robot in an known environment is usually supported by an a priori available 2-D map. Within a potentially changing real-world environment, the capability to modify and update an internal map is of vital interest to an autonomous mobile robot. Through the same capability, a robot can even learn a completely unknown environment and construct a map suitable for path planning and other purposes.

This paper presents a self-contained, real-time navigation system, build around a topological structured world model. Starting from a 2-D representation of all obstacle contours restricting the robot's motion, free motion space as well as obstacle space is subdivided through a black-and-white triangulation. By connecting adjacent elements of free motion space, a road-like representation of all viable paths within the environment is obtained. A two-level path planning scheme based on this representation is used for computing mission-specific trajectories, taking into account the three degrees of freedom and kinematic constraints of typical robot vehicles.

An algorithm for mapping a real-world environment accumulates geometrical knowledge, obtained from the environment through a 3-D range imaging device. Due to the unique structure of the world model, every version of the accumulated map is readily applicable for path planning purposes. The flexibility and high real-time performance of the navigation system is demonstrated by selected experiments with the mobile robot MACROBE operating within a laboratory environment.

1 Introduction

The information processing system of a mobile robot is concerned with basic tasks, such as:

- navigating within known or partly unknown environments using suitable maps,

- piloting the vehicle through preplanned paths, while avoiding collisions or evading unexpected obstacles,

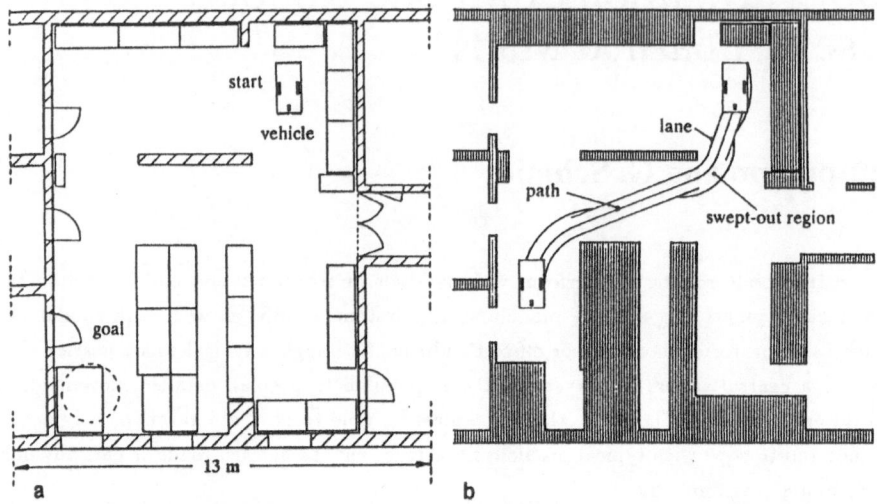

Figure 1: Typical environmental situation for a mobile robot: **a** floor plan of a laboratory; **b** pre-planned path for the indicated mission from start to goal

- monitoring and interpreting the environmental scene next to the vehicle with the support of a multisensor system.

Usually the navigation system of a mobile robot is initialized with a detailed map, representing most of the operation area of interest. Together with a description of the desired mission, the robot will be able to plan and execute a movement, which may be related to transportation of tools or materials within a factory or a laboratory.

Although most of the environment may be known, it is important for an autonomous operation, that a robot is able to update, modify and extend its internal map, whenever this becomes nessecary. As an extrapolation of this exploration and mapping capability a robot may learn and map a completely unknown environment by using its sensors and intelligent algorithms. Only few authors have addressed the problem of a sensor-based construction of a map suitable for navigation purposes [5, 6, 7, 8, 9, 10, 13, 24, 26].

The paper presents a self-contained, real-time navigation system for mobile robots, including the related algorithms for path planning within known or already explored environments and a mapping algorithm for constructing the required maps from 3-D range images.

Within a laboratory or factory the primary motion task (mission) of a mobile robot is to move from a starting location to a specified goal location (Fig. 1a). In this case, path planning is concerned with two major tasks: deciding where to go, in respect to the adjacent obstacles, i.e. finding a coherent region of free motion space (motion channel) connecting start and goal location and computing a feasible path (Fig. 1b) taking into account the vehicle's three degrees of freedom and typical kinematic constraints (section 3).

To achieve real-time performance for path planning, it is neccessary to construct a world model from a geometrical description of the robot's environment, e.g. a floor plan (Fig. 1a), revealing the topological structure of the environment by an unique black-and-white triangulation of both obstacle and free motion space (section 2).

It is possible to construct this type of a topologically structured world model by successively merging newly acquired geometrical knowledge to an existing world model, while preserving and updating ist internal black-and-white triangulation. Thus every version of the world model, acquired e.g. throughout the exploration of an unknown environment is immediately applicable for the presented path planning scheme (section 4)

The flexibility and high real-time performance of the algorithms for constructing the topologically structured world model as well as for path planning is shown by some experimental paths planned within a laboratory for a variety of different vehicle shapes and kinematics.

The effectiveness and correct operation of the presented mapping algorithm is demonstrated through an experiment with the mobile robot MACROBE. Based on a series of 30 range images a detailed 2-D map is constructed. The quality and accuracy of this map becomes evident through a comparision with the known floor plan (section 5).

2 A Topologically Structured World Model

Prior to the design of a navigation system a suitable structure of the underlying maps has to be fixed. Since maps are mainly used for path planning purposes, they should contain the following information:

- a description of all obstacles limiting the robot's free motion,

- a road-like representation of all viable paths within the environment and

- a representation of the available free motion space based on geometrical primitives like triangles.

2.1 Motion Oriented 2-D Geometrical World Model

Motion planning needs a good representation of free motion space [3, 2, 25, 23]. In rare cases, e.g. for an empty building or a well-structured factory plant, it is possible to use a direct representation, showing free motion space as distinct roads or corridors. In all other cases it is much simpler to describe the outer shape and dimensions of the obstacles within the environment thus generating an implicit representation of free motion space. The necessary information about the obstacles such as inventory or machinery can be extracted e.g. from a 3-D CAD database containing the factory layout or from 3-D range images [15, 16]. Due to the nature of mobile robots moving or walking on solid surfaces it is sufficient to map obstacles only within the range from the floor up to the robot's maximum height. The given information can be substantially reduced by projecting the outer contours of the obstacles onto the floor (2-D projection range images [15]). If necessary these contours are approximated by polygons, resulting in a set of vertices and pairwise connected edges. The amount of data for this representation depends on the required accuracy. Motion-relevant 3-D information, e.g. the height of the obstacles is packed into an attribute list associated with every obstacle edge. Some other geometrical information such as doors, slopes or docking sites can be represented by so-called virtual obstacles. The edges of these virtual obstacles can be crossed by the mobile robot only under certain circumstances, e.g. if "the door is open and the opening is higher than

2.0m". Also, virtual obstacles lock robots out from unexplored regions or from the working space of manipulators. Thus relabeling or deleting edges of virtual obstacles controls the motion space available to mobile robots. By exploring an unknown region, the associated virtual obstacle will be splitted into some real obstacles and the interconnecting free motion space. This representation technique can be easily extended, e.g. to larger buildings with several floors. Each floor is separately mapped containing appropriately modelled virtual obstacles for each staircase or elevator channel entering or leaving this floor. By pairwise connecting these virtual obstacles between adjacent floors, motion planning is extended over the entire building.

2.2 Reveiling the Topological Structure of an Obstacle-Clustered Environment

The major drawback of modelling the environment through a description of known obstacles is the poor knowledge about free motion space between these obstacles. By considering obstacle polygons only it is initially not possible to decide whether two obstacles are geometrical neighbours. To do so at least the space between the obstacles must be structured e.g. by simple geometrical elements such as triangles or quadrangles (raster elements). As the presented geometrical world model contains virtual obstacles which may turn to free motion space after further exploration it is advisable to structure the entire environment *including* obstacle space. Using structured motion space (raster or triangles) it is easily possible to derive motion channels, e.g. connecting known start and goal locations by applying a graph search on all elements labeled "free" (section 4).

There exist some techniques for uniquely structering an obstacle clustered plane [5, 6, 25]. By means of triangles as structural elements it is possible to represent exactly polygonal obstacles and the free space between them. Furthermore the number of elements required for a description of a given environment is significantly smaller than for a raster map. This fact becomes extremely important during graph search used in most motion planning schemes.

2.2.1 Computing a Delaunay Triangulation

The triangulation of polygons and the free space between them is generally *not* unique and it can be performed by several methods. Nevertheless, it is possible to force uniqueness by applying certain constraints to the resulting triangles. As we are interested in geometrical relations, it is sensible to group obstacle vertices in triples by minimizing the geometrical distance between them. In other words, we obtain triangles which do not contain *any* other vertex within their circumcircles. This kind of triangulation is known as Delaunay triangulation [1, 20]. Unfortunately, it can be computed only for the *vertices* of the given polygons. Therefore it is not guaranteed that all obstacle edges are identical with one of the edges of the computed triangles. In other words, the Delaunay triangulation does not inherently distinguish between obstacle *and* free motion space, however this deficiency can be easily repaired.

Instead of directly computing Delaunay triangulation it is more efficient to construct the straight-line dual, also known as Voronoi diagram [21, 19, 4, 17]. The Voronoi diagram consists of a set of Voronoi polygons, one for each given vertex and each Voronoi edge beeing part of the perpendicular bisector to a pair of vertices (Fig. 2a, b). A straight forward but not the most efficient algorithm $(O(N^2))$ works as follows:

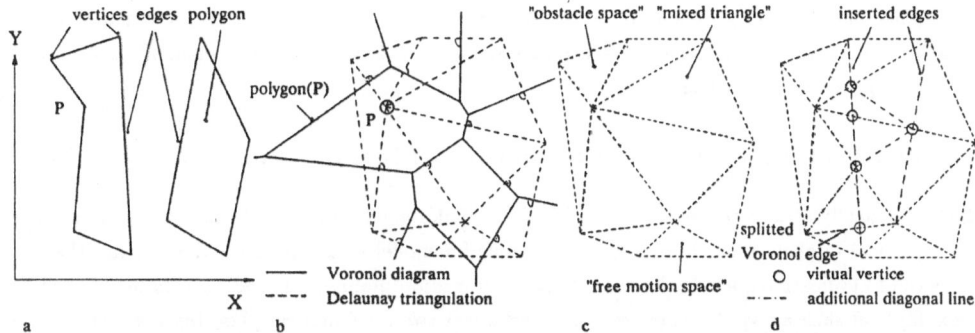

Figure 2: **a** obstacle configuration; **b** Voronoi diagram and Delaunay triangulation; **c** overlapping obstacle configuration and Delaunay triangulation; **d** black-and-white triangulation

> for each vertex
> compute the perpendicular bisectors to all other vertices and
> select the smallest polygon including the given vertex.

A more efficient algorithm $(O(N \log N))$ [21] computes the Voronoi diagram by recursively dividing the given set of vertices into two disjoint sets separated by a straight line. The two Voronoi diagrams for these sets are then merged. The recursion stops, if either set contains only two vertices, with the perpendicular bisector being the starting Voronoi diagram.

As already mentioned the straight-line dual of the Voronoi diagram is the Delaunay triangulation. It covers an area of the plane bounded by the convex hull of the given vertices. Thus the environment represented by this technique is always limited to a convex region, inhibiting a mobile robot from leaving the known world.

2.2.2 Converting the Delaunay Triangulation to a Black-and-White Triangulation

As the construction of the Delaunay triangulation is based on the vertices of the given obstacle polygons only, it may contain up to three different types of triangles (Fig. 2c): triangles representing "free motion space" and "obstacle space" as well as "mixed" triangles covering both obstacle and free motion space. Mixed triangles, however, can be separated along an obstacle edge into black (obstacle space) and white (free motion space) partitions. The principle of this partioning becomes obvious by overlapping the initially given obstacle configuration (geometrical world model) and the Delaunay triangulation. As shown in Fig. 2c there may exist some obstacle edges *not* corresponding with a triangle edge. By adding these edges to the triangulation, all mixed triangles are separated into black and white elements. The intersections between these added edges with the existing triangle edges are so-called virtual vertices and are required for unique represention of normal and splitted triangles. However, repeatedly cutting off a vertex from a triangle may result in a polygon with four or more vertices. To overcome potential problems with complex polygons, every quadrangle will be immediately divided into two triangles by inserting the shorter one of two diagonal lines (Fig. 2d). The final result of this procedure is a black-and-white triangulation representing uniformly both obstacle and free motion space. Due to the underlying Delaunay triangulation the free motion space is partioned into triangles extending from one side to other side of the corridor between adjacent obstacles. Furthermore these triangles are as large as possible, so only few elements are needed to describe large areas of white space.

3   Path Planning Based on Structured Motion Space

The topologically structured world model is primarily designed for fast and easy storage and retrieval of geometrical information. Besides the algorithms for model construction and updating there exist a variety of applications, especially various tasks of motion planning, based on this world model. As an example of such an application we developed a fast two-level motion planning algorithm suitable for various vehicle shapes and kinematical features. The algorithm computes a collision free path expressed by a series of steering commands for the mobile robot as well as a more descriptive path template useful for sensor-controlled movements.

3.1   Deriving Mission-Specific Motion Channels

Traditional path planning algorithms search the entire obstacle description or world model before they retrieve a vehicle-specific and somehow optimal path. A more challenging approach is the extraction of a motion channel connecting start and goal locations directly from the entire free motion space. This motion channel includes all available motion space between adjacent obstacles bounded by a polygon consisting of obstacle vertices. The edges of this polygon are either real obstacle edges or edges spanning between two different obstacles. Using topologically structured motion space, this motion channel is equivalent to a series of adjacent triangles.

The extraction of a motion channel is a simple graph search on all white triangles starting with the triangle containing the starting location and finishing at the goal triangle. Both, the start and goal triangles are found by the following approach:

> choose the obstacle vertex with the minimum distance to the given start or goal location and
> walk along the straight line connecting this vertex with the start or goal location from triangle to triangle and remember the last triangle visited.

Since the black-and-white triangulation exists only implicit within the Voronoi diagram, it is not possible to colour each triangle. Nevertheless a black triangle can only be entered by crossing an edge labeled "obstacle". Thus walking from one triangle to another is allowed if and if only the common edge is crossable by the robot. Assuming that both start and goal triangles are located within free motion space, it is impossible to enter obstacle space by using this rule.

For a given pair of start and goal locations, there may exist more than one motion channel reflecting the existence of topologically different paths. Thus, a performance criterion for the graph search is needed. It seems reasonable to weigh both an estimation for the final path length as well as the minimum path width compared to the vehicle's width. Within a single triangle the robot is assumed to move between the midpoints of pairwise adjacent edges (Fig. 3a). The width associated with such a path segment is computed as the perpendicular projection of the passed edges onto the

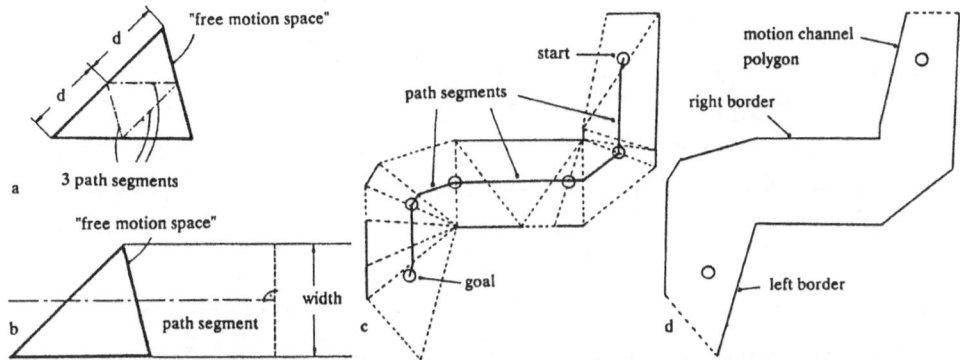

Figure 3: **a** path segment within a free motion space triangle; **b** width of path segment; **c** motion channel with merged path segments; **d** motion channel polygon

path segment (Fig. 3b). These path segments form a graph connecting all white triangles. The number of nodes within this connectivity graph, e.g. the midpoints of all crossable triangle edges can be significantly reduced by actually using only white triangles with three crossable edges and partially merging adjacent path segments (Fig. 3c). The final list of triangles connecting start and goal locations is collapsed to a polygon with, in respect to motion direction, a right and left border (Fig. 3d). This polygon is then forwarded to the vehicle-specific final path planning algorithm.

3.2 Vehicle-Specific Path Planning within a Motion Channel

Besides a coarse measure of the vehicle's width the extraction of a motion channel does not use any vehicle-specific details. However, the vehicle's shape, dimensions and its kinematical features, such as allowable turning radii and swept-out regions, must be considered during final path planning. Due to dynamic uncertainties of vehicle motion, it is advisable to keep a specified minimum clearance near obstacles during path execution. Also, the number of turns along the path should be reduced to a minimum. Since the motion channel contains all available free motion space between each pair of adjacent obstacles, searching for the shortest path completely included within the motion channel is a very attractive approach for path plannning. As shown in Fig. 4a, this path only changes direction at certain vertices of the motion channel polygon. These vertices are found by examing the inner angle of two adjacent edges; a vertex is critical if and if only the corresponding angle is greater than 180° (Fig. 4a). Passing a subset of these critical vertices on a circle suitable to the vehicle's behaviour results in a path consisting of circular arcs and straight lines (Fig. 4b, c). The path planning algorithm successively checks all critical vertices on both sides of the motion channel whether or not they belong to the final path. This check is based on the visibility graph within the motion channel polygon with the circular expanded obstacle vertices as nodes and tangential lines to each pair of circles as arcs. Since this modified visibility graph is never completely generated, there is no need for a comprehensive graph search. Thus the algorithm's performance is approximately proportional to the number of critical vertices within the motion channel. The resulting path can be compared to a rubber band connected between start and goal locations. The rubber band is

- 158 -

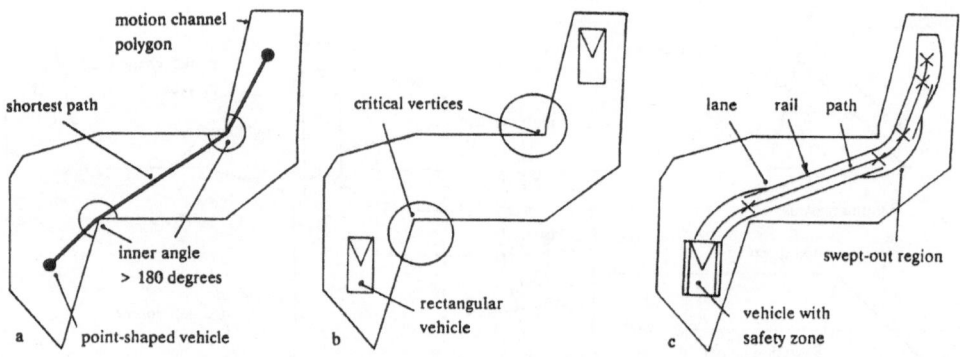

Figure 4: **a** motion channel with shortest path; **b** circular expansion of critical vertices; **c** final path

command	goal position			direction
	x	y	ψ	
	[mm]	[mm]	[°]	
Set Position	9561	11482	-90.0	
Move Absolute	8985	9289	-108.	forward
Move Absolute	4938	7521	-163.	forward
Move Absolute	3162	5128	-90.0	forward

Table 1: Steering commands specifying the preplanned path in Fig. 4

deflected on some of the critical vertices, expanded by circles, thus following the shortest path within the motion channel.

The radii of the circles used to expand the critical vertices are adapted according to the available motion space and should be as large as possible. The centers of these circles are usually located on the bisector of the turning angle with a programmable clearance for the inner rail of the final path (Fig. 4b). The swept-out region of the vehicle is modelled by a circle large enough to circumscribe all parts of the vehicle lying on the outer side of the turn. The path is collision-free if both rails and the swept-out circle never cross an obstacle edge (Fig. 4c). This is inherently guaranteed by the path planning algorithm for the straight parts and the inner rail at turns. Only the swept-out circles must be tested explicitly for collision-freeness.

The computed trajectory is converted to a series of steering commands, each specifying the vehicle's position and orientation at the end of each path segment consisting of a circular arc and an optional straight line segment (Table 1).

4 Constructing a World Model by Accumulation of Sensor Information

4.1 Acquisition of Geometrical Knowledge

Movements of a mobile robot are essentially restricted to parts of free space between the obstacles located within the robot's environment. Knowledge about the environment is usually incomplete,

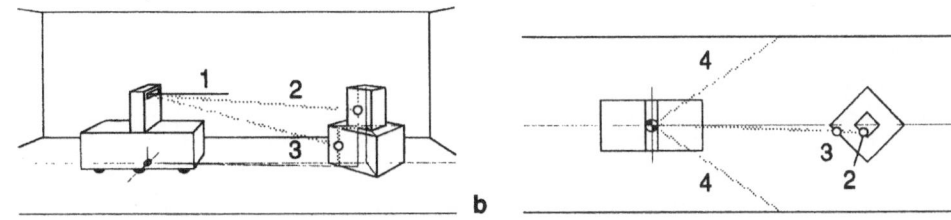

Figure 5: a acquisition of 3-D range data; b constructing a 2-D top view profile

thus acquisition of motion relevant geometrical information by means of a suitable sensor system becomes an extremely important task. The most significant sensor information is the (three dimensional) distance between the robot's current position and surrounding obstacles. This data can be directly obtained through a 3-D range imaging device. Taking into account that mobile robots are moving on floor, the initially 3-D range information can be substantially reduced to a 2-D description of usable free motion space. The related algorithm is based on the laws of geometrical projection as well as on heuristic rules for interpreting free motion space in case of missing or erroneous sensor information. The final result of this procedure is a polygon giving a restrictive description of visible free motion space.

Reducing 3-D Range Images to 2-D Top View Profiles

The 3-D range image provided by our laser range camera [15, 16] consists of an array of 81 × 41 individual range values derived from a spatial angle of 60° × 60°. As an alternative a similar geometrical description of the robot's local environment can be obtained from a 3-D multifunction radar system [18]. The amount of information delivered by these sensors is small if compared to video camera data. Nevertheless, further information reduction is possible as well as necessary to achieve real-time performance required for fast surveillance and online path planning tasks.

For this purpose the outer contours of the detected obstacles are projected onto the floor. Due to limitations within the vertical deflection angle, only obstacles ranging between the floor and the robot's maximum height (beam position 1, Fig. 5a) are visible to the sensor system.

At this level of information processing all transformations are done using a spheric coordinate system (elevation, azimuth and distance). After projecting all valid range values (pixels) onto the ground plane (elevation angle ≡ 0°) a polar coordinate system (horizontal direction and ground-level distance) is sufficient for further processing. Since only obstacle surfaces directly facing the sensor system are detectable, computation of the outer contours of the projected obstacles results in the selection of the pixel with the minimum ground-level distance among all valid pixels within a given horizontal direction (Fig. 5, beam position 3). Therefore only one pixel is retained for each of the 81 horizontal directions. Next the selected pixels are transformed into 2-D cartesian coordinates and collected within an array of 81 elements. Pixels with invalid range data are marked as "gap", *not* necessarily indicating visible free space. This array represents a type of top view of the environmental scene (top view profile, TVP). TVPs have to be examined according to the geometrical context of the detected pixels.

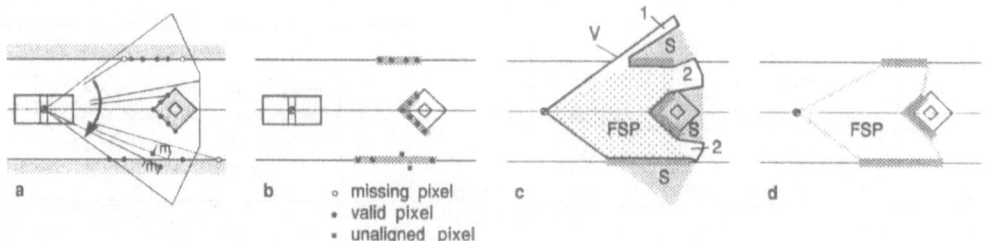

Figure 6: **a** 2-D top view profile; **b** segmented 2-D top view profile; **c** interpreting free motion space; **d** free space polygon (FSP)

As shown in Fig. 6a many pixels contain hits to the same obstacle surface or, within the top view of the scene, to the same obstacle edge. Due to the operating principles of range imaging devices, there may be two types of disturbances within the pixel-oriented top view profile (Fig. 6a): *missing pixels* carrying invalid range information ("gap") and *unaligned pixels* caused by the limited sensitivity and resolution of the sensor's signal processing systems.

Adjacent pixels can be grouped to a cluster representing the same obstacle surface, using the assumptions that all obstacle surfaces are smooth or even plane and there are no gaps smaller than a given minimum (e.g. 10–20cm) within the same surface.

Thus, small gaps (about 2–3 pixels wide) can be patched with the range values of their neighbour pixels. A series of pixels belongs to a common cluster, if and only if the relative distance value between two directly adjacent pixels is smaller than a given minimum (e.g. 10cm). Then each cluster can be replaced by a polygonal line segment or, in most cases, by a single straight line segment. The line segment is selected through best fitting all pixels within a given cluster. The resulting line segments i.e. the detected obstacle contours or obstacle edges are ordered with respect to the initial scan direction, forming a list of few geometrical elements called *segmented top view profile* (Fig. 6b).

Constructing Free Space Polygons

The segmented top view profiles are an obstacle-oriented interpretation of the available range information. With respect to path planning purposes a description of the available free motion space around the robot's position seems to be more useful. There exist essentially two methods for transforming the list of line segments to a polygon of free space bounded by actually detected obstacle edges and additionally virtual edges:

1. *visual free motion space* (V) is constructed from the complete measurement area of the sensor system within an unobstructed environment excluding only the invisible shadow regions (S in Fig. 6c), generated by the detected obstacle contours,

2. the more restrictive *interpreted free motion space* is constructed by subsequently connecting the vertices of two adjacent obstacle contours (with respect to the initially given scan direction) including the current sensor position as a virtual vertex (Fig. 6c, FSP).

The main drawback of method 1 is the fact, that solid obstacle edges may look transparent due to missing pixels (region 1, Fig. 6c), whereas with method 2 some of the actually detected free space

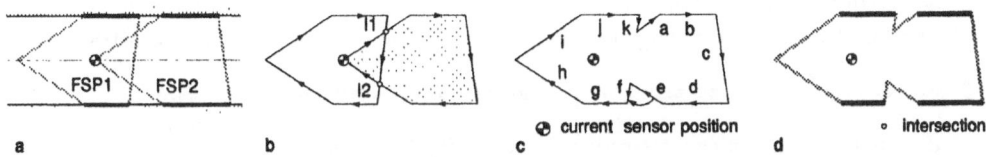

Figure 7: **a** measured free space polygons FSP₁, FSP₂; **b** computing intersections between FSP₁ and FSP₂; **c** walking along the boundary of merged free space; **d** newly generated map

(region 2, Fig. 6c) may be discarded. As a whole, the more restrictive method 2 is better suited for path planning and mapping purposes and will be used by our approach.

If there are no detectable obstacle contours on either side of the vehicle, the resulting free space polygon (Fig. 6d) may collapse to a narrow corridor in front of the robot. However, using some heuristic rules [13, 14] well-formed free space polygons (FSP) can be constructed for every given segmented top view profile.

4.2 Merging Visible Free Motion Space

Each free space polygon represents an area of known world. In other words, the boundary of the free space polygon resembles world's end. Exploration of an unknown environment proceeds by moving world's end into unknown regions. The main principle of mapping is to merge the available free space of two subsequent polygons, while conserving the outer boundaries of both polygons (the extended world's end) and deleting any obsolete virtual edges inside the known world.

The process of mapping starts with an initial free space polygon (FSP₁ in Fig. 7a). The available free space inside the polygon is marked by defining a sense of direction on the boundary (indicated by arrows in Fig. 7b). Walking clockwise on the boundary means, that the known world can be always found righthand. Merging a second polygon (FSP₂ in Fig. 7a) to the existing map (FSP₁) starts with the computation of all intersections between the current free space polygon and the existing map. Using the black-and-white triangulation computed for the existing map, the search for intersections can be done in a very efficient manner, by only examing those triangles directly crossed by the boundary of the newly acquired free space polygon (FSP₂).

These intersections are collected within an ordered list (I₁, I₂ in Fig. 7b) and classified whether the boundary of the free space polygon is leaving (type IO → I₁) or entering (type OI → I₂) known world. The classification is performed by walking along the boundary of the free space polygon and numbering all intersections passed. Starting at the current sensor position (always located within known world) an odd numbered intersection is of type IO whereas even numbered intersections are of type OI. There exists always an even number of intersections.

Next, the polygonal lines between every pair of two adjacent intersections must be rearranged to form the extended boundary of the known world. This is accomplished by the following algorithmic steps:

1. Remove an intersection (I₁ in Fig. 7b) from top of the list of intersections (starting intersection).

2. If the current intersection is of type IO then continue walking on the boundary of the free space polygon (otherwise on world's end) using the indicated direction until reaching the next

intersection (edges a–e, Fig. 7c, stopping at I_2, Fig. 7b) and remove that intersection from the list. All edges visited become part of the extended world's end and the newly constructed map.

3. Repeat step 2 until the starting intersection is reached again (I_1 in Fig. 7b).

4. If there exists remaining intersections (none in Fig. 7) within the list, then repeat step 1 to 4 until the list becomes empty.

All edges visited by the algorithm form new polygons, one polygon for every intersection removed from the list in step 1. The computed intersections are always vertices of these polygons. The remaining boundary segments between these intersections are discarded. Most of these discarded edges are virtual edges. However, it is possible to discard obstacle edges according to the algorithm. In this case, there is usually no loss of information, because the same obstacle edge is included within the current free space polygon as well as the current map.

After adding a new free space polygon, the final map includes at least one polygon (Fig. 7d). The outer boundary of the map (world's end) is marked in clockwise direction, meaning that the available free space is located inside the polygon.

This extended map (2-D geometrical world model) is again structured by a black-and-white triangulation as explained in section 2.2. In this case, the underlying Voronoi diagram must be recomputed only for the modified and newly added regions of known world, thus saving most of the computational efforts neccessary for constructing the topologically structured world model for the whole obstacle configuration.

Isolating Obstacles within Known World

If there exists more than one pair of intersections between the current free space polygon and the available map, more than one polygon may be generated by the mapping algorithm. These additional polygons are separated from the current world's end and form a piece of unknown world lying inside the available free space (ISO in Fig. 8f). Since the interior of such a polygon is obstacle space (unknown space), the boundary is marked in counterclockwise direction.

The construction of an isolated obstacle region is based on the general mapping algorithm as discussed in this section. Overlapping of two free space polygons (FSP_1, FSP_2 in Fig. 8c) drawn from the environment in Fig. 8a, b results in three pairs of intersections (I_1–I_6). Starting with intersection I_1, the algorithm generates first the new world's end (Fig. 8c). While walking on the new boundary of known world, intersections I_2–I_4 are removed from the list. After returning to intersection I_1, there remains an additional pair of intersections (I_5, I_6) on the list. Walking from intersection I_5 to I_6 according to step 2 of the algorithm (Fig. 8d) isolates a region bounded by one obstacle and several virtual edges. The isolated unknown region (ISO in Fig. 8f) covers both obstacle and unexplored free motion space. The amount of free motion space within such regions depends on the topological structure of the environment as well as on the selection of suitable sensor positions.

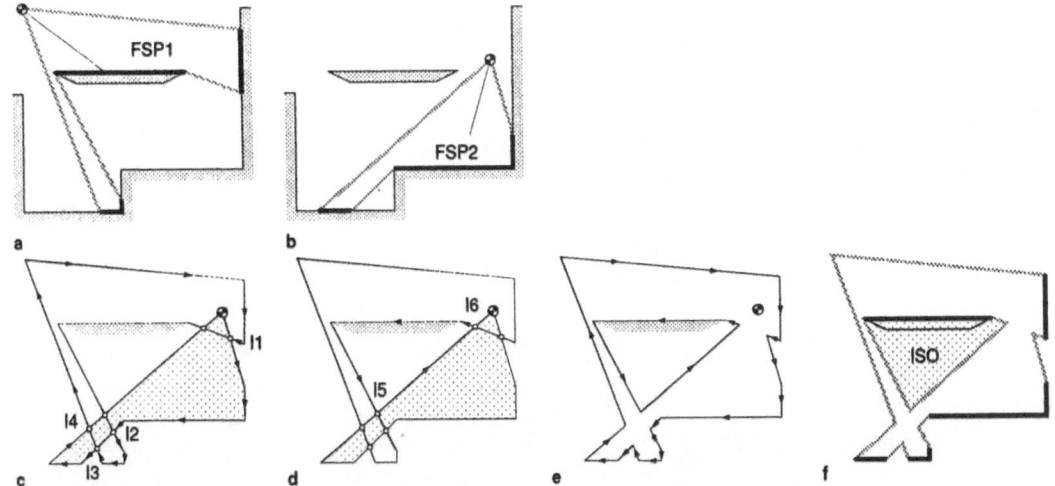

Figure 8: a measured free space polygon FSP$_1$; b measured free space polygon FSP$_2$; c walking along the boundary of known world; d walking along the boundary of an isolated unknown region; e discarding obsolete edges; f newly generated map including an isolated obstacle (ISO)

5 Experimental Results

The world model and the associated algorithms for construction of the black-and-white triangulation as well as the motion planning algorithm are implemented on a MicroVAX II using MODULA II language. To support the implementation of complex geometrical based algorithms, we developed an interactive simulation and debugging environment [27]. This tool is based on a high-performance window-manager, which allows the transparent overlapping of several windows containing e.g. the obstacle configuration and the Delaunay triangulation. The figures of all examples given in this paper were plotted with this development tool.

Path Planning for Various Types of Mobile Robots

Fig. 9 shows as an example of motion planning for mobile robots within our laboratory. The obstacle description (geometrical world model, Fig. 9a, b) presents in detail several rooms as well as typical laboratory equipment, such as tables etc. In the lower left corner a manipulation robot is indicated by its working space. The drawing of the Voronoi diagram (Fig. 9c) clearly shows the Voronoi polygons each associated with an obstacle vertex and gives an idea of the complexity of the corresponding data model. As shown in Fig. 9d the black-and-white triangulation consists of some large triangles covering most of the free motion space (white space). The set of small triangles are located within or near the walls of the laboratory. The connectivity graph for white space (Fig. 9e) was drawn for a vehicle width of 1m; this is the width of our full-scale mobile robot MACROBE including a sufficient safety clearance. This graph can be considered as a coarse road map indicating clearly all topologically different paths within the laboratory. Using this graph the mission-specific path planning algorithm derives a motion channel (Fig. 9f) for the mission shown in Fig. 9a. All vehicle-specific path planning considering the vehicle's shape and kinematic features is done within this

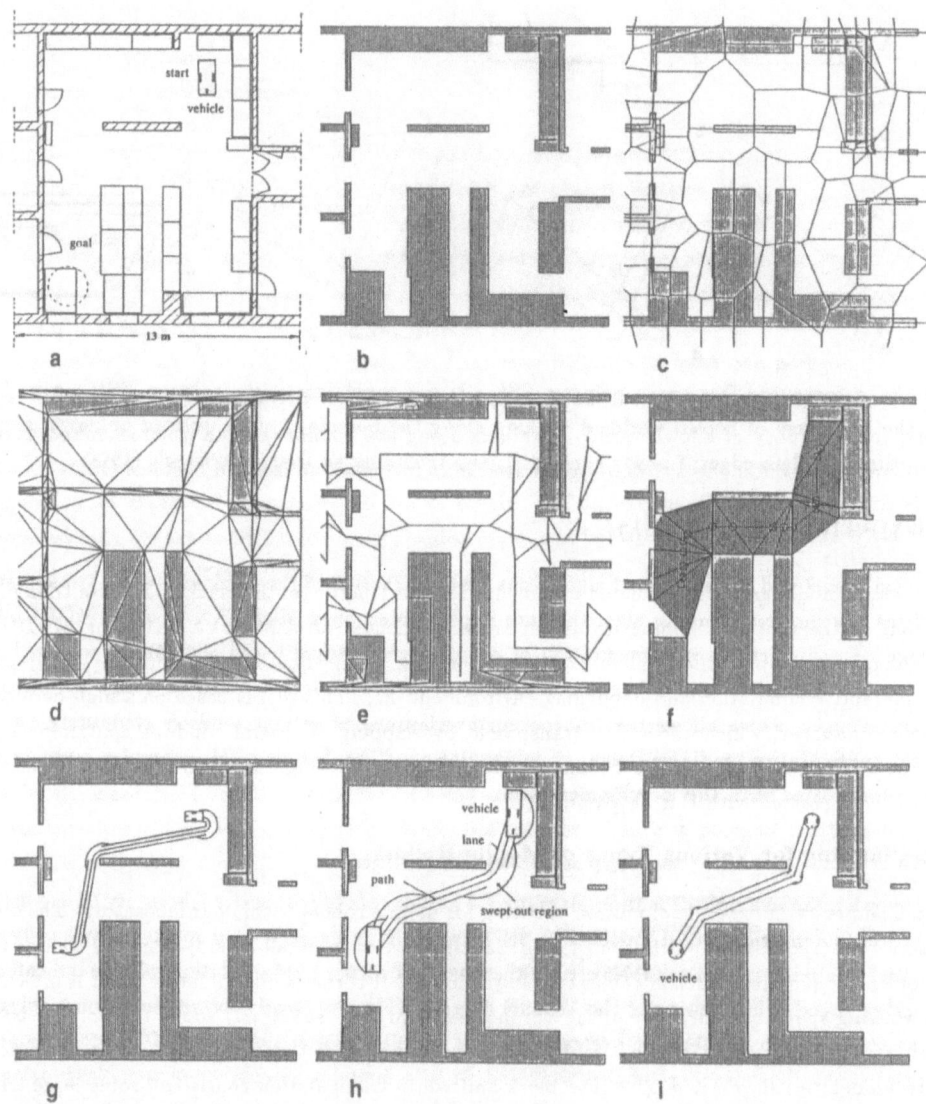

Figure 9: Steps of motion planning: **a** top view of laboratory; **b** obstacle configuration; **c** Voronoi diagram; **d** black-and-white triangulation; **e** connectivity graph; **f** motion channel; **g–i** final paths: **g** small and agile vehicle; **h** long and slim vehicle; **i** cylindrical vehicle turning in place

	Laboratory	City-Plan	Pilot Plant
vertices	87	327	845
added edges	12	29	53
added vertices	41	72	168
computation time	[s]	[s]	[s]
Voronoi-Diagram	7.84	36.26	82.31
split mixed triangles	0.61	1.08	2.43
derive motion channel	0.41	3.51	8.08
final path planning	0.10	0.19	0.37

Table 2: Computation times for motion planning in various environments

motion channel. Therefore, only this step is to be repeated for various vehicle types. Figs. 9g, h show the computed final paths for rectangular shaped vehicles without the ability to turn in place. Notice the turning radii adapted to the vehicle's dimension and the large swept-out region for the long vehicle used in Fig. 9h. Using a cylindrical shaped vehicle able to turn in place the final path essentially consists of straight lines (Fig. 9i). All final paths are computed by the *same* algorithm by simple adaption of some vehicle-specific parameters.

Table 2 gives an overview of computational times associated with typical real environments. A major time portion is consumed by the generation of the topologically structured world model. This is due to the fact that computing the Voronoi diagram is an $O(N \log N)$ problem. However, this computation is only done during initialization or after major changes within the robot's environment. Deriving a mission-specific motion channel being actually a $O(N^2)$ graph search, takes only a few seconds, while the vehicle-specific path planning runs under 1s. Compared with the typical speed of an experimental indoor mobile robot of about 1m/s, these computation times prove to be sufficiently short. Using a MicroVAX 3 system all computations are done 3.7 faster.

Mapping a Laboratory Environment

The effectiveness and correct operation of the presented algorithm is demonstrated by the mapping of a real-world laboratory environment. The mobile robot MACROBE travels on a preplanned path through a narrow corridor, enters a laboratory room, turns around a pillar and moves back to its starting location (Fig. 10f). During execution of this movement, the 3-D laser range camera takes 30 pictures which are subsequently processed to form a map (Fig. 10e). Since the horizontal viewing angle of the laser camera is only 60°, it proves to be rather difficult to detect openings within the walls on either side of the vehicle (Fig. 10a).

When entering the laboratory room, a large area of the environment lying inside the turn, remains unexplored due to the limited viewing angle of the sensor system (Fig. 10b). The same phenonemon can be observed when moving around the pillar in the middle of the room (Fig. 10d). Before leaving the laboratory room, the pillar is isolated as an obstacle within the current map and the shadow regions near the entrance are now fully explored (Fig. 10e). Travelling back to the starting location adds only little information to the map.

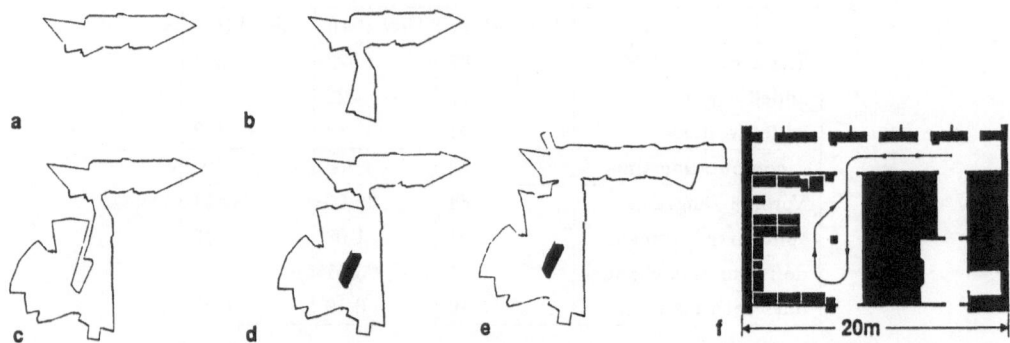

Figure 10: **a** moving along the corridor; **b** entering the laboratory room; **c** detecting the pillar; d isolating the pillar; **e** final map; **f** floor plan of laboratory including preplanned path

The final map (Fig. 10e) is a direct result of the mapping algorithm. Its construction relies heavily on the quality of MACROBE's position measurement system (odometric dead-reckoning system) and the precision of range data provided by the 3-D laser range camera. The quality of the final map becomes evident by comparing the map to the known floor plan of our laboratory (Fig. 10e, f).

The presented mapping algorithm is part of the navigation system of the experimental mobile robot MACROBE [11, 12, 27]. It is implemented using MODULA II and needs less than 1s for merging a new free polygon into the current map. The preparation of the collected map data for path planning purposes (constructing the topologically structured world model) requires about 1–20s, depending on the number of free space polygons added. Compared to the vehicle's average speed during exploration (0.2m/s) and to the travelled distance (about 40m) most of the mapping process is finished when returning to the starting location.

6   Conclusions and Outlook

The presented navigation system for mobile robots includes three major components and the related algorithms:

- a topologically structured world model containing a detailed geometrical description of the robot's environment,

- a two-level path planning system for navigating the robot within known regions of the world and

- a mapping system, enabling the robot to accumulate geometrical information provided by 3-D range imaging devices and to autonomously learn an inknown environment.

The topologically structured world model is a flexible and powerful tool for managing and manipulating geometrical information including sensor data. Although primarily designed as a 2-D world model all motion-relevant 3-D information including different floors within buildings can be modeled by properly attributing existing geometrical elements. By structuring both obstacle *and* free motion space through triangles the world model is readily applicable to all sorts of motion planning. Furthermore, obstacles representing unknown regions can be explored later, thus turning black to white space without any special treatment within the path planning algorithms. The multilevel approach for managing geometrical information as well as the two-level motion planning algorithm save processing time by executing costly algorithms only when necessary. Especially the final path planning algorithm runs so fast compared to the vehicle's speed, so that it can be repeatedly used during path execution, thus adapting the preplanned path to the actual environmental situation provided by the robot's sensor system.

The presented mapping algorithm is a fast and general approach for constructing 2-D maps using 3-D range data as for example provided by a laser range camera. By incrementally maintaining a topologically structured world model during the mapping process, every version of the accumulated map is immediately applicable to path planning [2, 3, 22, 12]. Starting with the initially available 3-D range image (81 × 41 pixels) the sensor information is consequently reduced up to a highly abstracted polygonal description of the environment. Before removing redundant obstacle data, this information can be used to improve e.g. the accuracy of the robot's position estimates, giving the mapping process a reasonable robustness against unavoidable position errors (due to dead-reckoning). The algorithm can be easily extended for dealing with time variant environments and is well suited for indoor operations.

An implementation of the mapping algorithm was tested within an exploration and mapping experiment using the mobile robot MACROBE and its 3-D laser range camera. While travelling along a preplanned path 30 range images are collected to build a map of a rather complex indoor environment.

Next major research steps will be concerned with implementation and test of exploration strategies, enabling a mobile robot to autonomously *explore and map* unknown indoor environments. Some of these exploration strategies are an adaption and reimplementation of earlier research results [7, 8, 10].

Acknowledgement

The work reported in this paper was supported by the Deutsche Forschungsgemeinschaft, as part of an interdisciplinary research project on "Information Processing in Autonomous Mobile Robots" (SFB 331).

References

[1] Bowyer, A., "Computing Dirichlet Tessellations", *Computer J. 24*, 1981, pp. 162–166.

[2] R.A. Brooks, T. Lozano-Pérez, "A Subdivision Algorithm in Configuration Space for Findpath with Rotation" *8th IJCAI, Karlsruhe*, 1983, pp. 799–806.

[3] R.A. Brooks, "Solving the Find-Path Problem by Good Representation of Free Space", *IEEE Trans. Sys., Man a. Cyb. 13*, 1983, pp. 190–197.

[4] Chazelle, B. and Edelsbrunner, H., *An Improved Algorithm for Constructing kth-Order Voronoi-Diagrams*, IEEE Trans. on Comp. 36 (1987), 1349–1354.

[5] A. Elfes, "A Sonar-Based Mapping and Navigation System", *Proc. IEEE Int. Conf. on Rob. a. Aut.*, 1986, pp. 1151–1156.

[6] A. Elfes, "Sonar-Based Real-World Mapping and Navigation", *IEEE J. RA-3*, 1987, pp. 1151–1156.

[7] F. Freyberger, P. Kampmann, G. Karl, G. Schmidt, "MICROBE - Ein autonomes mobiles Robotersystem", *VDI-Z 127*, 1985, pp. 231–236.

[8] F. Freyberger, P. Kampmann, G. Schmidt, "Ein wissensgestütztes Navigationsverfahren für autonome mobile Roboter", *Robotersysteme 2*, 1986, pp. 149–161.

[9] G. Giralt, R. Chatila, M. Vaisset, "An Integrated Navigation and Motion Control System for Autonomous Multisensory Mobile Robots", *C. Robotics Research*, Brady and Paul, Eds., MIT Press, 1983, pp. 191–214.

[10] P. Kampmann, F. Freyberger, G. Karl, G. Schmidt, "Real-Time Knowledge Acquisition and Control of an Experimental Autonomous Vehicle", *IAS-1, An Int. Conf.*, Amsterdam, 1986, pp. 294–307. in Hertzberger, L.O. (ed.) *Intell. Auton. Systems, An Int. Conf.*, Elsevier Science Pub. Amsterdam, 1986, pp. 294–307.

[11] P. Kampmann, G. Schmidt, "Topologisch strukturierte Geometriewissensbasis und globale Bahnplanung für den autonom mobilen Roboter MACROBE", *Robotersysteme 5*, 1989, pp. 149–160.

[12] P. Kampmann, G. Schmidt, "Multilevel Motion Planning for Mobile Robots Based on a Topologically Structured World Model", *IAS-2, An Int. Conf.*, Amsterdam, 1989, pp. 241–252.

[13] F. Freyberger, P. Kampmann, G. Schmidt, "Constructing Maps for Indoor Navigation of a Mobile Robot by Using an Active 3D Range Imaging Device", *IEEE Int. Work. on Intel. Robots a. Systems*, Tsuchiura, 1990, pp. 143–148.

[14] P. Kampmann, G. Schmidt, "Mapping of Indoor Environments for Robot Navigation Based on 3D Range Images", *Proc. IEEE Int. Conf. on Rob. a. Aut.*, Sacramento, California, April 7-12, 1991, in print.

[15] G.K. Schmidt, G. Karl, "3D Laser Range Camera for Mobile Robot Motion Control", *IEEE Int. Work. on Intel. Robots a. Systems*, Tokyo, 1988, pp. 605–610.

[16] G. Karl, G. Schmidt, "3D-Laser Range Camera for mobile Robot Applications, Sensor Model based Preprocessing and Motion oriented feature extraction", *Proc. IFAC, Symp. on Robot Cont.*, Karlsruhe, 1988, pp. 70.1–70.7.

[17] Kirkpatrick, D., "Effizient Computation of Continous Skeletons", *Proc. 20th IEEE Ann. Symp. Found. Computer Science*, 1979, pp. 18-27.

[18] Lange, M. and Detlefsen, J., "94 GHz 3D-Imaging Radar for Sensorbased Locomotion", *Proc. Int. Microwave Symp. IEEE MTT-S (Long Beach)*, 1989, pp. 1091–1094.

[19] Lee, D., T. and Drysdale, R., "Generalization of Voronoi Diagrams in the Plane", *SIAM J. Comput. 10*, 1981, pp. 73–87.

[20] Lee, D., T. and Preparata, F., "Computational Geometry - A Survey", *IEEE Trans. on Comp. 33*, 1984, pp. 1071–1101.

[21] Lee, D., T., "Two-Dimensional Voronoi Diagrams in the L_p-Metric", *J. ACM 27*, 1980, 604–618.

[22] T. Lozano-Pérez, M.A. Wesley, "An Algorithm for Planning Collision Free Paths Among Polyhedral Obstacles", *Commun. ACM 22*, 1979, pp. 560–570.

[23] T. Lozano-Pérez, "Spatial Planning, A Configuration Approach", *IEEE Tra. Comput. 32*, 1983, pp. 108–120.

[24] M.J. Mataric, "Environment Learning Using a Distributed Representation", *IEEE Conf. on Robotics and Automation*, Cincinnati, 1990, pp. 402–406.

[25] A.C-C. Meng, "Free Space Modeling and Geometric Motion Planning Under Unexpected Obstacles", *IEEE Conf. on RA*, Philadelphia, 1988, pp. 82–87.

[26] U. Raschke, J. Borenstein, "A Comparison of Grid-type Map-building Techniques by Index of Performance", *IEEE Conf. on Robotics and Automation*, Cincinnati, 1990, pp. 1828–1832.

[27] K. Robl, F. Freyberger, P. Kampmann, G. Schmidt, "Simulation Tools for the Development of Autonomous Mobile Vehicles", *Intern. Adv. Rob. Progr., 1st Workshop on Manipul., Sens. a. Steps Towards Mobil.*, Kernforschungszentrum Karlsruhe, 1987, pp. 211–233.

A Recursive Control Structure for Mobile Robots

E. Badreddin

ABSTRACT

A generalized control structure for shaping the behavior of a mobile robot is described. The proposed structure possesses many desirable properties such as recursiveness, nestedness of the behavior band-width and predefined interface of the behavior levels. It uses different system representations at the different levels of abstractions. A general formal description as well as the block-diagram realization of the different levels are given. The concept of analog gates is suggeseted to combine the individual behaviors into a desirable overall-behavior. The realization of several levels has been successfully validated on our mobile robot under realistic environmental conditions.

1. INTRODUCTION

T wo control architectures are common in the design of mobile robots today. The so called vertical/behavior decomposition e.g. [1],[7],[3] also sometimes called the "subsumption architecture" and, earlier, the horizontal/functional decomposition e.g. [8], [9].

We believe, however that, due to the fact that the problem is "two dimensional", one cannot strictly follow one decomposition scheme while completely ignoring the other. While the behavior decomposition is helpful during the design stage of the control strategy, the functional decomposition, which is also required at every behavior level, is important in the realization phase (soft- and hardware). In our work [6] we propose an alternative architecture, called the "Recursive Nested Behavior Control", RNBC, (Fig.1) which possesses many desirable properties for robot control. This architecture is paired with the the so called "functional-unit" decomposition which emphasizes the realization aspects of the soft- and hardware to achieve an asynchronous processing in a robust and efficient way. In contrast to the work of [2], the RNBC consists of generally non-linear systems and uses multiple system representations at different levels of abstraction.

In this paper, we shall discuss the "anatomy" of the RNBC proposing a realization for each of the behavior levels, briefly explaining its function and showing the interface to other levels. Also, the problem of combining different behaviors into an overall-behavior is tackled using the analog-gate concept. Finally some experiments are presented to show the fruitfulness of the RNBC for the behavior levels which have been implemented so far.

2. RECURSIVE BEHAVIOR-SHAPING

The block-diagram of the RNBC is shown in Fig.1 The main properties of the RNBC can be summarized as follows:

1) it is strictly *recursive*. Interactions are made only between the i-th and (i+1)-th levels. The restrictions imposed by recursiveness is over rewarded by the ability to specify well-defined software and hardware interfaces in the "work-package" of each project member. This is essential in projects with many co-workers of different experience and back-ground.

2) the behavior levels are *nested*. The behaviors of broader band-width (reflexive behaviors) occupy the inner-most "loops" and those with narrower band-width (intellectual behaviors) are settled in the outer-most "loops". The nestedness property allows the estimation of system band-width and, thus, the speed of reaction to environmental changes. It also endows the system inherent robustness.

3) the RNBC is a *bottom-up* approach *i.e.* it provides gradually increasing sophistication and an operational prototype through out all development stages. The importance of this property is immanent since the experience of the developers increase with time and the motivation is greater as they bring the robot to "act" at any level and watching "Him or Her" getting smarter the more they invest in the development.

4) the different behavior levels use different system descriptions according to the *level of abstraction*. For example, differential/difference equations are used for dynamics, axis-level and robot-level control while a graph is used for path-planning and a push-down automaton for monitoring and reasoning.

5) the RNBC possesses a *fixed structure* unlike e.g. the "subsumption architecture" [7], which inherently changes the topology of the "behavior network" due to the inhibition/suppression interactions. Stability of variable structure systems are difficult to investigate let alone to guarantee. The stability problem may appear to be of less importance when only few levels with limited, predictable behaviors communicating over a simple topology of interconnections are considered. It becomes, however, more complex as the number of behavior-levels grows and the connectivity increases as a result of higher sophistication of the overall-behavior.

6) each behavior level, except the lowest and the highest, interconnects to the lower and the higher levels through *feedforward and feedback* respectively. Recall that feedforward can be effectively used for disturbance cancellation using suitable sensors, and thus increasing the apparent band-width, while feedback can improve the stationary response of the loop.

7) *modelling is distributed* over all levels. The models used in each level are usually sparse since they model system-dynamics needed only on that particular level.

8) *no explicit sensor-fusion* is necessary. Sensors supply information on the environment to the spot where they are needed in the level of concern. If the same sensor-data are used on different levels, they are fed-forward or fed-back through the interface and are consequently delayed. For the purpose of obtaining the largest possible band-width, it is, therefore, recommended to inject the sensor-data in the lowest-level they are required and then passing them through the feed-forward interconnections to higher-levels, eventually after an appropriate preprocessing.

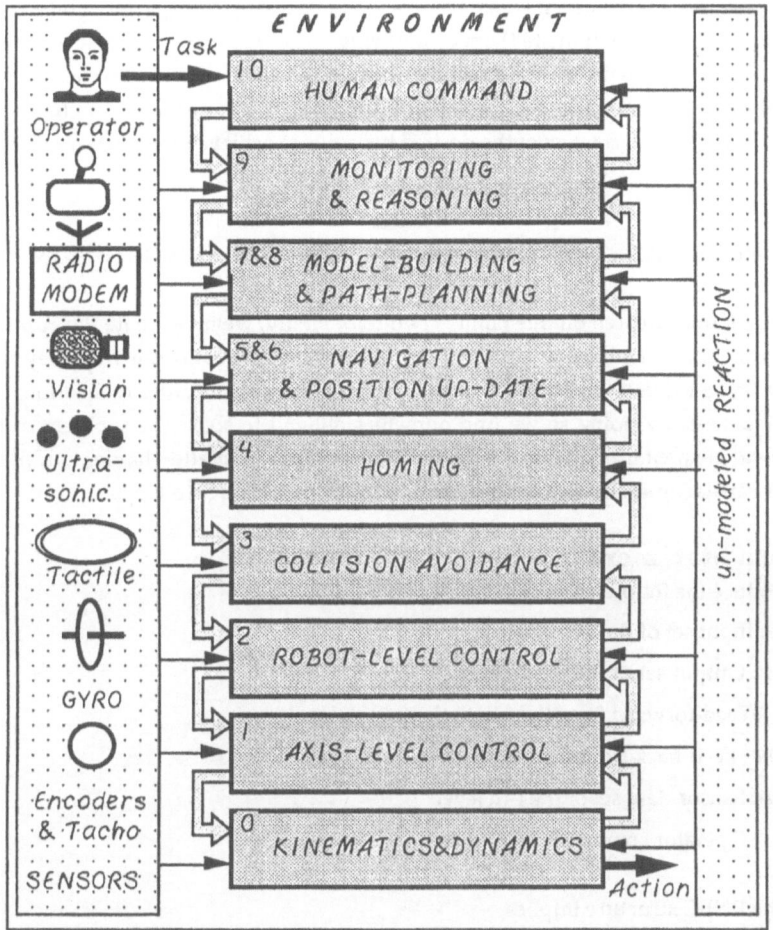

Fig.1 The RNBC of RAMSIS

Basically, the architecture can be viewed as a traditional nested control where the behavior levels implement the feedforward and feedback compensators. The main contribution can then be regarded as the conception of *feedforward and feedback compensations as behavior levels* and, thereby, using *different system descriptions within the same structure*. This poses, however, several problems to be solved for any successful realization:

- *Interfacing* different system descriptions,
- Treatment of *goal conflicts* between several behavior-levels,
- Appropriate *composition* of the individual behaviors to produce an acceptable overall behavior which achieves the goal. In the next section we propose the use of
- Stability investigations of the control system especially under *environmental feedback*.

This implies the stringent requirements on the robustness of the control system since the

robot will always act upon and its environment, and eventually changing it, whilst reacting to disturbances trying to eliminate them by applying an appropriate control strategy,
• Due to the recursive nature of the structure, the permanent challenge is to successively implement a higher level under the *constraints* imposed by the lower ones, and finally,
• Detailed *on-line implementation* of the desired behavior to achieve the necessary response.

3. PROBLEM STATEMENT AND FUNCTIONAL DESCRIPTION

The aim of our research on this point is two-fold. Firstly, we investigate the applicability of different system-descriptions at different levels of abstraction and the way they interact. Secondly, we try to validate the RNBC structure by proposing a particular realization, with no claim on uniqueness. By doing so we find ourselves obliged to solve the problem of how to combine several elementary behaviors such that the resulting overall-behavior completes the specified task satisfactorily. In the subsequent discussion, please refer to Fig.2.

3.1 Problem Statement
We, first,introduce the following *conventions*:

$IP_i \subseteq IP$: Input set of the i-th level.

$OP_i \subseteq OP$: Output set of the i-th level.

$FF_i \subseteq FF$: Feed-forward from the i-th to the (i-1)-th level .

$FB_i \subseteq FB$: Feed-back from the i-th to the (i+1)-th level.

$SD_i \subseteq SD$: Sensor-data set of the i-th level.

$OC_i \subseteq OC$: Operator command applicable on the i-th level.

Notice that the RNBC structure implies,

$IP_i = FB_{i+1} \cup FF_{i-1} \cup SD_i \cup OC_i$ and,

$OP_i = FB_i \cup FF_i$,

where $\cup$, is the set union-operator.
The i-th level can then be, formally, described by the function (mapping):

$f_i : IP_i \rightarrow OP_i$

The *problem statement* can now be made:
Find f_i such that the composition function,

$g_i : IP_i \rightarrow OP_0$

$g_i = f_i \circ f_{i-1} \circ f_{i-2} \circ f_{i-3} \circ ... f_0$, realizes the desired behavior on a particular level.

Notice also that due to the recursive property,

$g_i = f_i \circ g_{i-1}$

The sets used are listed below in the order of their appearance in the description:

$\mathcal{M}$: { Mechanical moments/torques [N.m] }

A : { Linear accelerations [m/sec^2] }

Ω : { Angular velocities [rad/sec] }

$\mathcal{F}$: { Mechanical force [N] }

$\mathcal{U}$: { Electric potential [Volts] }

Γ: { Angular accelerations [rad/sec^2] }

S : { Boolean status for processes, actuators, sensors, modem, operation-state, etc.[true/false] }

Φ: { Angular displacement [rad] }

$\mathcal{V}$: { Linear velocities [m/sec] }

Λ : { Linear displacement [m] }

Ξ : { Environment's features: walls, edges, corners, windows, doors, gangways, co-operative landmarks, surface-texture, temp., tactile reaction-force, color}

N : { Graphs representing the operation-space topology }

Σ: { Commands: go, stop, turn-right/left, keep-heading, look-for-free-space, dock, home , open/close-door, activate effector, pause-for-perception, wait , send/receive-message}

Z : { Mission: go-home, charge-batteries, run-simulation }

T: { Task: clean-floor, transport-pay-load, carry-object-to-person, hause-guard}

3.2 Functional Description

A formal description of the mappings each behavior level performs is difficult and can be very complicated, but for the lowest levels. For sake of clarity, we shall therefore, give a functional description for each level. We shall briefly describe f_i together with $\mathcal{FF}_i$, $\mathcal{FB}_i$, SD_i and OC_i. We cannot claim, however, that neither IP_i or OP_i described here are exhaustive let alone that g_i is a one-to-one function. The block-diagrams for each level, shown in Fig.2, are self-explanatory to a large extent.

3.2.1 Kinematics and Dynamics:

This is the robot itself as a multi-body with a certain number of degrees-of-freedom, which are reduced to three in our robot.

IP_0 = { Axes torques : $\mathcal{M}_i \in \mathcal{M}$}

$\mathcal{FB}_0$ = {}

$\mathcal{FF}_0$ = {Axes angular-accelerations: $\alpha_l \in A$, Axes-angular velocities: $\omega_l \in \Omega$}

SD_0 = {}.

OC_0 = { Muscular force: $f_m \in \mathcal{F}$}.

3.2.2 Axis-level Control:

This level establishes the, classical, control of each axis and produces the drive torques.

$\mathcal{FB}_1$ = { Axes torques : $\mathcal{M}_i \in \mathcal{M}$ }

$\mathcal{FF}_1$ = { Axes angular-accelerations: $\alpha_l \in \Gamma$, Axes angular-velocities: $\omega_l \in \Omega$, Status: $s_1 \in S$}

SD_1 = { Current sensors: $I \in \Gamma$, Encoders: $E \in \Phi$, Pneumatic-pressure sensors:$P \in \mathcal{F}$}.

OC_1 = { Axes-joy-stick voltages: $u_{ref} \in \mathcal{U}$}.

3.2.3 Robot-level Control:

At this level, the forward and inverse kinematics will be used to establish the speed (evtl. also the acceleration) control of the robot as a whole.

$\mathcal{FB}_2$ = {Axes ref. angular-accelerations: $\alpha_{ref} \in \Gamma$, Axes ref. angular-velocities: $\omega_{ref} \in \Omega$ }

$\mathcal{FF}_2$ = { Robot linear and angular-accelerations: $a \in A$, $\alpha \in \Gamma$, Robot linear and angular-velocities: $v \in V$, $\varphi \in \Omega$, Status: $s_2 \in S$ }

$\mathcal{SD}_2$ = { Gyro: $\psi \in \Omega$, Accelerometer: $a \in A$ }

$\mathcal{OC}_2$ = { Robot -joy-stick voltages: $\mathcal{U}_{ref} \in \mathcal{U}$ }.

3.2.4 Collision Avoidance:

Being now able to control the body-velocities, we require further that the robot whilst moving, at whatever speed it may have, should actively avoid collision with stationary or moving objects. We do so by using a repulsion function which takes into account robot velocities, direction of motion and sensing and computation delays to produce a reflexive behavior. The repulsion is composed of a potential-field produced by the ultra-sonic sensor-array and the tactile force/moment acting on the robot body. The conflict between reaching the goal and avoiding collision is treated by the × operation which decreases the repulsion caused by the potential-field as the robot gets closer to the goal until finally it reaches zero at the home-(goal) position. The tactile repulsion is kept "alive" to prevent the robot from reaching a forbidden goal-position which may cause collision.

$\mathcal{FB}_3$ = { Robot ref. linear and angular-accelerations: $a_{ref} \in A$, $\alpha_{ref} \in \Gamma$, Robot ref. linear and angular-velocities: $v_{ref} \in V$, $\varphi_{ref} \in \Omega$}

$\mathcal{FF}_3$ = {Robot linear and angular-velocities: $v \in V$, $\varphi \in \Omega$, Robot repulsion-linear and angular-velocities: $v_{rep} \in V$, $\varphi_{rep} \in \Omega$, Status: $s_3 \in S$}

$\mathcal{SD}_3$ = { Range from ultra-sonic array : $\mathcal{R}_i \in \Lambda$, Linear and angular displacement of tactile-ring: $\mathcal{D}_{xy} \in \Lambda$, $\mathcal{D}_\phi \in \Phi$}

$\mathcal{OC}_3$ = {Robot-joy-stick voltages: $\mathcal{U}_{ref} \in \mathcal{U}$, Proximate: $\mathcal{P}_{op} \in \Lambda$ }.

3.2.5 Homing:

This is, basically, the position-loop. A self-tuning controller is used to overcome the difficulties encountered due to the non-holonomic constraints. Notice that we do not implicitly build a trajectory generator which satisfies the kinematic constraints but, instead, adjust the gains according to robot velocities and position. This gain tuner is the most crucial block in this level.

$\mathcal{FB}_4$ = { Command for robot linear and angular-velocities: $v_{cmd} \in V$, $\varphi_{cmd} \in \Omega$, Robot linear and angular position: $xy \in \Lambda$, $\phi\psi \in \Phi$}

$\mathcal{FF}_4$ = { Command for robot linear and angular-velocities: $v_{cmd} \in V$, $\varphi_{cmd} \in \Omega$, Robot linear and angular position: $xy \in \Lambda$, $\phi\psi \in \Phi$, Status: $s_4 \in S$}

$\mathcal{SD}_4$ = {}

$\mathcal{OC}_4$ = { Goal position: $xy_{goal} \in \Lambda$, $\phi_{goal} \in \Phi$}.

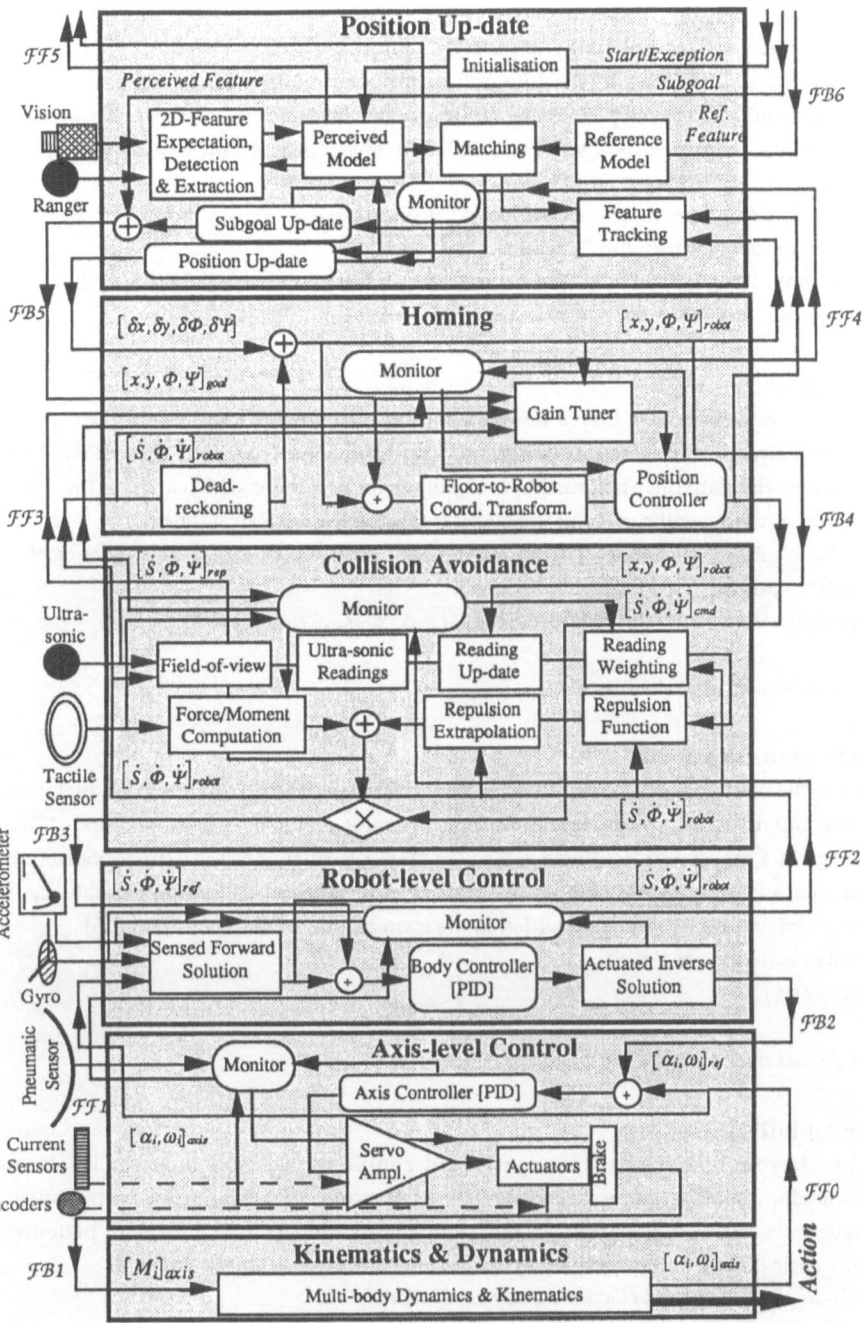

Fig.2-a Details of the Behavior Levels

3.2.6 Position Up-date:

The kinematic data, used for position computations, includes errors such as drift, bias and slippage which accumulate due to the integration property of the dead-reckonning..This imposes a serious limitation to the positioning accuracy. This level is responsible for acquiring a sparse model of few features of known position in the environment and matching them to obtain the correct, relative, robot-position.

$\mathcal{FB}_5$ = { Position up-date: $\delta_{xy} \in \Lambda$, $\delta\phi\psi \in \Phi$, Goal position: $xy_{goal} \in \Lambda$, $\phi_{goal} \in \Phi$}

$\mathcal{FF}_5$ = {Perceived model-features: $\mathcal{F}_e \in \Xi$, Status: $s_5 \in S$}

$\mathcal{SD}_5$ = { Range from ultra-sonic sensor-array: $\mathcal{R}_i \in \Lambda$, Two-dimensional visual features: $\mathcal{VF}_e \in \Xi$ }

OC_5 = { Goal position: $xy_{goal} \in \Lambda$, $\phi_{goal} \in \Phi$}

3.2.7 Navigation:

The navigation level will build a fictive line-of-sight between the goal and the current robot position and will always track this line generating sub-goals when the sight is blocked. This strategy combines well with the previous levels to build a local planner which attracts the robot to the goal whilst avoiding collision until the target is reachable over a straight-line.

$\mathcal{FB}_6$ = { Ref. feature: $\mathcal{F}_{eref} \in \Xi$, Sub-goal linear and angular position: $xy_{subgoal} \in \Lambda$, $\phi_{subgoal} \in \Phi$, Flag for Start/Exception: $\mathcal{Flg} \in S$}

$\mathcal{FF}_6$ = { Subgoal linear and angular position: $xy_{subgoal} \in \Lambda$, $\phi_{subgoal} \in \Phi$, Status: $s_6 \in S$}

$\mathcal{SD}_6$ = {}

OC_6 = { Goal position: $xy_{goal} \in \Lambda$, $\phi_{goal} \in \Phi$}.

3.2.8 Path-planning:

Until this level, the operator had to enter the floor co-ordinate of the goal. The way which lead to the goal was unknown. Now it is desired that the operator enters only the goal "node" in a known topology of a graph representing a gang-way map. We further require the robot to find a cost-optimal path leading to the goal. At this level of sophistication, a learning capability is considered mandatory and is, therefore, added in the form of an associative memory [5].

$\mathcal{FB}_7$ = { Goal-list: $xy_{list} \in \Lambda$, $\phi_{list} \in \Phi$}

$\mathcal{FF}_7$ = { Status: $s_7 \in S$}

$\mathcal{SD}_7$ = {}

OC_7 = { Start / Goal nodes: $\mathcal{N}_s \in N$, $\mathcal{N}_g \in N$}.

3.2.9 Model-building:

The robot had to rely on the relatively sparse and simple models in the lower levels. A complete 3D-model was never necessary. Now, however, more sophistication in the treatment of complex events is required. In particular, the generation of the 2-D model for path-planning and laying the basis for reasoning and goal synthesis have become an emergent need.

$\mathcal{FB}_8$ = { 2D-Model : $\mathcal{M}2d \in \Xi$, Start/Goal nodes: $\mathcal{N}_s \in N$, $\mathcal{N}_g \in N$ }

$\mathcal{FF}_8$ = { Start/Goal node: $\mathcal{N}_s \in N$, $\mathcal{N}_g \in N$, Status: $s_8 \in S$ }

$\mathcal{SD}_8$ = { Vision: $2\_\mathcal{D} \in \Lambda$, Ranger: $1\_\mathcal{D} \in \Lambda$ }

OC_8 = { Command sequence: $Cmd_{seq} \in \Sigma$}.

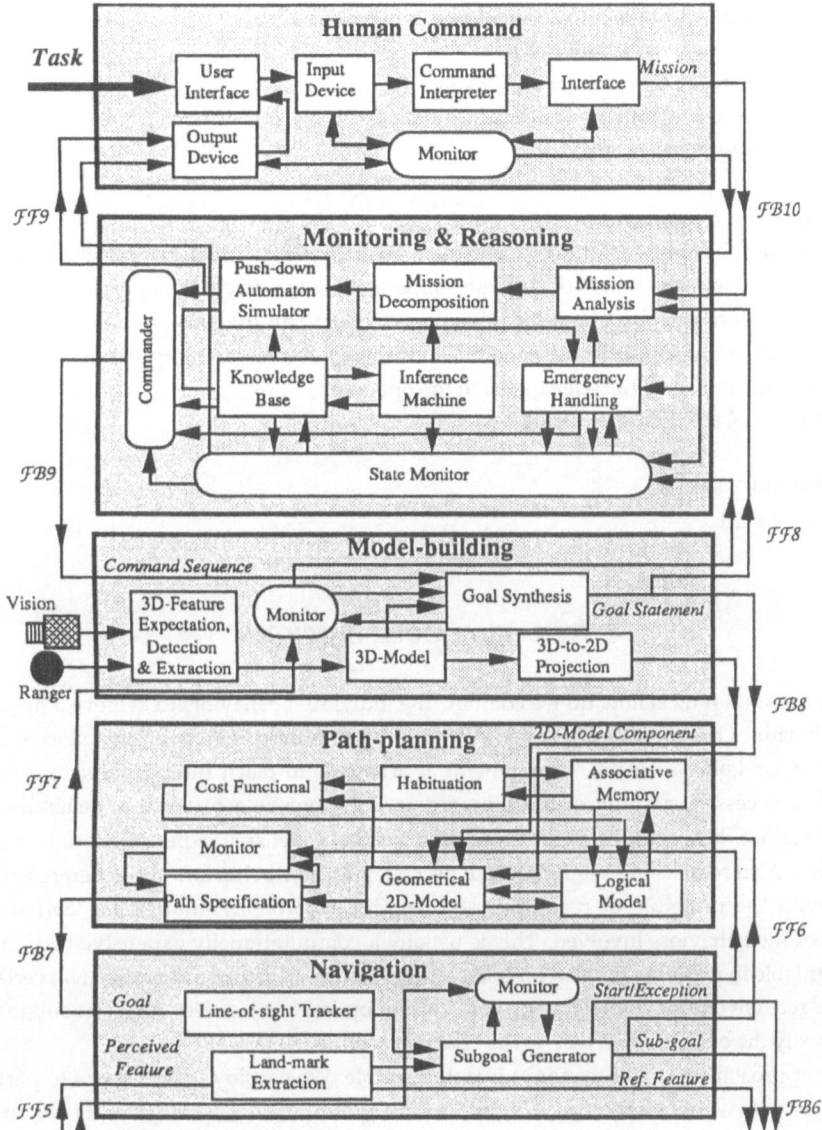

Fig.2-b Details of the Behavior Levels (cont.)

3.2.10 Monitoring & Reasoning:

Each of the previous levels has its own local monitor which observes the state of that level, prevents the stage from reaching some forbidden states and , further, sends the status to the higher level. The same is now to be established for the whole machine. Further, the mission will be analyzed and decomposed into sub-missions whose feasibility is first checked by running a simulation and then cut-down into a sequence of commands.

$\mathcal{FB}_9 = \{$ Command sequence: $Cmd_{seq} \in \Sigma \}$

$\mathcal{FF}_9 = \{$Start/Goal node: $\mathcal{N}_s \in N$, $\mathcal{N}_g \in N$, Status: $s_9 \in S$ $\}$

$\mathcal{SD}_8 = \{$ Vision: $2\_\mathcal{D} \in \Xi$, Ranger: $1\_\mathcal{D} \in \Lambda \}$

$OC_8 = \{$Command sequence: $Cmd_{seq} \in \Sigma \}$.

3.2.11 Human Command:

This level enables the operator to interface to the robot according to the behavior-level progress. In its most primitive form, where the operator muscular force is used, the operator interface is trivial. The interface remains simple also when a joy-stick, on the axis or the robot level, is incorporated. The most sophisticated version will include a command interpreter and some sort of visual, acoustic or even a force feed-back to the operator.

$\mathcal{FB}_{10} = \{$ Mission : $\zeta \in Z$, Status: $s_{10} \in S$ $\}$

$\mathcal{FF}_{10} = \{\}$

$\mathcal{SD}_{10} = \{$ Operator's senses $\}$

$OC_{10} = \{$ Task: $\mathcal{T} \in T\}$.

4. BEHAVIOR COMPOSITION

The question now is how do we combine the individual behaviors to achieve a desirable overall behavior of the robot at any level?. We shall discuss this problem briefly, since it is the subject of an on going research. One general approach is to teach the behavior to a *neural network*. The necessary amount of teaching required to achieve a reasonable generalization capability cannot, however, be determined in advance. Extensive experimentation will be unavoidable. Also re-teaching might become necessary as new behaviors are generated which conflicts with the principle of recursiveness. Another rigorous method is to *"optimize" a functional* of the behaviors involved. This is usually a computationally expensive task that is rarely affordable in real-time with an on-board energy source. Again, this strategy will certainly violate the recursiveness principle, since the "sum of optimal trajectories between subgoals is not necessarily the optimal trajectory to the final goal".

The approach we currently adopt is rather simple. We employ logical-gates to perform operations on sets whose elements are of the Boolean space, and *Analogical-gates* (Fig.3) on sets whose elements are continuous functions in time, primarily velocity commands. Using these two types of gates, one may construct *circuits* that process the individual behaviors to produce a desired "plausible" overall behavior. Notice, however, that the same prerequisites of skill and experience required to successfully build conventional circuits of any kind are also required here. Currently, we do not have any quantitative performance criterion for which the behavior can be optimized. The only requirement is that the robot "gracefully" reaches the goal, no matter how. In the following section we present some of these gates, briefly explaining their main properties.

4.1 Analogical Gates

These analogical gates are of two kinds: symmetric and asymmetric. The symmetric gates perform operations similar to their logical counter parts such as union (OR) and intersection (AND) while the asymmetric ones implement "prevalence" operations. Symbols used to represent these gates are shown in fig.3. Their functions are briefly discussed below.

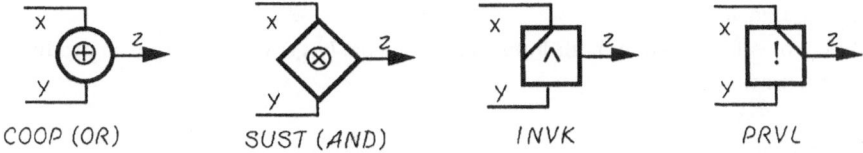

$$COOP (OR) \qquad SUST (AND) \qquad INVK \qquad PRVL$$

Fig.3 Analogical Gates

The names and symbols given to these gates are arbitrary and are only intended as a convention for their operation. Further, we shall assume the existence of the inputs and their *negations* which is equivalent to changing the sign.

Throughout the following treatment we use an exponential function in two variables x and y,

$$\wp(x,y) = e^{-\left(\frac{ax^2+bxy}{x^2+y^2}\right)} \text{ and } x,y \in \mathbf{R}$$

4.1.1 Symmetric Gates

The *symmetric gates* implement commutative operations corresponding to set union and intersection.

1) Co-operate (Analogical-OR) gate:

$$z = x \oplus y = x \, \wp(y,x) + y \, \wp(x,y)$$

with a=1.02889 and b=e^{-a} =0.3574.

Functional Description:

Both inputs contribute to the output in relation to their magnitudes. In the absence of one input, the output linearly tracks the other.

Major Functional Characteristics:
- $x \oplus y = y \oplus x$
- $cx \oplus cy = c(y \oplus x), c \in \mathbf{R}$
- $x \oplus x = x$
- $x \oplus 0 = x$
- $x \oplus -x = 0$
- $\min(x,y) \le x \oplus y \le \max(x,y)$
- $\frac{\partial z}{\partial x}\big|_{x=0, \forall y} = 0, \frac{\partial z}{\partial y}\big|_{y=0, \forall x} = 0$

2) Sustain (Analogical-AND) gate:

$$z = x \otimes y = x \left[1 - \wp(y,x) \right] + y \left[1 - \wp(x,y) \right]$$

with a=2.28466 and b=e^{-a} -1 =-0.89817.

Functional Description:

The output grows fastest when both inputs simultaneously grow. No output is produced if either input is zero.

Major Functional Characteristics:

- $x \otimes y = y \otimes x$
- $cx \otimes cy = c(y \otimes x)$, c$\in$ R
- $x \otimes x = x$
- $x \otimes 0 = 0$
- $x \otimes -x = 0$
- $\min(x,y) \le x \otimes y \le \max(x,y)$
- $\frac{\partial z}{\partial x}\big|_{x=0,\forall y} = 0$, $\frac{\partial z}{\partial y}\big|_{y=0,\forall x} = 0$

4.1.2 Asymmetric Gates

The *asymmetric gates* implement operations for which one port is given more competence than the other.

3) Invoke gate:

$$z = x \wedge y = x \, \wp(y,x) + y \left[1 - \wp(x,y) \right]$$

with a=2.284466 and b=1- e^{-a} = 0.89817.

Functional Description:

As the x-input grows, the share of the y-input to the output is increased. The absence of the x-input inhibits the output. In the absence of the y-input, the x-input is linearly passed to the output.

Major Functional Characteristics:

- $x \wedge y \ne y \wedge x$
- $cx \wedge cy = c(x \wedge y)$, c$\in$ R
- $x \wedge x = x$
- $x \wedge 0 = x$
- $0 \wedge y = 0$
- $x \wedge -x = 0$
- $-x \wedge x = 0$
- $\min(x,y) \le x \wedge y \le \max(x,y)$

4) <u>Prevail gate:</u>

$$z = x!y = x \, \wp(y,x) + y \, \wp(x^3,y)$$
with a=1.02889 and b=e^{-a} =0.3574.

Functional Description:
The x-port is assigned an exceptional prevalence over the y-port. The latter is put-through directly to the output as long as the former is absent. However, once the input at the prevalent port it strongly dominates the output.
Major Functional Characteristics:
- x!y ≠ y!x
- x!0 = x
- 0!y = y
- x!-y= -x!y
- min(x,y) ≤ x!y ≤ max(x,y)

4.1.3 Behavior Circuits and General Remarks

An example for a behavior circuit using analogical gates is shown in fig.4. This circuit is currently implemented in our mobile robot. The main purpose is to make the robot avoid collision with obstacles while underway to the goal. Once reached the goal, the robot will not repel away unless it feels a tactile force. The way it takes to "home" is considerably influenced by the free-space it traverses as long as it is still far from the goal position, but the influence gets weaker as it moves closer to the goal. A docking position is given a higher priority to other goals. The operator is allowed to adjust the velocity profile of the trajectory using the joystick. We do not have a "recipe" for how to build such circuits. Intuition, experience and experimentation are all important "ingredients" in a successful design. As mentioned earlier, behavior composition is a very lively research topic in mobile robots. Our on-going research aims at further exploring the properties of analogical gates and behavior circuits. Experiments conducted with this simple circuit are almost always, but not exclusively, satisfactory.

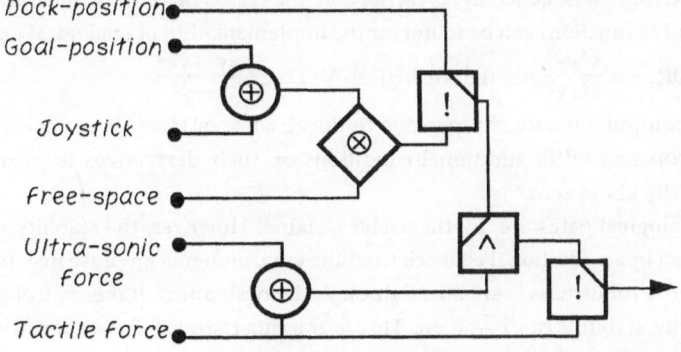

Fig.4 Behavior Circuit

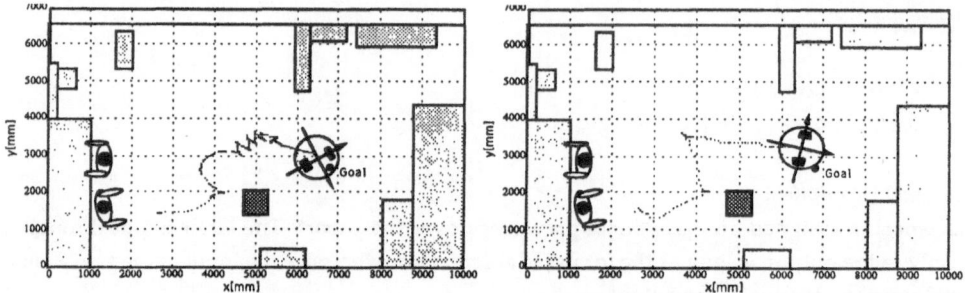

Fig. 5 Experiments without&with the behavior circuit

Recorded data for an experiment in which the robot is to reach a goal whilst avoiding an obstacle are shown in Fig.5. The figure to the left shows the trajectory when the behavior circuit is not used. The output of homing and collision-avoidance are simply added. The figure to the right shows the same situation but with the behavior circuit employed. Notice the absence of the oscillations encountered in the first experiment and the effect of the free space behind the robot on the first portion of the trajectory. Neither the joystick nor the Docking procedure was engaged. In both experiments, the ultrasonic repulsion force was sufficient to prevent the robot from touching the obstacle and engage the tactile repulsion. Further improvement on this trajectory can be obtained by overlaying a velocity profile and/or specifying subgoals.. Nevertheless, these experiments show that this simple goal conflict of reaching a goal-position and simultaneously avoiding collision is feasible even with few primitive behaviors when properly augmented.

Finally, before concluding this section, some general notes are due:

1) Let "o" denote any of the above operations, $\oplus, \otimes, \wedge$ and !, then:

• $z=xoy$ always exists $\Rightarrow$ o is well defined,

• for $x,y \in V$ where $V=\{v \mid v=[v_{min}, v_{max}], v \in \mathbf{R}\}$, $z=(xoy) \in V \Rightarrow V$ is closed under o,

• for where $x \in X$ and $y \in Y$, $X, Y \subseteq V$, $z=xoy$ is unique for $X \times Y$

$\therefore$ "o" is a *binary operation*.

2) The operations performed by the above analogical-gates are not associative. This limitation does not represent a serious drawback to this approach since the order in which the operations are performed is, generally, prescribed by the recursion of the control structure.

3) Alternative functions can be found for the implementation of analogical gates , e.g., for the Analogical-OR : $z = \dfrac{x^3+y^3}{x^2+y^2}$ and the Analogical-AND : $z = \dfrac{xy^2+yx^2}{x^2+y^2}$.

This version is computationally cheaper due to the absence of the exponential function. The functions we proposed fulfill additional conditions on their derivatives to provide smooth transition especially about zero.

4) The analogical gates are, by themselves, stable. However, the stability of the entire system remains an open question. Feedback from the environment can cause instability in some cases. Although the robustness is enhanced through the nestedness of the control structure, no guarantees for the stability can be given. This is the most crucial question in the design of controller for mobile robots in particular due to the dynamically changing environment.

5. CONCLUSIONS AND FINAL REMARKS

In this paper we discussed an alternative control structure for mobile robots which can be realized recursively and possesses some important desirable properties such as inherent robustness, nestedness, and distributed modelling. It endorses the significance of combining different system descriptions according to the level requirements. Furthermore, it offers a basis for tailoring the behavior of a mobile robot to the customer requirement, who might not be interested in ultimate autonomy. Another advantage is that this "bottom-up" approach always produces a functioning prototype from which more experience is attained and some important problems for the further design are exposed and treated. One of the most important issues, namely behavior composition and treatment of goal-conflict has been addressed by the so called "Analogical gate" concept to perform binary operations for a pair of behaviors. One may then build a circuit of these gates to implement the desired behavior. The recursive structure described in this paper has been validated in realistic environment up to the homing level on our mobile robot RAMSIS [4] using multiple processors running asynchronously. Path-planning is implemented on a host-computer but not yet integrated in the robot system. The work in inertial/odometric navigation is progressing, although not yet completed. Position-update, including initial position estimation using ultrasonic images is the subject of current experimentation. Recently, We began the realization of the state monitor as part of the "Monitoring and Reasoning" level. A last point we would like to point out is the stability and robustness investigations of mobile robot control. This major problem has not been sufficiently treated neither in our work nor in the literature to the best of our knowledge.

REFERENCES

[1] Albus, J.S., Brains, Behavior, and Robotics, BYTE Books, McGraw-Hill,1981.

[2] Antoulas,A.C., "On Recursiveness and Related Topics in Linear Systems", IEEE Trans. Automatic Control,Vol. AC-31,No.12,1986, pp.1121-1134.

[3] Arkin,R.C.,"Motor Schema Based Navigation for a Mobile Robot", Proc.IEEE Int. Conf. on Robotics and Automation, 1986, pp.264-271

[4] Badreddin,E.,"RAMSIS Concept and Specifications", Proc.19th Int. Symp. on Allied Techn.&Automation (ISATA), Monte Carlo1988,pp.103-121.

[5] Badreddin,E.,"Associative Memory Implementation in Path-Planning for Mobile Robots", Proc. IEEE Int. Conf. on Robotics and Automation, Cincinnati-Ohio, May 1990, pp.14-19.

[6] Badreddin,E.,"Recursive Nested Behavior Control Structure for Mobile Robots", Proc. IAS-2 Int. Conf. on Intelligent Autonomous Systems, Amsterdam, Dec.1989, pp.586-596.

[7] Brooks,R.A.,"A Robust Programming Scheme for a Mobile Robot", NATO ASI Series, Vol. F29,Languagees for Sensor-Based Control in Robotics, Springer-Verlag, 1987,pp.509-522.

[8] Crowley,J.L.,"Navigation for an Intelligent Mobile Robot", IEEE Journal of Robotics and Automation, Vol. RA-1,No.1,1985,pp.31-41

[9] Moravec,H.P.,"The Stanford Cart and the CMU Rover",Proc.IEEE,71,July 1983,pp.872-884.

Real-Time Control in an Autonomous Mobile Robot

T. Knieriemen and E. von Puttkamer

Abstract:

Information processing in an Autonomous Mobile Robot is done by a control system, which has to incorporate sensor data processing and action control tasks on an appropriate computer architecture. The software and hardware related questions of the organization of the control system are subject of this paper.

To cope with the real-time requirements here an orthogonal control structure with cascaded control loops is presented, combining aspects of both, functional and behavioural oriented control structures. The underlying computer architecture is organized as a distributed system following the task oriented decomposition of the control structure. The data to be transmitted between processing units are those exchanged between modules of the control structure. For the communication structure three design alternatives are discussed: a decentralized system and a centralized one, both with broadcasting mechanisms, as well as a centralized system with point-to-point communication.

Contents:

1. Introduction

Autonomous mobile robots (AMR) are systems which can move and perform useful operations without external support or human intervention over substantial time intervals. The ability to operate in an unknown or partly known environment is essential for an AMR to be considered fully autonomous. According to their locomotion system AMRs are classified in underwater, water, land, air and space robots. They need a sophisticated perception system, which allows to maintain a world model of the operating environment. Because of the inaccuracy and uncertainty of the perception system this model may be incomplete and some parts may be inconsistent, but at least it must be consistent in such a way that the robot can record its movements, determine its position and recognize obstacles in a reliable manner. Besides sensor data processing, several planning and navigational capabilities like obstacle avoidance, exploration of unknown environment or path planning in partly known environment must be provided.

The entire information processing in an AMR is done by a control system, which has to coordinate action control and sensor data processing tasks on an appropriate computer architecture under real-time conditions. Design and realization of such a control system requires hard- and software related questions to be solved in a harmonized manner. That means that the organization of the software modules, the so called control structure, must be designed with respect to the intended hardware organization of a multiprocessor architecture.

The general information processing structure of an AMR may be considered to consist of two distinct subtasks:
- **action control** (planning, execution and monitoring of a given task)
- **sensor data processing** (abstraction, fusion and mapping of sensor information)

Figure 1 shows both subtasks of the control system with their correspondence to the peripheral AMR-components (MMI: man-machine-interface, locomotion system and sensor system). The vertical triangle represents action control and describes the top-down partition of a particular task (specified by the operator) to a set of subtasks and finally to a sequence of commands which affect the motors of the locomotion system. The horizontal triangle represents sensor data processing and describes the bottom-up construction of complex information structures from raw sensor data. This sensor data processing delivers more and more abstract descriptions of the environment (maps with geometric, topologic or semantic information) which are used for planning and monitoring of system actions. That means that the action commands are derived from a given task and controlled by the results of the sensor data processing. Influenced by the actions of the system, which are the results of command execution, the actual environment is changing. Finally the reaction of the environment is recognized by the perception system, so that a closed loop is formed.

After this overview of the general organization of an AMR and its control system the following sections will provide a more detailed discussion of both aspects (soft- and hardware) of the control system with emphasis to real-time and harmonized hard-/software structures. Section 2 explains essential tasks of the control structure in context with known approaches for their organization. The real-time requirements which are considered in section 3 lead us to-

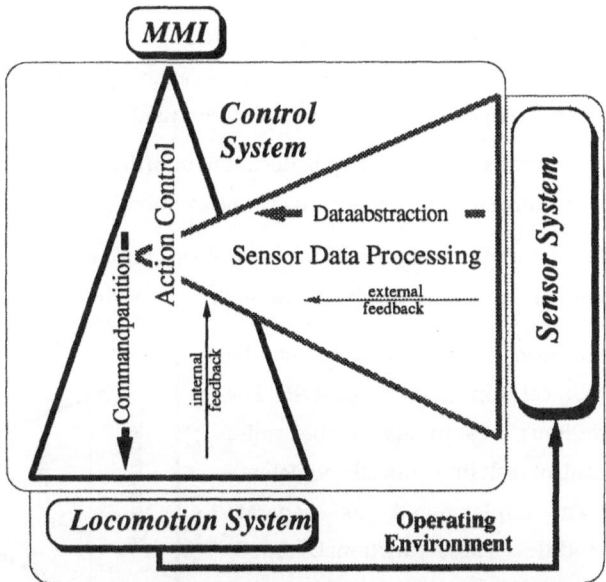

Figure 1: General information processing structure of an AMR

wards the orthogonal decomposition of the control structure described in section 4. Finally the underlaying computer architecture which fits to the suggested control structure is discussed in section 5.

2. Control-Structure of an AMR

In order to consider a mobile robot to be fully autonomous its control structure has to handle the aspects of a changing world and therefore to provide the following capabilities [1]:

- collision avoidance (obstacle detection together with appropriate evasive movements)
- position estimation (internal, external or combined methods)
- world modelling (generation and modification of internal maps describing the environment)
- global and local path planning (route, course and track generation)

These essential capabilities of an AMR must be realized in the control structure in such a way, that a robust, modular and expandable control structure with minimized computational redundancy can be provided:

- Robustness is necessary to cope with inaccuracies and uncertainties of perception as well as with hardware errors like randomly occurring communication failures.
- Modularity is very important for parallel development of several subtasks and separate testability of single components.

- Expandability is a useful feature during system configuration. It allows to build up the entire system stepwise from low-level capabilities like obstacle avoidance to high-level capabilities like reasoning about the environment.

- Minimized redundancy is desirable in order to reduce quantitative aspects like computing resources, system size, power consumption, system costs etc.

For the implementation of the control structure two completely different approaches are well known: functional and behavioural decomposition.

Functional decomposition is the classical top-down approach. The designer has to divide the entire control task into subtasks which are then implemented as separate modules. These functional components work together in a processing chain of vertical slices (see figure 2) through which information flows from the sensors to the actuators. As

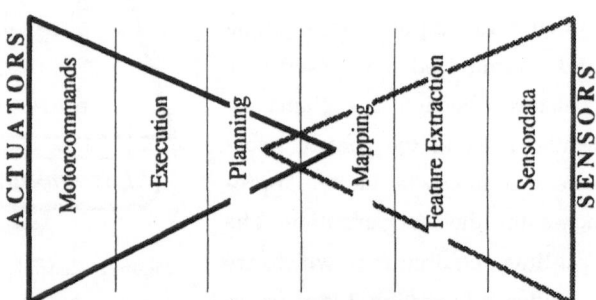

Figure 2: Horizontal decomposition of the control structure into functional modules

in figure 1 the underlaying triangles illustrate the degree of data abstraction or command partition for each slice. In this control architecture an instance of each slice must be built in order to run the robot at all. A failure of a single functional module results in failing of the complete system. This approach is also known as horizontal decomposition (into vertical slices) [2] or analytic approach [3].

Behavioural decomposition uses an approach, which allows the designer to combine the desired system behaviour by bottom-up adding basic competences like avoid obstacles, wander, explore etc. (see figure 3). Each level of competence realizes an individual connection between some kind of sensor data and actuator commands. These schemes of behaviour run in parallel and if one doesn't work only that particular competence is lost; the lower rest of the

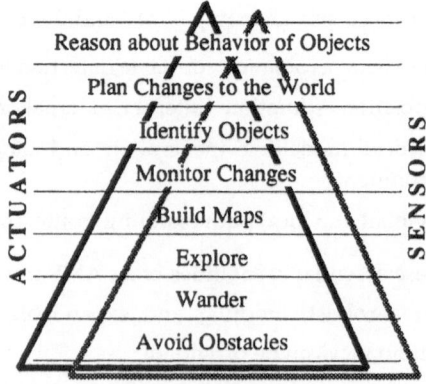

Figure 3: Vertical decomposition of the control structure based on task-achieving behaviours

system carries on doing something sensible. At a time only one control layer actively influences the systems overall behaviour by suppressing or subsuming the behaviour of other control layers. As mentioned above the underlaying triangles illustrate the degree of data abstraction or command partition for each layer. This approach is also called vertical decomposition (into layers of competence) [2] or synthetic approach [3].

The control structure used within the MOBOT-III project at Kaiserslautern University is a combination of both approaches. Before this so called orthogonal decomposition is being described in section 4, a closer look to the real-time requirements of an AMR is provided in section 3.

3. Real-time requirements: where they come from

Information processing under real-time conditions requires, that the result of processing is being delivered within predefined time constraints. These constraints should not be given by the performance of the computer system but by the demands of the application.

In case of an AMR the real-time requirements to the control system are strongly influenced by the following circumstances:

- moving system (velocity, braking action, kinematic)
- restricted perception possibilities (sensor capabilities)
- changing environment (static and dynamic aspects)

The influence of these features on the real-time constraints of an AMR are discussed exemplary for obstacle avoidance.

In order to have long time intervals for information processing, an obstacle should be detected as early as possible. This desire is being restricted, first by the limited maximum range of the sensors and second by suddenly occurring obstacles (e.g. moving objects at a crossing which were hidden by the static environment). The second case requires, that a minimum distance to a mobile object (MO_{min}) must be specified. For a moving object an appropriate evasive movement can only be realized if the object doesn't occur within the distance MO_{min}. As a result of this C_{min} the minimum distance of collision is being defined as:

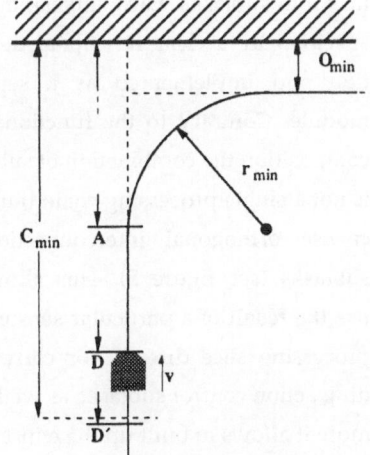

Figure 4: Real-time parameters for obstacle avoidance

$$C_{min} := min \{maximum\ range\ of\ sensors,\ MO_{min}\}$$

Considering the highest acceleration and angle velocity as well as the possible radius of steering and the delay of stopping, the stopping distance and the smallest radius of evasive movement can be estimated for a given velocity of the mobile robot. Figure 4 illustrates these real-time parameters of obstacle avoidance: For O_{min} a minimum security distance to objects of the environment and r_{min} the smallest steering radius at a given velocity v an evasive movement must be started at A. Let C_{min} be the minimum obstacle detection distance and $(D'- D)/v$ the time between two distance measurements then at D an obstacle may be detected. Therefore the remaining time for obstacle detection and planning of suitable movements can be estimated by:

$$\Delta t = (D - A) / v$$

Similar considerations are possible for the stopping distance.

4. Orthogonal decomposition with cascaded control loops

To cope with the real-time requirements of an AMR here an orthogonal control structure with cascaded control loops is suggested, combining aspects of both, functional and behavioural oriented control structures. Each control loop realizes a specific task like motor control or obstacle avoidance.

Just as the functional decomposition a top-down design approach is possible, which has to divide the entire control task into subtasks. Each subtask may be specified by a clear functional description and implemented as a separate module. Contrary to the functional decomposition the combination of subtasks is not a single processing chain but a slicewise orthogonal interconnection of subtasks (see figure 5). This allows to use the result of a particular sensor data processing slice directly for corresponding action control subtasks as well as in the subsequent sensor data processing slice. Furthermore it allows to built up the entire control system incrementally, starting with low-level tasks like motor control and finishing with high-level tasks like reasoning about the behaviour of objects. This characteristic is one of the main advantages of behavioural decomposition and from our point of view it is the only possible approach to realize such a complex control task.

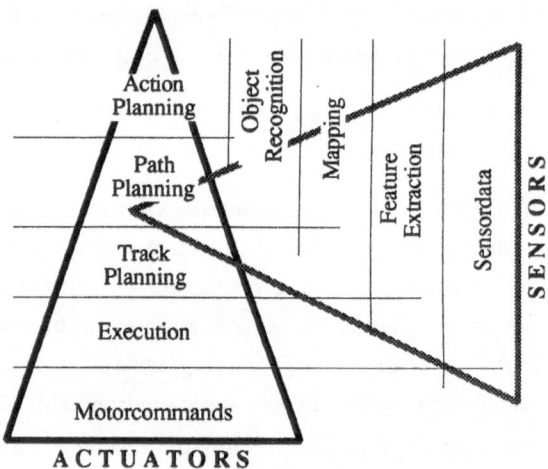

Figure 5: Orthogonal decomposition of the control structure into separate tasks

A closer look to the features of the orthogonal approach is given by the following overview of the MOBOT-III control structure, which is an example of an orthogonal decomposition and being realized at Kaiserslautern University.

Figure 6 shows the actual components of the control structure. On one hand it consists of sensor data processing components (IPE - Internal Position Estimation, SMG - Segment Map Generation, PFE - Primary Feature Extraction, CCG - Correlator Composer Geographer and TOPO - Topographer) which process raw sensor data from internal and external sensors to composed models in a bottom-up manner. On the other hand a hierarchical control structure is composed of five levels of control. Only the upper three ones are influenced by the environment situation, which means that they use the results from the world modelling components for their individual tasks. The components and interfaces of the entire control structure are described in [1] with emphasis on the sensor data processing part.

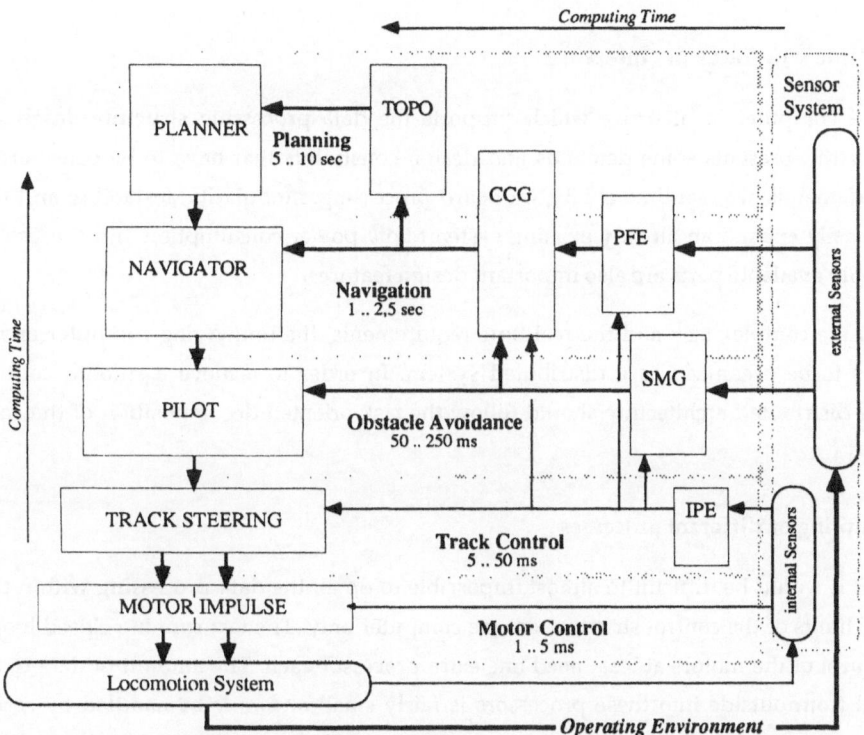

Figure 6: Cascaded control loops of MOBOT-IIIs control structure

Within this paper not the functionality but the combination of particular components is being considered. As seen in figure 6 there are five cascaded control loops. With the exception of motor control, all control loops consist of components for a particular control task and also of sensor data processing components, which process sensor data at a certain degree of abstraction (geometric, topologic and semantic representations). In each control loop (e.g. *obstacle avoidance*) the sensor data processing components deliver a representation (*a map with distance information to the surrounding obstacles*) which in context to a given task (*drive to a specific positionvector*) causes the action control component (*pilot*) to generate commands (*subvectors which consider local obstacles on the straight on course to the specified positionvector*) for the underlaying control loop (*track steering*) [1]. Seen as closed control loops, the sensor data processing components produce the controlled variable to be correlated to the reference input inside the action control component, which generates the manipulated variable.

The cascaded control structure allows all control loops to affect the overall system behaviour at any time. This allows a range of system behaviours to be derived from a limited number of basic functions.

5. Real-time computer architecture

Choosing a computer architecture which supports the data processing structure described above, an AMR presents some demands and design constraints that have to be considered. The basic demands are real-time ability, on board processing, modularity, testability and the ability to easily expand an already existing system. Low power consumption and the use of commercially available parts are also important design features.

Because of the complex task and the real-time requirements, the underlying computer architecture has to be organized as a distributed system. In order to achieve a smooth control system the distributed architecture should follow the task oriented decomposition of the control structure.

5.1 Decoupling of different processes

In an AMR it would be difficult to almost impossible to do all the data processing within the given time limits of the control structure on one computer only. The fast running closed loops for the control of the motors at least need one extra processor each. The amount of data to be transferred from outside into these processors is fairly small and may be handled by serial links. The remaining mass of data processing should be divided onto different processors too. A possible solution could be to attach a separate processor to each main module of the control structure.

The real-time requirements of the control structure specify hard time limits for generation and distribution of data packages being exchanged between processing elements. The distributed computer architecture must guarantee these limits and allow an effective decoupling of the different modules in order to remain the system manageable and testable.

The data to be transferred between processors correspond to the arrows between the modules of the control structure (compare figure 6). They indicate high-level variables, which are directly transferred to and from memory of the respective processing units.

The timing requirements for each processor can be read off the control structure too: a module running on a processor has to produce its data within a fraction of the given time interval for all modules inside the particular control loop. The transport of data has to be done within fixed time limits by the communication system interconnecting the processors.

Distributing the different tasks of the control structure to different processors gives an effective decoupling and holds the system manageable. Each module has a well defined data interface and precise time limits. It may be handled therefore by different programmers. As long as the intended functionality of a module remains preserved and its time limits are met, algorithms may be altered or changed at will; the exchanged data structures are the invariants of the system, forming the interface between modules.

As seen from the different processors, they should address high level variables (records, arrays, lists, ...) in their respective programs via access routines, which causes the underlying data transfer to be transparent. The difficult task of debugging could be made easier if the transport system were able to monitor the exchanged data. Regarding the underlying computer architecture these demands show the main area of investigations: the communication system.

5.2 Communication system requirements

There are three basic requirements the communication system between the processors has to meet: tight time limits, transparent transport of higher level variables and fault tolerance. They are discussed in the following sections.

5.2.1 Cyclic processing

Because data transmitted in the system may be seen as measured and control variables in closed loop systems sampled digitally, they have to be produced and transported cyclically.

All data types being fixed at compile time, the packet lengths and the bandwidth required may be calculated in advance. This leaves the scheduling to be done. In order to simplify this task, the sampling rates in the different control loops should be adjusted to be in a power-of-two relation to each other.

The transport capacity should be such, that within the shortest time slot not only those urgent (short) packages may be transported, that need a high frequency sampling but also the longest package, so that no package needs to be splitted with all the additional handling overhead.

These considerations show, that data transport is fixed in its timing behaviour throughout the system.

5.2.2 Transparent transport of higher level variables

As seen from the modules residing inside a processor they access higher level variables via access procedures, the transport system being transparent to them.

Aside from data management invisible to the outside, access procedures deliver a pointer to a variable accessed (call by reference), the program then being free to read multiply or to write respectively. Another access to the same variable via the access procedure is handled as beginning of a new cycle in the data processing loop of the module. Whenever a variable is calculated the program puts a time stamp to this variable to show the consumer the time of production.

An access procedure and the transport system share three buffers for a variable: the part reading a variable does this from one buffer, the part writing new values has two buffers to write into alternately in order to keep a relatively fresh buffer in stock if the reader wants to read new data, thereby automatically freeing a buffer for writing new data into. The buffers are a shared resource between two independent parties: an access procedure and the transport system. Their critical regions have to be managed.

In order to decouple the different modules from each other, access to a variable in a module must not be influenced by the timing behaviour of other modules. This excludes blackboard systems for communication [6], as traffic jams cannot be excluded because of unpredictable access to variables from inside the modules. Changing a module may thus produce different timing behaviour in others.

5.2.3 Fault tolerance

In a cyclic data transport scheme all data are transported in this manner, so the transport system always runs at full capacity. There is no worst case to be handled [7]. There will be failures in data transport, mainly transient failures of the kind, that once in a while a bit gets lost.

As long as the transport system can detect such a flaw by parity or CRC check the corresponding data package (a variable) may be signed as false and dropped. The data processing modules can detect this via the time stamp [7]: they get old data instead of a new one and can interpolate a future value. In such a sampled system one sampling point may get lost without doing harm as long as the sampling period is short enough compared to the period of the highest system inherent frequency. If more than one datum fails in succession, than an emergency situation must be stated to the supervisor of the control structure, which in the worst case could stop the robot for further inspection.

By this simple scheme the system is fault tolerant against transient failures. At any time any data out of n data transported in a cycle may be wrong without doing harm. This is not allowed for two data packages in succession. As a result of this the control system satisfies a kind of (n-1)-security.

5.3 Design alternatives

For the communication structure three design alternatives are discussed: a decentralized system and a centralized one, both with broadcasting mechanisms, as well as a centralized system with point-to-point communication. They are described in the following sections.

5.3.1 Decentralized communication system
The most natural organisation of a communication system would be to interconnect processors directly, following the control structure: where data are to be transported, an interconnection is formed.

As long as the different communication links from and to any processor are less than 5 a transputer system might be build [8]. The transputers must be equipped with the communication software to transmit the necessary data. Synchronisation of data transfer in the needed cyclic manner may turn out to be a problem. Moreover this solution does not decouple data processing and data transport: they are done on the same processor, and data transport time is lost time for running the algorithm to produce data.

Using separate communication controllers inside each processing unit and performing all data transport on a common bus as physical medium, one gets a solution like figure 7. Inside a processing unit the communication controller and an access procedure share the 3 buffers for each variable. The communication controller has to organize the data transport necessary for its

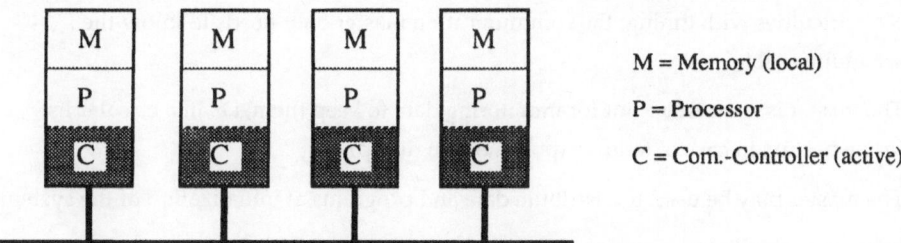

Figure 7:   Configuration with active controllers for broadcast communication

host processor.This can be done either by point-to-point communication or by broadcast. In the second case every communication controller monitors the communication line to pick up data intended for its host processor. The data are addressed not for a processor but by the system wide known name of a variable.

In this configuration communication controllers operate as active emitters. They are sending regularly the data their host processors have produced, and need not to bother about the recipients [4]. This broadcast scheme copes favourably with the allowed rare losses of data packages.

There is a synchronisation problem to be solved: each communication controller must know its time slot to send with respect to a system wide known clock. So one processor needs to keep a clock distributing a time signal in regular intervals for synchronisation.

5.3.2 Centralized communication system

The amount of hardware in each processor may be made smaller in a system with one central communication controller, regularly polling all processors to transmit their respective data. They broadcast their data to all and each communication controller picks up its relevant data from the bus, thus operating in a passive manner. A possible solution of a centralized system is shown in figure 8.

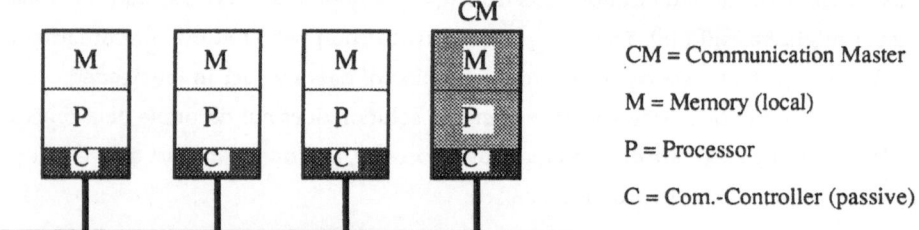

CM = Communication Master

M = Memory (local)

P = Processor

C = Com.-Controller (passive)

Figure 8: Configuration with a master module responsible for broadcast communication

This type of centralized system has the following advantages

- No difficulties with timing; the communication master only needs to know the scheduling table.
- The master is a natural point for monitoring data to keep them, i.e. in a circular list, for inspection in case of failures (post mortem dump).
- The master may be used to distribute data and programs at initialization of the system.
- It forms naturally the interconnection to the outside world for debugging.

There are disadvantages too:

- Any failure in the communication master stops the whole system.
- There is still a need for a kind of passive communication controller in each processing unit.

This latter disadvantage may be overcome in a system like fig. 9, a centralized communication system with point-to-point communication.

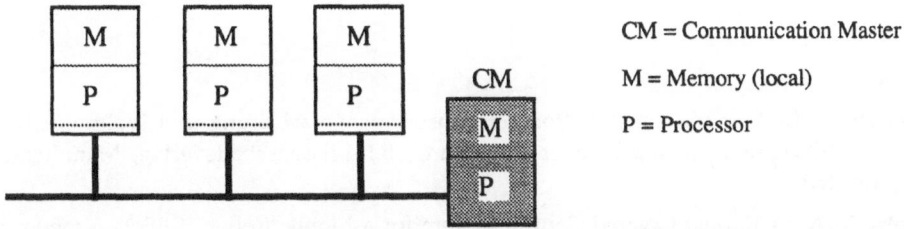

CM = Communication Master

M = Memory (local)

P = Processor

Figure 9: Configuration with a master module responsible for point-to-point communication

There is one master to address all processors, taking up data from buffers either via a DMA or from buffers in dual ported memories, and distributing data into buffers the same way. This approach is supported by commercially available VME-bus-systems for 68000 based computers, using dual ported memories for parallel access via host processor and VME-bus.

The disadvantage of this scheme is a lot more data to be transported. Each variable has to be transported at least twice: from the producer to the communication controller and from here to all consumers in succession. The time slot is the same as for the communication scheme with separate controllers. Using a bus system with 32 bit words gives the needed bandwidth.

6. Conclusion

The design alternatives discussed above are being investigated within the MOBOT-III project at Kaiserslautern University in three stages of development: The first design was a decentralized solution with communication controllers at each processor, the R-bus-system [4]. It used a serial communication link with independent controllers. Then this scheme was altered into a system with a central communication master, polling the communication controllers in each processor to broadcast their data, the S-bus-system [5]. Both systems had rather small bandwidth and needed specially developed hardware.

Current work is being concentrated on the third alternative: a VME-bus-system with 68000 processors and dual-ported memories with one computer as communication master. This master is used as system supervisor and debugging link too.

A newly developed operating system kernel allows the code compatible transfer of programs, being implemented on Apple Macintosh computers, to the VME-bus-system via the communication master. Programs may thus be developed and debugged inside a rather comfortable simulation environment running on Mac-II-systems [9]; debugged programs are downloaded into the VME-bus-system on the robot.

REFERENCES:

[1] Knieriemen, T.: Autonome mobile Roboter: Sensordatenverarbeitung und Weltmodellie-rung zur Navigation in unbekannter Umgebung; B.I.-Wissenschaftsverlag, Mannheim, being printed

[2] Brooks, R. A.: A Robust Layered Control System for a Mobile Robot, IEEE Transactions on Robotics and Automation, Vol. RA-2, No. 1, 1986, pp. 14-23

[3] Nehmzow, U.; Hallam, J.; Smithers, T.: Really Useful Robots, Second International Conference on Intelligent Autonomous Systems, Amsterdam, The Netherlands, 1989, pp. 284-293

[4] Hinkel, R.; Knieriemen, T.; von Puttkamer, E.: An Application for a Distributed Computer Architecture - Realtime Data Processing in an Autonomous Mobile Robot, IEEE International Conference on Distributed Computing Systems, San Jose, California, 1988, pp. 410-417

[5] Weiß, G.: Entwurf und Implementierung eines Multiprozessor-Kommunikationssystems für Echtzeitanwendungen, Diploma-Thesis, Kaiserslautern University, Computer Science Department, 1991

[6] Harmon, S. Y.; Bianchini, G. L.; Pinz, B. E.: Sensor Data Fusion through a Distributed Blackboard, IEEE International Conference on Robotics and Automation, Vol. 3, San Francisco, California, 1986, pp. 1449-1454

[7] Kopetz, H. et. al.: Distributed Fault Tolerant Real-Time Systems: The Mars Approach, IEEE Micro Magazine, Feb. 1989, pp. 25-40

[8] Einstein, J.R., Barhen, J.: A Virtual Time Operating System for a Hypercube in Robotics Applications, Intelligent Robots and Computer Vision, SPIE Vol. 726, 1986, pp. 424-430

[9] Knieriemen, T.; von Puttkamer, E.; Trieb R.: 3d7 - A 3D Simulation Environment for Autonomous System Design, IAS-2, Intelligent Autonomous Systems Conference, Amsterdam, Netherland, 1989, pp. 434-440

Advanced Sensor-Guided Motion Control for Mobile Robots

G. Pritschow and J. Heller

The application of mobile robots in flexible manufacturing systems requires suitable methods for navigation and path planning. In this paper, a path-planning method, considering kinematic and dynamic constraints of mobile robots, will be presented. The real-time calculation of the nominal values is based on teach-in programming and on navigation with landmarks. An important feature is the hardware independence, so that the developed modules can be used as standard elements in mobile robots.

1   INTRODUCTION

Mobile robots without a physical guideline require suitable sensors for navigation, communication and collision avoidance. Looking at mobile robots in production environments, the tasks of the multi-sensor navigation system are: following a pre-programmed path and recognizing possible obstacles /1/. The task of the path planning is the real time generation of nominal values for the drives. With respect to an optimal generation of nominal values the kinematic and dynamic constraints of the used mobile robots have to be taken into account.

These are, for example, the maximum curvature while going around curves or the maximal velocity or acceleration of the drives. These so-called hardware-dependent modules are closely related to the hardware-independent path-planning modules. The user should be able to define the three-dimensional path (position (x,y) and orientation (γ)) and the velocity profile independent from each other. The path will be divided into single segments which are represented by an analytical function with the parameter s. The velocity profile defines the time dependence of the parameter s /2/. The consideration of steadiness in the time behaviour as well as the geometrical values lead to nominal paths, which are steady up to the third derivative with respect to time, and up to the second derivative with respect to the parameter s. Thus, there are no jumps in the acceleration profile.

Like in conventional robot controls the Cartesian nominal values will be transformed into the machine coordinates. In contrast to this, the measured machine coordinates can be transformed into Cartesian coordinates in order to realize a closed-loop path control in addition to the closed-loop position control /3/. With the presented method, future nominal values will be compared with the dynamics of the system so that an optimal path-planning can be achieved. Similar methods are known from Olomski /6/ where the constraints lead to velocity changes. It depends on the quality of the algorithms in which way these changes are taken into consideration.

With mobile robots, the path planning is closely related to the sensors for navigation. Since the robots have to follow a pre-programmed path, it is not necessary to scan the surrounding continuously. In the following, a procedure will be presented which is based on dead reckoning, inertial measurements (gyroscope) and discrete path correction with the aid of landmarks. In contrast to the navigation in Evans /4/, the absolute position of the landmarks has not to be known so that there is an additional increase of flexibility.

2 ANALYSIS OF NAVIGATION METHODS FOR MOBILE ROBOTS IN A PRODUCTION ENVIRONMENT

The navigation system is an essential task while introducing mobile robots without a physical guideline. Since a few years, there have been a lot of research activities in Europe, Japan and USA on the topic of mobile robots. A classification of the navigation methods will be shown in <u>Figure 2.1</u> /5/.

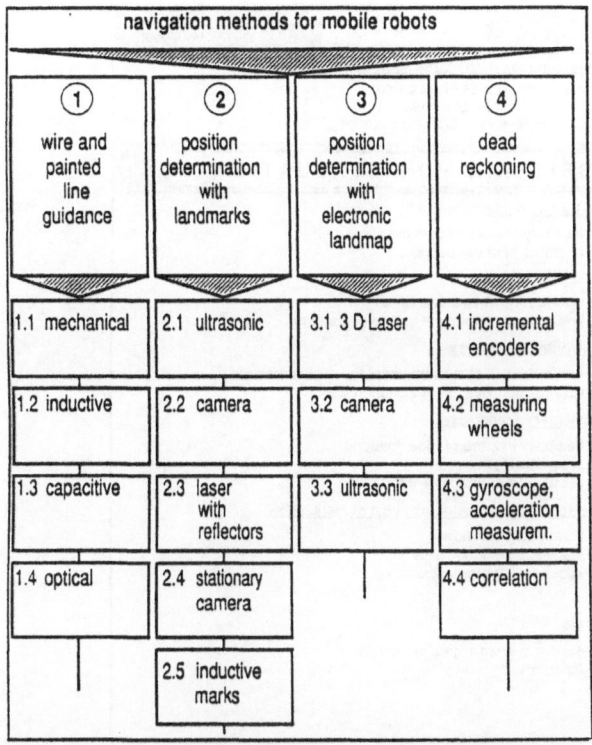

<u>Figure 2.1</u>: Classification of navigation methods for mobile systems

The methods in group 2, 3 and 4 will be considered next. Almost all research activities use dead reckoning. Incremental encoders, gyroscopes and sensors for acceleration measurements are being used, depending on the desired accuracy. Velocity measurements with correlation are also known. The use of these sensors will increase the accuracy of the dead reckoning but since there are still errors due to drift, limited resolution etc., the single use of dead reckoning is not possible for navigation. Therefore, some kind of absolute position detection is necessary. However, there is no need for continuously scanning the surrounding, so that the methods in group 2 are the most interesting to the described application. In dependence of the desired measuring values and the specific advantages and disadvantages, one of the systems will be used. Considering a vehicle movement in a surface, the three degrees of freedom (x,y,γ) have to be determined. This has to be done not in absolute Cartesian coordinates but with respect to the coordinate system of the landmark (<u>Figure 2.2</u>).

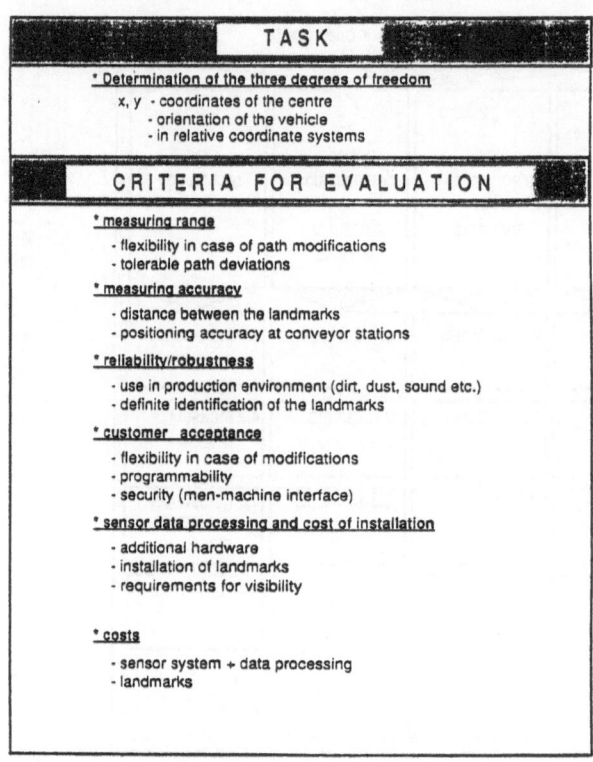

<u>Figure 2.2</u>: Requirements on navigation systems

Figure 2.2 also shows the criteria in order to evaluate the different systems. A qualitative evaluation is shown in Figure 2.3.

sensor / criteria	2.1	2.2	2.3	2.4	2.5
measuring range	+	+	+	+ +	O
measuring accuracy	+	+	+	O	O
amount of data processing	O	+	O	+ +	+
reliability	O	+	O	O	+
customer acceptance	O	+	O	− −	+ +
expense of stationary installations	+	O	O	+ +	O
expense of installation in robot	−	+	+	− −	+
costs	O	+	+	+ +	+

(++) very good or high   (−) very bad or low

Figure 2.3: Evaluation of navigation methods for discrete referencing

From today's point of view, the methods 2.2, 2.3 and 2.5 are the most appropriate ones. In the following, the navigation with landmarks whose absolute position is not necessary will be discussed.

2.1 Discrete Referencing with Landmarks

The troublesome measuring of the landmarks is needless if a Cartesian coordinate system R_i is assigned to a landmark i, Figure 2.4.

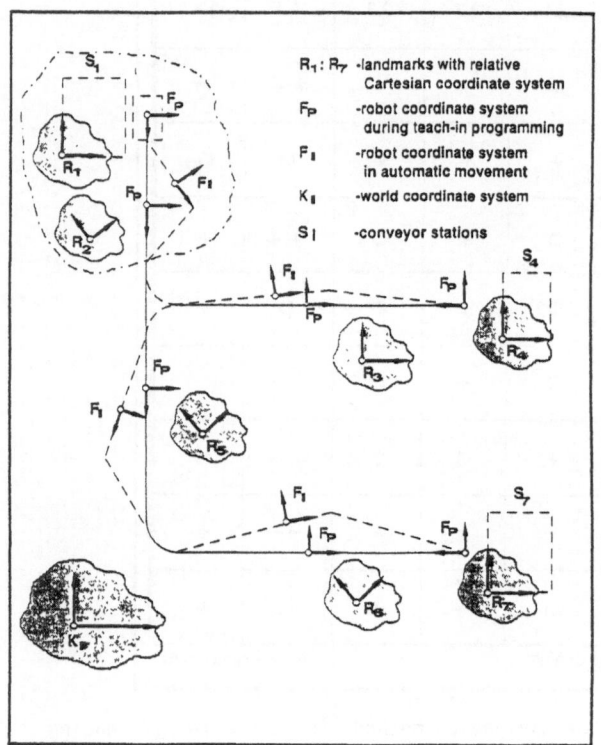

Figure 2.4: World modelling using relative Cartesian coordinates

During the programming phase, the teach points are stored with respect to the last relative coordinate system. This means, that the data of the nominal path between R_1 and R_2 are related to R_1, between R_2 and R_3 to R_2 and so on. In the automatic movement, the sensor system will be activated when the vehicle approaches a landmark. The real position and orientation of the vehicle will then be measured. Considering nominal, real and future values, a path correction will be calculated. However, the vehicle is moving so that there is a time difference between the actual measurement and the beginning of the correction. This difference will also be determined and taken into consideration. Finally, the overall structure of a kinematically independent mobile robot control can be achieved (Figure 2.5).

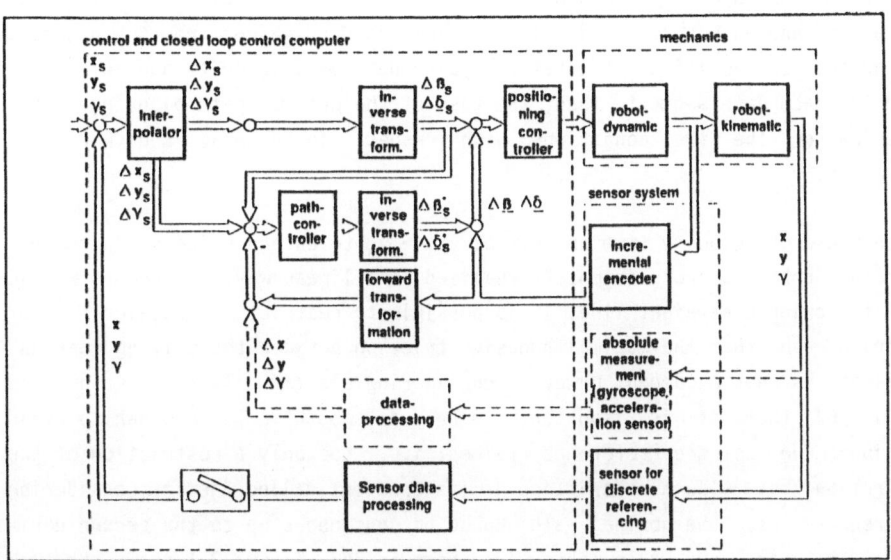

Figure 2.5: The structure of a kinematically independent mobile robot control

The Cartesian coordinates are represented by x, y, γ and the machine coordinates by β_1, β_2, δ.

3   OPTIMAL PATH PLANNING FOR MOBILE ROBOTS

The input values of the path planning are on the one side the desired movements given from the sequence control, e.g. straight lines or curved lines to the aimed point, and on the other side the on-line measured values from the discrete referencing. In addition to that, the user defines a velocity and an acceleration for each path segment. Thus, the task of the optimal path planning is the connection of the two conditions with respect to dynamic and kinematic constraints.

The increase of velocity from an initial value v_a to an end value v_e in today's commercial robot controls is mostly realized in a linear way. In case of mobile robots following a straight line, it is possible to restrict the acceleration of the drives so that the maximal adhesive friction between the driving gear and the floor is not exceeded. However, considering elastic drives it is shown in Olomski /6/ that it is important to have a "smooth" path in order to avoid vibrations due to the reference values; i.e., not only a restriction of the acceleration but also a restriction of its gradient called "jerk". Considering this requirement, the nominal path should be continuous up to the second derivative. The necessary algorithms are shown in /6/. Figure 3.1 shows the time behaviour of jerk r, acceleration a, velocity v and the path parameter s.

Beginning with their initial values, all the parameters are led to an end value while considering the limiting values. The integration of the nominal velocity into each sensing time will lead to a nominal value for the parameter s.

In addition to path continuity with respect to time, the nominal path in the x-y surface should also be continual with respect to the parameter s. An optimal path with straight lines and curved segments has a continuous progress of the curvature, and therefore no jump in the centrifugal force. If the nominal path is given as a function of the path parameter s the requirements for the x and y coordinates are as follows:

$$x_{(s^+)} = x_{(s^-)} \quad ; \quad y_{(s^+)} = y_{(s^-)} \tag{3.1}$$

$$\frac{\partial x}{\partial s}_{(s^+)} = \frac{\partial x}{\partial s}_{(s^-)} \quad ; \quad \frac{\partial y}{\partial s}_{(s^+)} = \frac{\partial y}{\partial s}_{(s^-)} \tag{3.2}$$

$$\frac{\partial^2 x}{\partial s^2}_{(s^+)} = \frac{\partial^2 x}{\partial s^2}_{(s^-)} \quad ; \quad \frac{\partial^2 y}{\partial s^2}_{(s^+)} = \frac{\partial^2 y}{\partial s^2}_{(s^-)} \tag{3.3}$$

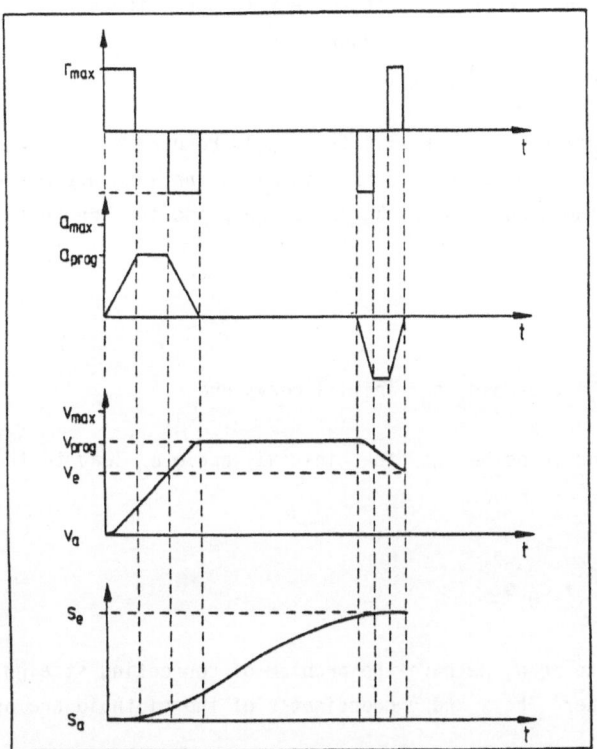

<u>Figure 3.1</u>: Time behaviour of jerk r, acceleration a, velocity v and path
parameter s

Besides these requirements there are other constraints like centre distance,
maximal speed of the steering angle or maximal centrifugal force which will determine the path velocity as well as the curvature c. For each nominal point of
the path it is necessary that

$$c(s) \leq c_{max} \qquad\qquad\qquad (3.4)$$

During the real-time generation of the nominal path straight line segments are
connected by curved segments. For this purpose clothoids will be used since they
optimally fulfill the following needs /7/, /8/:

- continual path
- linear and continual curvature with respect to parameter s
- continual change of the radial acceleration
- low expense of calculation.

A clothoid is a planar curve which will frequently be used in high-way design. The curvature $c(s,k)$ of the curve is proportional to the length s measured from the origin of the clothoid where both the curvature c and the length s are equal to zero:

$$c(s,k) = k \, s + c_0 \qquad (3.5)$$

k: parameter of the clothoid, c_0: initial curvature

Integrating (3.5) and considering the initial angle ϕ_0 lead to the tangent $\phi(s,k)$ of the path:

$$\phi(s,k) = \phi_0 + \frac{k}{2} s^2 + c_0 \, s \qquad (3.6)$$

If c_0 and ϕ_0 are set to zero, we have the problem of connecting straight line segments with each other. The x and y coordinates of the clothoid are as follows:

$$x(s,k) = \int_0^s \cos\left(\frac{k \, \tilde{s}^2}{2}\right) d\tilde{s} \qquad (3.7)$$

$$y(s,k) = \int_0^s \sin\left(\frac{k \, \tilde{s}^2}{2}\right) d\tilde{s} \qquad (3.8)$$

The determination of the integrals (3.7) and (3.8) cannot easily be done. However, if the integrals are expanded into a series and the single form of the sum is integrated, it follows:

$$x(s,k) = \sum_{i=0}^{\infty} (-1)^i \frac{s}{(2i)!(4i+1)} \left(\frac{k}{2} s^2\right)^{2i} \qquad (3.9)$$

$$y(s,k) = \sum_{i=0}^{\infty} (-1)^i \frac{s}{(2i)!(4i+3)} \left(\frac{k}{2} s^2 \right)^{2i+1} \qquad (3.10)$$

In order to connect two directed lines g_i, g_{i+1}, a pair of clothoids will be used, Figure 3.2.

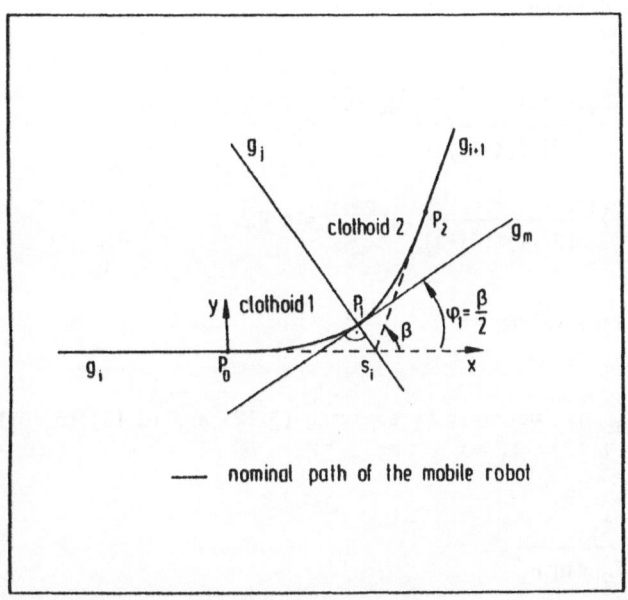

Figure 3.2: Connecting two directed lines g_i, g_{i+1} with a pair of clothoids

Clothoid 1 starts at point P_0 with initial curvature $c_0 = 0$ and initial tangent $\phi_0 = 0$. The curvature increases proportionally to the length s up to point P_1 where $s = s_1$ and $c = c_1$. Then, the curvature decreases from c_1 to zero at point P_2 of the clothoid 2. It follows that the angle of the tangent at P_1 is equal to $\phi_1 = \beta/2$, and for the parameter of the clothoid follows: $s_1 = s_2$ and $k_1 = k_2$. This means that the clothoids are similar. If P_0, g_i and g_{i+1} are given, the point P_2 is also defined. This has to be considered when the connecting points P_0 and P_2 are chosen from the path-planning method.

The directed line ϕ_i is used to determine the parameter s_1 and k_1, while g_i is defined as follows:

$$
g_i = \begin{bmatrix} S_{ix} \\ S_{iy} \end{bmatrix} + t \begin{bmatrix} -\sin \beta/2 \\ \cos \beta/2 \end{bmatrix} \tag{3.11}
$$

By introducing $\beta = 2 \phi_1 = k\, s_1^2$ into (3.9) and (3.10), the x,y coordinates of point $P_1\ (x_1,\ y_1)$ are defined:

$$
x_1 = s \sum_{i=0}^{\infty} (-1)^i \frac{1}{(2i)!(4i+1)} \phi_1^{2i} = s\, F_1 \tag{3.12}
$$

$$
y_1 = s \sum_{i=0}^{\infty} (-1)^i \frac{1}{(2i+1)!(4i+3)} \phi_1^{2i+1} = s\, F_2 \tag{3.13}
$$

$F_1,\ F_2$ are constant values

The equation for s_1 is obtained by applying (3.12) and (3.13) to (3.11) and eliminating parameter t:

$$
s_1 = \frac{S_{ix} \cos \phi_1}{F_1 \cos \phi_1 + F_2 \sin \phi_1} \tag{3.14}
$$

with $S_{iy} = 0$ in the coordinate system with its origin in P_0 and x-direction in direction of g_i. This leads to

$$
k_1 = \frac{\beta}{s_1^2} \tag{3.15}
$$

With k_1 and s_1, the pair of clothoids for connecting the directed lines g_i and g_{i+1} are defined.

If the maximal curvature c_{max} is exceeded, the path-planning algorithms will reduce the velocity or choose other connecting points P_0, P_2, if possible.

4   THE MOBILE ROBOT CALLED FLEXL

The mobile robot "Flexl", <u>Figure 4.1</u>, has been developed at the Institute of Control Technology for Machine Tools and Manufacturing Systems, ISW. It is used as an automated- guided vehicle, AGV, in a production environment.

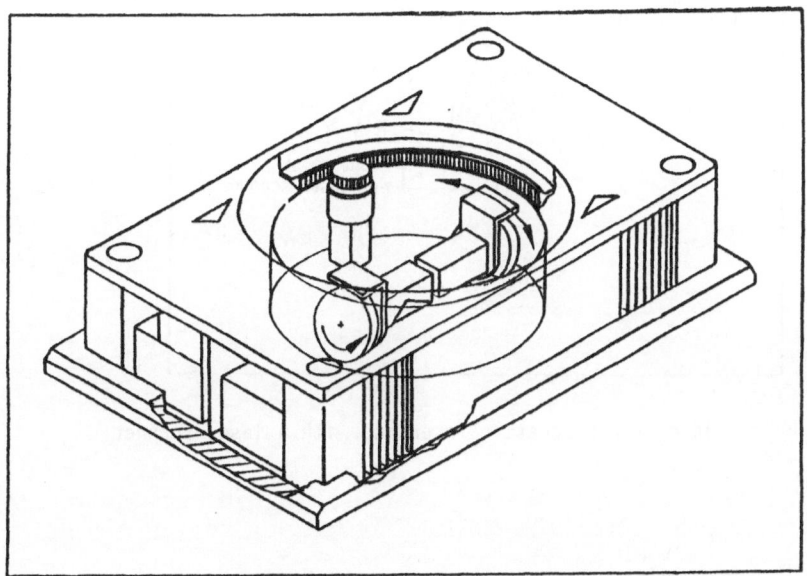

<u>Figure 4.1</u>: The mobile robot called FLEXL

A laser scanner with, reflectors in the surrounding will be used for navigation. In order to position it with an accuracy of ± 1,5 mm at conveyor stations, ultrasonic sensors are being used.

<u>Figure 4.2</u> shows the principle of the discrete referencing while the vehicle is passing a reflector. The three angles α_1, α_2, α_3 are measured. Thus, the Cartesian coordinates can be calculated within the current Cartesian coordinate system, shown in <u>Figure 4.2</u>. The comparison between the measured value and the nominal value at the reference point leads to the calculation of a correction path.

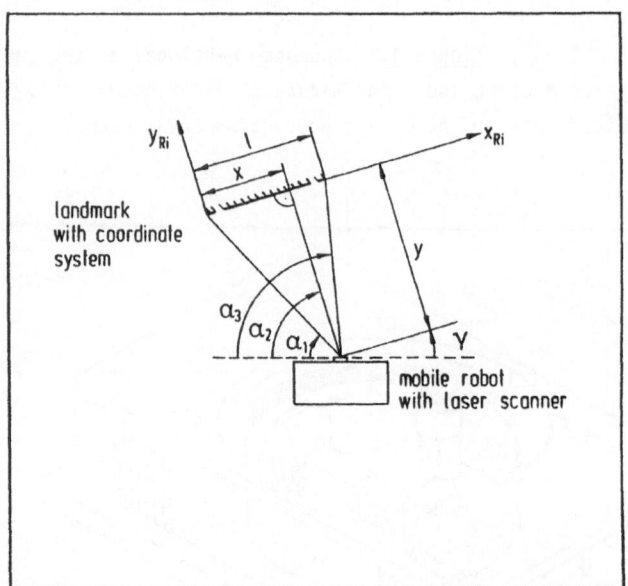

Figure 4.2: Principle of the discrete referencing with a laser scanner

5 OUTLOOK

The presented path-planning method for mobile robots allows an efficient use of different kinematics. This makes it possible to configurate systems with auto-mated-guided vehicles quickly and flexibly. Looking at applications in production environment, future research has to be done in the field of insensitive sensors for navigation and collision avoidance.

Thus, the reliability of a plant will increase and the advantages of a wireless-guided AGVS will become even more important.

6 LITERATURE

/1/ Pritschow, G.;   FTS: Flexibel navigieren ohne Leitdraht.
    Heller, J.   F + H, Fördern und Heben, 40 (1990) Nr. 11.

/2/ Bruhn, H.;   Grundlagen der analytischen Bewegungsplanung
    Ersü, E.   für Roboter.
                          VDI Berichte: 598. Düsseldorf: VDI Verlag, 1986.
                          S. 461 - 472.

/3/ Jantzer, M.   Bahnverhalten und Regelung fahrerloser Trans-
                         portsysteme ohne Spurbindung.
                         ISW-Bericht 82. Berlin, Heidelberg, New York,
                         London, Paris, Tokyo: Springer Verlag 1990.

/4/ Evans, D.F.   Non-wire guidance: System flexibility is what
                         counts.
                         Proceedings of the 6th International Conference
                         on AGVS, Brussels, 1988.

/5/ Heller, J.   Steuerungsstrukturen und Navigationskonzepte.
                         In: Seminar "Neue Entwicklungen bei fahrerlosen
                         Transportfahrzeugen", ISW, Oktober 1990.

/6/ Olomski, J.   Bahnplanung und Bahnführung von Industrie-
                         robotern.
                         Fortschritte der Robotik 4.
                         Friedr. Vieweg & Sohn, Braunschweig/Wiesbaden.

/7/ Bronstein, I.N;   Taschenbuch der Mathematik.
    Semendjajew, K.A.   20. Auflage. Teubner Verlag, Leipzig.

/8/ Makino, H.   Clothoidal interpolation - A new tool for
                         high-speed continuous path control.
                         Annals of the CIRP, Vol. 37/1, 1988.

Knowledge-Bases and Computer Architectures

Reactive Planning
- A Model of Knowledge-Based Real-Time Planning -

J. Dorn and G. Hommel

Abstract

Second generation expert systems are characterized by model based knowledge representation. In most cases these models are object-centered structures with constraints between them. In many technical applications, processes and events are more important than objects. The changes in physical or chemical processes have to be represented adequately.

This paper concentrates on planning systems for technical processes. Traditional knowledge-based planners do not support real-time planning. They are constructed to solve problems sequentially regardless of temporal constraints of the application. A plan can only be carried out after having been constructed. It is impossible to mix the process of planning with plan execution. These planners do not consider that the real world is continuously changing even during planning. We describe a model which considers all of these requirements, and we call this model of knowledge-based real-time planning reactive planning.

1 Motivation

Second generation expert systems are characterized by model based knowledge representation. In most cases these models are object structures with constraints between objects or between properties of the objects. In many technical applications, processes and events are more important than objects. The changes in physical or chemical processes have to be represented adequately. One important feature hereby is the representation of the quantitative and qualitative notion of time.

1.1 Knowledge-based Systems for Technical Applications

In order to control technical applications such as airplanes or power plants, thousands of signals have to be identified and combined so that consequences of a given situation can be derived. Redundant components for enhancement of safety and reliability result in additional complexity. For autonomous mobile robots, complex programs have to be developed in order to explain and to react to all occurrences in the robot's environment. This complexity cannot be handled sufficiently with traditional control methods. New methods are necessary for the growing complexity in process control applications.

Knowledge-based methods are investigated in order to find a solution to that problem. There are two different methods of incorporating knowledge-based methods into process control applications. A knowledge-based system can generate controlling programs written in a procedural language. This approach is taken e.g. in a programming system for a CIM environment [1] where Ada programs are generated. The second attempt proposed here is based on the idea that the controlling program itself uses knowledge-based methods. The advantage of the first attempt is that in most cases the program will be faster and more efficient. The disadvantage is that events in the application which the programmer had not considered can not be handled as easily as in the second approach.

The possibility of automatic knowledge acquisition and learning motivates the application of knowledge-based methods further. For an autonomous mobile robot working in different unknown environments, modification of available knowledge is absolutely necessary. In process control applications the system has to perform continuously. An interruption in order to install new software components is often undesirable. The possibility of changing the knowledge during run-time has to be considered in the specification of the control system's software. This is a natural process for knowledge-based systems. Kowalski writes:

> We argue that computer programs would be more often correct and more easily improved and modified if their logic and control aspects were identified and separated in the program text. [2]

1.2 Requirements of Real-Time Applications

First we will outline some of the important characteristics of real-time applications that distinguish them from other applications. These and further problems are discussed in more detail in [3].

- Sequential and concurrent activities

In real-time applications different activities occur concurrently (e.g. several autonomous robots are moving) and sometimes a strict sequence is necessary (e.g. in the process of assembling parts). In a knowledge-based system this has to be described explicitly.

- Timeliness

If an event occurs in an application, often a reaction has to be executed within a specific deadline. The controlling program has to guarantee this time limit. If this is not the case, the control over the technical process is lost and could result in great damage. To fulfil this requirement it is necessary that planning takes place as early as possible and some reasonable action can be taken before the specified time expires.

- Nonmonotonicity

Real technical processes are not static. We cannot suspend the real process until the planning process is finished (e.g. we cannot stop the flying aircraft as though it were a simulation model). During planning time the process will change its state. In a planning process we have to foresee and to consider this changing environment. The planner has to be "self-conscious".

- Asynchronous events

In technical applications a controlling component has to manage asynchronous events and has to worry about them. A typical asynchronous event might be a failure in the application. The controlling program has to be interruptible.

- Continuous execution

Technical applications are running for long time periods without intermission. Programs controlling such applications have to be designed so that they are not terminating.

2 Deficiencies in Traditional Planners

Looking into well-known knowledge-based planners we discover some deficiencies with regard to real-time applications. STRIPS [4], one of the early systems, can be used to control a real robot, a typical real-time application. But STRIPS does not handle real-time problems. The planning process and the execution of the robot task are performed sequentially. During the planning process it is assumed that the environment will not change.

The program controls a robot that is capable of solving the following tasks in an environment described in figure 1:

1. Turn on the light switch (a special version of the "Monkey and Bananas" problem).

2. Push three boxes together (a special version of the "Blocks World" problem).

3. Go to a location in another room

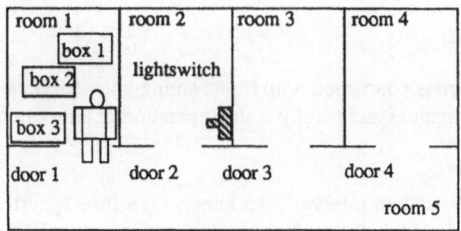

Figure 1: The STRIPS-World

The efforts of the following years are directed towards improving the planning process itself. So reduces nonlinear planning [5] the complexity from exponential to polynomial cost. Hierarchical planning [6] also results in a reduction of complexity. With constraint posting [7] there seems to be an elegant representation for further improvements. In [8] a planning model is described for temporally rich domains. This model comes closest to our requirements. But there are no possibilities for coping with hard real-time problems.

To illustrate this we will pose three tasks in the STRIPS-world that cannot be described and solved in the quoted systems. The only system we are familiar with that would also solve those tasks is described in [9].

2.1 A Goal State Described Using Intervals

It is 2 pm. The first task requires that the robot is in room 2 between 3 pm and 3.15 pm in order to perform some actions. Reasoning about time intervals is feasible using interval logic. With TIMELOGIC [10] a planner based on interval logic exists. This planner considers temporal relations but not temporal quantities. It would not take a great deal of effort to extend it with this feature.

If the robot needs between 4 and 6 minutes to carry out the plan, it could begin moving to room 2 at 2.54 pm.

2.2 Knowledge About Other Goals

Assuming that it is now 2.05 pm, a new task is added: The robot shall stay in room 4 from 3.40 pm to 3.55 pm. To plan this task, the planner has to know that the robot should be in room 2 before 3.15 pm. The second plan then will consist of the move from room 2 to room 4 at 3.33 pm. The planners mentioned above are not capable of handling this problem because the old goals are no longer part of the knowledge base. After a plan has been generated the goals are discarded.

2.3 Knowledge About Already Planned Actions

Assuming that it is now 2.10 pm, another task is added: A box has to be placed in room 3 at 3.30 pm. To find a good solution the planner must have access to the previously generated plans. The new plan results in a modification of both plans. In the first plan the robot only had to pass room 5. In the modified plan the robot carries the box to the front of door 2 before it completes the rest of the first task. Afterwards the robot leaves room 2, lifts the box, and carries it into room 3. From there the robot moves to room 4 to perform the second task. This task, as well, would not be solved by the quoted planners because they do not have access to the previously generated plans. Knowledge concerning already planned actions is not stored in the knowledge base.

These problems exist because knowledge-based planners are constructed like theorem provers. The planner gets a goal and proves it. As a side effect of the proof some actions are generated that have to be executed in the real world in order to reach the goal. The prover has access to a knowledge base describing the environment but not goals and plans of actors in the environment.

3 Basic Representation

The first aspect described here is concerned with representing knowledge to overcome the problems mentioned above. The most important aspect hereby is the separation of temporal and causal knowledge.

3.1 Intervals

The whole representation is based on intervals. An interval is a time specification consisting of four attributes: name, *start*, *end*, and *duration*. Start, end, and duration are granulation intervals.

A *granulation interval* is the smallest possible interval that can be measured in a system. It corresponds to the resolution of a clock and is a discrete number. After every increment of this clock the time of one granulation interval has elapsed. The real interval in the application differs in general from the interval in our representation (see figure 2).

The largest deviation of our interval from the real interval in the technical process can be one *granulation* at both sides of the interval. In technical applications the duration of an granulation interval is typically in the range of milliseconds.

Often exact limits of an interval cannot be specified . Therefore a window around the interval is defined. In this case an interval begins within a *beginning window* and ends in an *ending window*. The specification of an interval attribute may consist of a fixed value or a range of possible values.

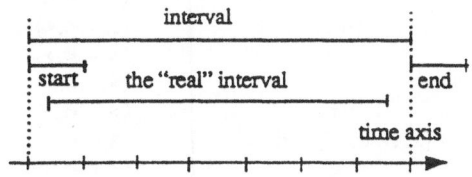

Figure 2: Deviation of the real interval

The following example states that the interval i_1 starts at the time of granulation interval 3 and ends between the granulation intervals 6 and 9. Therefore the duration is a time between 3 and 6 granulation intervals.

$\langle i_1, 3, 6..9, 3..6 \rangle$

The possible values of a granulation interval range from 0 to ∞. If no values are known we describe this with the window: $0..\infty$.

3.2 Constraints Between Intervals

On these intervals we define *interval constraints* comparable to Allen´s relations [11]. We assume that the reader is familiar with these temporal constraints. Constraints between two intervals are defined on the ending granulation intervals of both intervals. A constraint is described by a set of four constraints on granulation intervals.

$I_1 \langle B_1, B_2, B_3, B_4 \rangle I_2$

This is illustrated by figure 3.

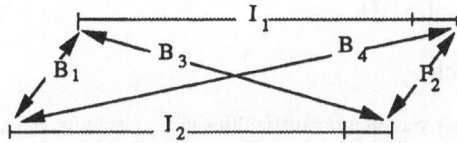

Figure 3: Internal Representation

The internal representation is given by the following axiom:

$[\ I_1 \langle B_1, B_2, B_3, B_4 \rangle I_2\ ] \leftrightarrow$
  $\text{constraint}(\text{start}(I_1), \text{start}(I_2)) = B_1 \wedge$
  $\text{constraint}(\text{end}(I_1), \text{end}(I_2)) = B_2 \wedge$
  $\text{constraint}(\text{start}(I_1), \text{end}(I_2)) = B_3 \wedge$
  $\text{constraint}(\text{end}(I_1), \text{start}(I_2)) = B_4$

We introduce some new basic relations that are convenient for simplicity in planning. These correspond with disjunctions of Allen´s relations. These and the original relations of Allen are defined as shown in figure 4.

Name	Allen	Representation
meets	m	$\langle <,<,<,= \rangle$
before	<	$\langle <,<,<,< \rangle$
sequence	{< m}	$\langle \leq,<,<,< \rangle$
equal	=	$\langle =,=,<,> \rangle$
starts	s	$\langle =,<,<,> \rangle$
finishes	f	$\langle >,=,<,> \rangle$
in	{s, si, d, di, f, fi, =}	$\langle \geq,\leq,<,> \rangle$
overlaps	o	$\langle <,<,<,> \rangle$
starts-with	{<, m, o, fi, di}	$\langle <,\geq,<,\geq \rangle$
follows	{m, o, s, d}	$\langle -,<,<,\geq \rangle$
intersects	{o, oi, d, di, s, si, f, fi, =}	$\langle -,-,<,\geq \rangle$

Figure 4: Definition of interval constraints

For the propagation of these constraints we have implemented a more efficient strategy based on sequence graphs [12].

3.3 Occurrences

One important aspect of the planning environment are *occurrences*. We distinguish between the prototype of an occurrence and the occurrence itself. An occurrence always takes place at a specified interval. This is specified:

Occurrence-Type @ Interval

This means that an occurrence of type *Occurrence-Type* will take place at the specified interval. Formally, an occurrence is a proposition that is constrained to be true during a specified interval.

Different occurrence types are distinguished: facts, objects, constraints, events, processes, and actions. This is described in more detail in [16].

4 The Modal Approach

To represent causality and to overcome the difficulties in representing and accessing goals and actions we define a modal interpretation based on the modal system S_4 [13].

4.1 Possible and Necessary Propositions

The basic operator of this logic is $\Diamond p$ which states that the proposition p is *possible*. The *necessity* of a proposition $\Box p$ is defined using the $\Diamond$ operator:

$$\Box p =_{def} \neg \Diamond \neg p$$

We use the following definition to define strict implication:

$$p \Rightarrow q =_{def} \Box (p \rightarrow q)$$

The necessity of a proposition means that this proposition not only is true but is enforced to be true and perhaps some actions have to be taken in order to make it come true (e.g. a goal in a planning process).

4.2 Always and Sometimes Valid Occurrences

Using interval logic and those modal definitions we can specify special temporal propositions. In general, goals are not true at the time they are specified. But we desire that there will be a time in the future when they will be true. We specify that there will be an interval I_2 in the future in which the proposition goal @ I_1 will be true:

$$\text{true(goal @ } I_1) @ I_2.$$

We distinguish between propositions that are always true and those that might be true:

$$\text{always}(X @ I_1) =_{def} \Box \text{true}(X @ I_1) @ I_2$$
$$\text{perhaps}(X @ I_1) =_{def} \Diamond \text{true}(X @ I_1) @ I_2$$

That an occurrence X is always valid means: It is always true that the occurrence X will hold at interval I_1.

4.3 Attainable and Demanded Goals

The specification of an additional goal during the planning process is also an occurrence. This occurrence is qualified with the attribute "demanded". For a demanded occurrence we have to prove that it is attainable.

$$\text{attainable}(X@I_1) =_{def} \Diamond \text{ true}(X @ I_1) @ I_2 \wedge now <= I_2$$

An occurrence is attainable if we find a plan so that it would be necessary that the goal occurs if the plan would be performed. The definition of attainability is comparable to the definition of necessity.

$$\text{demanded}(X @ I) =_{def} \neg \text{attainable}(\neg(X @ I))$$

4.4 Feasible and Scheduled Actions

Feasible actions are actions that are performable. This means that all preconditions of such an action can be matched with the state description so that the action can be performed.

$$\text{scheduled}(X @ I) =_{def} \neg \text{feasible}(\neg(X @ I))$$

A scheduled action has to be feasible.

5 Planning With Scripts

Scripts, a structured knowledge representation mechanism developed by Schank [14], are event-oriented. Opposed to the object-oriented view of most of the other well-known representations, the basic idea of this concept is that an environment can be described with prototypal stories. As in a frame-based system there are slots for describing the single events (in our terminology occurrences) in the story, one or several agents of a story, and some props (objects that are concerned with the story).

Important for the planning process are two further slots – the entry conditions and the results. In backward-chained planning the results of different scripts are matched with goals and subgoals in the planning process. Then the entry conditions of the matched script form new subgoals if not yet satisfied.

Scripts may also be used in a forward-chained planning process. Starting from a state description the entry condition determines which script is applicable.

Two additional slots are for interval and causal constraints. With the interval constraints we can constrain the temporal relations of the occurrences of one script. The causal constraints are for planning purposes to support the planner.

5.1 The Architecture of the Script System

A central part in a script based system is the script planner that receives a task specified by the user of the system. The planner tries to find a sequence of scripts, the plan. This plan would accomplish the given task provided that all actions could be executed without failure. The planner has access to the global object-oriented knowledge base. The planner selects a sequence of scripts from the knowledge base and passes it to the script interpreter which is responsible for their execution. The script supervisor compares information from sensors and from the interpreter and detects whether a given situation is erroneous. If a failure should occur, the supervisor stops the interpretation of one or more scripts. Another of the script supervisor's tasks is the verification of the object-oriented knowledge base.

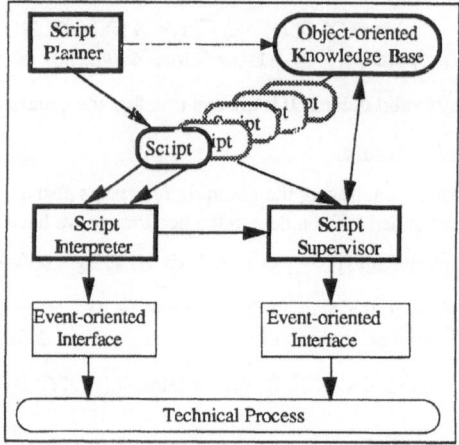

Figure 5: The Architecture of the Script System

5.2 Definition of Scripts

A script contains a prototype of a story. Various instances of one script can be generated by the script planner. These instances of a script are called *tracks*. Various tracks differ in the values of variables for objects and other properties which are concerned with the story. A specific track is the result of matching a script with objects from the knowledge base.

To illustrate a script we give a graphical description of a typical track in figure 6 and the same track in a PROLOG based list notation as it is used in the script based planner in figure 7.

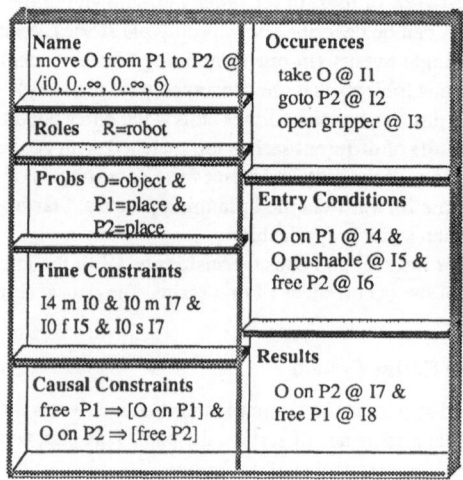

Figure 6: Graphical Description of a Track

Slots	Filler
Name	`[move O from P1 to P2 @` `(i0,0..∞,0..∞,6),`
Roles	`[role(R, robot)],`
Probs	`[prob(O, block),` `prob(P1, place),` `prob(P2, place)],`
Occurrences	`[take O @ I1,` `goto P2 @ I2,` `open_gripper @ I3],`
Entry Conditions	`[O on P1 @ I4,` `pushable(O) @ I5,` `free P2 @ I6],`
Results	`[O on P2 @ I8,` `free P1 @ I7],`
Temporal Constraints	`[I4 m I0, I0 m I7,` `I0 f I5,I0 s I7],`
Causal Constraints	`[free P1 ⇒ [O on P1],` `O on P2 ⇒ [free P2]]`

Figure 7: Description of a Track in list notation

5.3 Concurrency and Scripts

Events in a robotic environment can occur concurrently. To cope with this there are two possibilities: Concurrent events can be described inside a script or we can allow scripts to describe concurrent stories.

Allowing concurrent events and/or scripts we have to consider the classical problems of synchronization and communication. If the events described in different scripts are not completely independent, communication between scripts is essential. This can be done via the global knowledge base. In this case the access to the global database has to be synchronized in order to avoid the well-known read-write or write-write conflicts.

This can be specified in the script either using specific entry conditions or time intervals that describe when the actions are allowed to occur. In the latter case the script planner assumes the responsibility for correct synchronization.

5.4 Planning

This section describes how the planner works and how the knowledge base is modified during the planning process.

Knowledge-based planning is based on a search for relevant operators. Operators describe changes in the environment caused by actions. As mentioned above, our operators are scripts. The entry conditions correspond with the precondition formula and the results with the add formula and delete list in STRIPS.

Given a goal, all scripts that contain a result which can be matched with the goal form a conflict set. This conflict set has to be reduced by knowledge-based strategies until only one script remains. This script is scheduled in order to solve the goal. But to execute this script we have to assure that the script is attainable. To prove this we have to show that the entry conditions of the script are holding.

The planning process uses a least-commitment-strategy based on a constraint-oriented approach. This means that the temporal relation between intervals and the causal relation between occurrences is constrained.

If there are several conjunctive goals, we use a form of nonlinear search for relevant operators. Like Chapman [5] we use the concept of incomplete plans. If we plan an action we look for a place in this incomplete plan.

Suppose we have the planning environment stated in figure 1. The environment before planning is described by three occurrences. We see them in the knowledge base attributed by the intervals i_1 - i_3.

To solve the three tasks specified in chapter 2, four operators are needed to describe the possible actions: GOTO, GOTHRU, TAKE, and PUT_DOWN. The semantics of these operators should be obvious.

After having planned for the three tasks, the knowledge base is changed as shown in figure 8. The first granulation interval is set to 2 pm. The duration of one granulation interval is set to 1 minute.

```
IN(ROBOT, ROOM₁) @ ⟨i₁, 1, 52..56, 51..55⟩,
IN(BOX₁, ROOM₁) @ ⟨i₂, 1, 52..57, 52..57⟩,
OFF(LIGHTSWITCH) @ ⟨i₃, 1, 0..∞, 0..∞⟩,
scheduled(GOTO(BOX₁)) @ ⟨i₁₃, 45..53, 48..54, 1..3⟩,
scheduled(TAKE(BOX₁)) @ ⟨i₁₄, 48..54, 49..55, 1⟩,
scheduled(GOTO(DOOR₁)) @ ⟨i₄, 49..55, 52..56, 1..3⟩,
scheduled(GOTHRU(ROOM₁, ROOM₅, DOOR₁)) @ ⟨i₅, 52..56, 54..57, 1..2⟩,
scheduled(GOTO(DOOR₂)) @ ⟨i₆, 54..57, 57..58, 1..3⟩,
scheduled(PUT_DOWN(BOX₁)) @ ⟨i₁₅, 57..58, 58..59, 1⟩,
scheduled(GOTHRU(ROOM₅, ROOM₂, DOOR₂)) @ ⟨i₇, 58..59,60 ,1..2⟩,
scheduled(GOTHRU(ROOM₂, ROOM₅, DOOR₂)) @ ⟨i₉, 82..86,(84..87,1..2⟩,
scheduled(TAKE(BOX₁)) @ ⟨i₁₆, 84..87, 84..87, 1⟩,
scheduled(GOTO(DOOR₃)) @ ⟨i₁₇, 84..87, 87..88, 1..3⟩,
scheduled(GOTHRU(ROOM₅, ROOM₃, DOOR₃)) @ ⟨i₁₈, 87..88, 89,1..2⟩,
scheduled(PUT_DOWN(BOX₁)) @ ⟨i₁₉, 89, 90, 1⟩,
scheduled(GOTHRU(ROOM₃, ROOM₅, DOOR₃)) @ ⟨i₂₁, 93..97,95..98,1..2⟩,
scheduled(GOTO(DOOR₄)) @ ⟨i₁₀, 95..98, 98..99, 1..3⟩,
scheduled(GOTHRU(ROOM₅, ROOM₄, DOOR₄)) @ ⟨i₁₁, 98..99,100 ,1..2⟩,
demanded(IN(ROBOT, ROOM₂) @ ⟨i₈, 60, 75, 15⟩,
demanded(IN(ROBOT, ROOM₄) @ ⟨i₁₂, 100, 115, 15⟩,
demanded(IN(BOX₁, ROOM₃) @ ⟨i₂₀, 90, 0..∞, 0..∞⟩,
i₁₃ m i₁₄ m i₄ m i₅ m i₆ m i₁₅ m i₇ m i₈,
i₈ < i₉,
i₉ m i₁₆ m i₁₇ m i₁₈ m i₁₉ m i₂₀,
i₂₀ m i₂₁ m i₁₀ m i₁₁ m i₁₂
```

Figure 8: The Knowledge Base

In figure 8 the occurrences are classified and sorted by time; usually the planner would not sort them. The sequence of the occurrences is defined by the interval relations. The relation m in the example means that the two intervals meet and the relation < means that the first interval precedes the second.

6 Conclusion

We have presented a new mechanism for real-time planning which overcomes some of the deficiencies in knowledge-based planning. One big problem remaining is the speed of execution. For real-time applications with short deadlines this model cannot be used without specific hardware support because the overhead for knowledge-based techniques is still too large.

How this hardware support can be provided realizing an event-oriented interface is described in [15]. This interface maps all data from the technical process to occurrences. It also helps the knowledge-based system in guaranteeing time limits. The concept was tested in a hard real-time environment. Despite some deficiencies we think that this approach is very important for further developments in artificial intelligence applications for real-time systems. The full description of this system is published in [16].

Future work will concentrate on building tools to validate these ideas in a broader field of technical applications. One of those applications will probably be a scheduling problem in an Austrian steel-making plant.

References

[1] *Programme für kooperierende Maschinen in der Fertigung.*
H. Czech, R. G. Herrtwich, G. Hommel, R. Sasse, A.-P. Winkler.
Technischer Bericht 89-11, TU Berlin, Fachbereich 20, 1989.

[2] Algorithm = Logic + Control.
Robert Kowalski.
Communications of the ACM, Vol. 22, No. 7, 1979, pp. 424-436.

[3] Real-Time Knowledge-Based Systems.
Laffey, Cox, Schmidt, Kao, Read.
AI Magazine, Spring 1988, pp. 27-45.

[4] STRIPS: A New Approach to the Application of Theorem Proving to Problem Solving.
Richard E. Fikes, Nils J. Nilsson.
Artificial Intelligence 2, (1971),
pp. 189-208.

[5] Planning for Conjunctive Goals.
David Chapman.
Artificial Intelligence 32, (1987) pp. 333-377.

[6] *A Structure for Plans and Behavior.*
Earl D. Sacerdoti.
Elsevier Computer Science Library, NY 1977.

[7] Planning with Constraints.
Mark Stefik.
Artificial Intelligence 16, (1981)
pp. 111-140.

[8] A Formal Logic of Plans in Temporally Rich Domains.
Richard Pelavin, James F. Allen.
Proceedings of the IEEE, Vol. 74, No 10, Oct 1986, pp. 1364-1382.

[9] Reactive Reasoning.
Michael P. Georgeff, Amy L. Lansky.
Proceedings of the 11th IJCAI, 1989, pp. 677 – 682

[10] Planning Using a Temporal World Model.
J. F. Allen, J. A. Koomen.
Proceedings of the 8th IJCAI, 1983, pp. 741-747.

[11] Maintaining Knowledge About Temporal Intervals.
James F. Allen.
Communications of the ACM, Vol. 26, No 11, pp. 823-843.

[12] TimEx – A Tool for Interval-Based Representation in Technical Applications.
Jürgen Dorn.
Proceedings of the Second Conference on Tools for AI, Washington DC, 1990.

[13] Strict implication. An emendation.
C.I. Lewis.
Journal of Philosophy 17, (1920), pp. 300-302.

[14] *Scripts, Plans, Goals and Understanding*.
Roger C. Schank, Robert P. Abelson,
Erlbaum, Hillsdale

[15] Der Ereignis-Prozessor,
(The event-processor).
Jürgen Dorn.
Proceedings WIMPEL '88,
B.G.Teubner-Verlag, 1988, pp. 205-219.

[16] *Wissensbasierte Echtzeitplanung*, (Knowledge-Based Real-Time Planning).
Jürgen Dorn. Ph.D. TU Berlin,
published by Vieweg-Verlag 1989.

An Object-Oriented Knowledge Base for CIM Environments

J. Schweiger and H.-J. Siegert

Abstract

The multi-user knowledge base developed for CIM environments is based on an object-oriented, hierarchical approach which supports multiple inheritance. In addition, an active component is integrated which automatically distributes changing data to user tasks in a blackboard-like manner. Furthermore, the structure of the knowledge base can be modified dynamically during run-time. Program products of the knowledge base, uniquely interfacing RAM structures as well as permanent database layers, have been built for various hardware and software platforms. The knowledge base administrator is able to modify the structure and the data in the knowledge base with an interactive graphics interface providing monitoring and browsing. User processes access the data in the knowledge base with a program interface which supports several programming languages. The use of the knowledge base in applications like image interpretation, navigation and task planning showed that modelling the data in the knowledge base provides permanent, redundancy-free storage of complex structured data with comfortable access. This paper deals with the concepts of the knowledge base and the results obtained through the applications mentioned above.

1 Introduction

Computer-Integrated Manufacturing (CIM) designates the integration of all computer applications in the field of manufacturing. In it each computer application uses a lot of data and knowledge coming from other applications and produces a lot of data and knowledge which is relevant to other applications. Therefore, special facilities are necessary to store common data and to support the exchange of data between the single applications. The modelling concepts of classical databases do not enable us to do this [Dadam et al. 87, Meyfarth 90]. One central point of the Real-Time and Robotics Group of the Institut für Informatik at the Technische Universität München is the development of a distributed real-time knowledge base for all the information needed to run a factory autonomously. The knowledge base is intended to be the central medium for storing and distributing knowledge and data from different application processes in the field of manufacturing. The work is part of the joint research project Sonderforschungsbereich 331 'Information Processing Concepts in autonomous, mobile Robot Systems'.

This paper presents the results of our research and shows open problems which are to be solved in future work. In section 2 the basic concepts of the knowledge base developed so far are explained. Section 3 describes the architecture and the interfaces as well as access times of the implemented program products of the knowledge base. Section 4 deals with several applications, including the program products of the knowledge base, which have been done in cooperation with other projects of the SFB 331. In section 5 our goal of distributing data into several local knowledge bases with an automatic exchange of data between all of them is discussed. Finally, section 6 gives a description of our future work.

2 Basic Concepts

The modelling concepts of the knowledge base as shown in figure 1 are founded on an object-oriented approach. The basic structure in the knowledge base is therefore the object. Objects can be devided into prototype objects, also called *classes*, and individual objects, also named *instances* [Bocionek 87]. Each class defines a set of instances from which each describes properties of a real-world entity. In figure 1 the classes *autonomous-syst, locomotion-syst, manipulation-syst* and *mobile-robots* are shown. The class *mobile-robots* has one instance named *mb-1*.

The properties of a real-world entity are described as *attributes*. The attributes are classified into *class, instance* and *administration attributes*. Class attributes describe properties whose values are equal in all instances of the class. In figure 1 the class *locomotion-syst* has one class attribute named *service-range*. Instance attributes express properties which may have different values in every instance of the class. So in figure 1 the instance attribute *manip-loc* has the value *approach* in the instance *mb-1*. Administration attributes contain information about which instances are an element of the class, and about the inheritance of attributes (in figure 1 all classes have the administration attributes: *instances, superclasses, subclasses*). Administration attributes are hidden attributes which cannot be accessed directly by the user. A class inherits all the attributes of its superclasses, which means that the attributes defined in the superclasses are added to the ones defined by the class considered. In the same way the subclasses of a class inherit all attributes of that class.

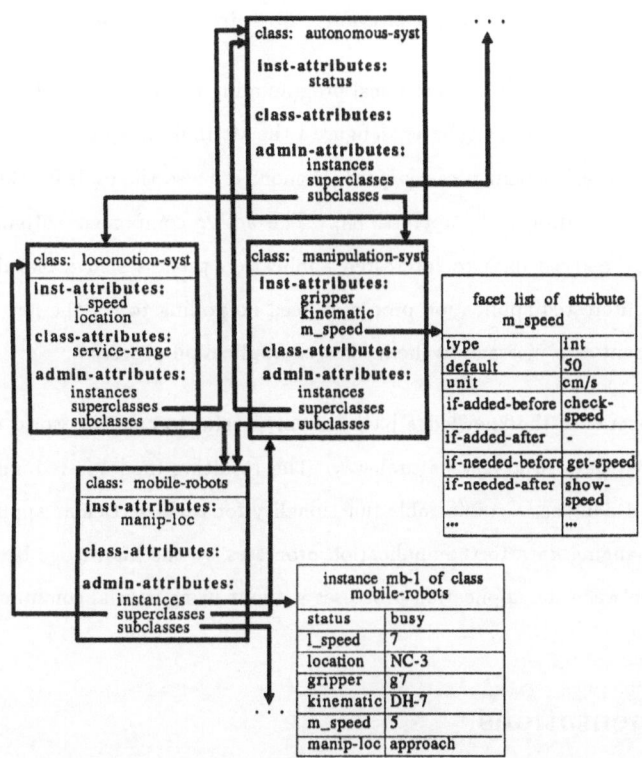

Figure 1: Basic modelling concepts of the knowledge base

The special case in which a class has at the most one superclass is called *single inheritance*. In contrast, if a class can have several superclasses, we speak of *multiple inheritance*. If single inheritance is given, superclasses are generalizations, whereas subclasses are specializations of a considered class. Figure 1 shows an example of multiple inheritance. The class *mobile-robots* inherits all attributes from the classes *autonomous-syst, locomotion-syst* and *manipulation-syst* (multiple inheritance). The class *locomotion-syst* is a specialization of the class *autonomous-syst* (single inheritance).

Each attribute of a class is described by a set of facets. Some are standard facets for all attributes and contain information like the type or the default value of the attribute. The user can define additional facets freely. In figure 1 the attribute *m_speed* of the class *manipulation-syst* has the facets *type, default, unit* and so on. Some special facets include the identifier of an attached procedure, called *demon*, which is executed if the attribute is used in a certain way. For instance, if the attribute is modified, the demon of the facet *if-added- before* is executed before the attribute value is changed. In the same way there are facets for *if-added-after, if-needed-before* and *if-needed-after* demons.

Demons are written by the user in classical programming languages and allow the storage of procedural knowledge in the knowledge base. In figure 1 the attribute *m_speed* of the class *manipulation-syst* has the if-added-before demon *check-speed.* Demons are also the basis for the active component of the knowledge base [Bocionek/Meyfarth 88]. The active component automatically distributes changing, specially marked data to interested application processes in a blackboard-like manner. This means that interested application processes need no polling to get the new data at once. The changing data is sent as a message to the interested application processes.

The basic concepts of the knowledge base are very useful to create a world model dealing with complex structured data in a very natural way. The resulting model is redundancy-free and provides, by means of demons, a comfortable functionality for each particular application. The active distribution of changing data to the application processes by the knowledge base allows a uniform communication between the application processes without using special communication facilities.

3   Implementations

The knowledge base with all the concepts mentioned above is implemented according to the architecture introduced in figure 2. In this implementation the objects of the knowledge base can be handled only by means of a set of access services. In order to offer multi-user access, the knowledge base consists of a knowledge base kernel process and a remote service interface linked to every application process (in figure 2 technical process 1, ..., technical process n, graphics shell, monitor and browser). The remote service interface currently supports the programming languages C, Pascal and Modula2. If a service of the remote service interface is called from an application procedure, the request is sent as a message via a communication interface to the knowledge base kernel process where it is executed. The communication interface operates on a ports interface to perform interprocess communication.

The result of a request is again sent back as a message to the corresponding application process in the same way.

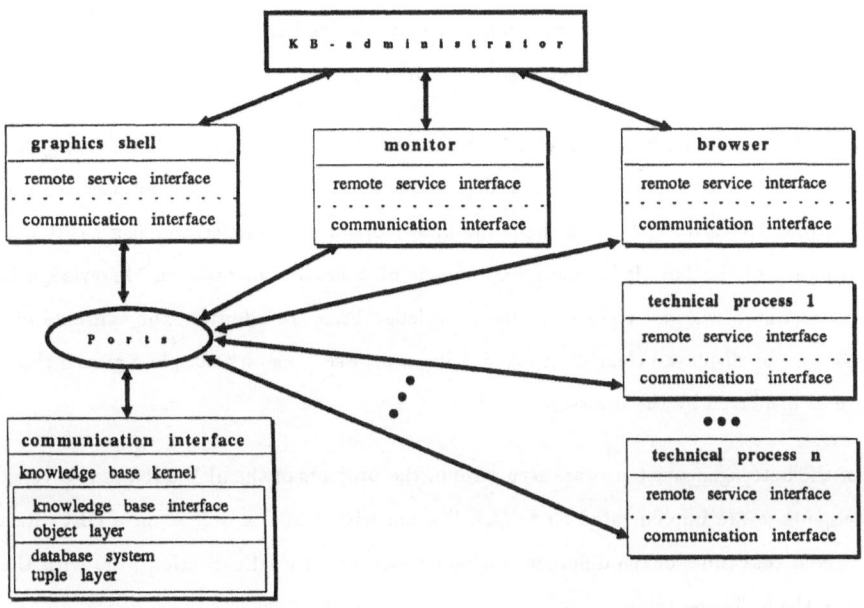

Figure 2: Architecture of the program product

The knowledge base kernel keeps the objects of the knowledge base in a relational database system. As the mapping between the objects and the database tuple is fixed and therefore the powerful functionality of a query language is not needed, the knowledge base kernel uses the tuple layer of the database system to handle the data in the database. The object layer of the knowledge base kernel thus maps every knowledge base request into a sequence of tuple layer operations of the underlying relational database system. The mapping is explained in detail in [Meyfarth 88].

The knowledge base kernel uses a well-defined tuple layer interface, so that it is possible to use the same kernel with different database systems. The current implementation of the knowledge base kernel can be configured both with the commercial database system TRANSBASE [TransAction 89] and with a RAM-resident database system developed by us [Schuster 90]. In contrast to the RAM-resident database system, which keeps the tuple as an AVL-tree resident in the main memory, the TRANSBASE database system stores every tuple permanently in a B^*-tree on the disk.

The TRANSBASE version is able to store huge amounts of data securely over a long time. One problem here is the large access time required by data on the disk. The second version of the knowledge base kernel stores the objects and structures in a RAM-resident database system, which in turn keeps the data in an AVL-tree in the main memory. This version of the knowledge base kernel thus provides short access times, but it is not suitable for handling huge amounts of data. Furthermore, data can be lost in a possible system crash.

Several tools have been built on top of the remote service interface, in order to support the knowledge base administrator . A graphics and a vt100 shell allow us to call all services of the knowledge base interactively. With these shells the knowledge base administrator can easily design the object structure of the knowledge base. By means of a flexible monitor the knowledge base administrator can watch relevant changes in the knowledge base. In addition, the contents of the knowledge base can be displayed clearly arranged using a browser. As an example, figure 5 shows a class structure as displayed by the browser.

Because of the heterogeneous hardware structure in the projects of the SFB 331 the two versions of the knowledge base are implemented on a VAX-Station with VMS as well as on a DEC-Station with Ultrix. The access times of the different versions measured on a DEC-Station 3100 with Ultrix are presented in the following table.

service	TRANSBASE KB	RAM-resident KB
read	450 ms	15 ms
write	1500 ms	15 ms

The different implementations are supplemented with a detailed manual [Bocionek et al. 90] to complete packages, the so-called program products of the knowledge base. The program products are used by other projects of the SFB 331 [Lainer 90, Wurstbauer 91, Kogler 90] and in some other applications, for instance [Fischer 89, Forster 89, Kern 90]. The final goal is to integrate all projects by connecting them to an active, distributed, real-time knowledge base system providing disk-based as well as RAM-based storage of data.

4   Applications

In this section some applications using the program products of the knowledge base are presented. The applications were developed in cooperation with other projects of the SFB 331.

4.1 Image Interpretation

In the first application the knowledge base is incorporated in the image interpretation system of SFB 331 project B5 [SFB-Bericht 88], which interprets given images by comparing them with given models of the real world. The image interpretation system is based on a blackboard architecture. In general, a blackboard architecture consists of a global database called the blackboard and of logically independent procedures named knowledge sources which act on the data in the blackboard. The knowledge sources relate only to the blackboard, but there is no straight communication among themselves. A blackboard architecture is designed to bring many different sorts of knowledge to bear on a single problem.

The special blackboard architecture of the image interpretation system as shown in figure 3 contains a blackboard and four knowledge sources. The first knowledge source creates the models out of CAD descriptions of the real world and stores them in the blackboard. The second knowledge source is an image processing system which prepares the given images to be compared. The third 'grouper' knowledge source combines partial structures of a given image (as prepared by the second knowledge source) into higher structures and stores them in the blackboard. Finally, the fourth knowledge source controls the actions of the other knowledge sources.

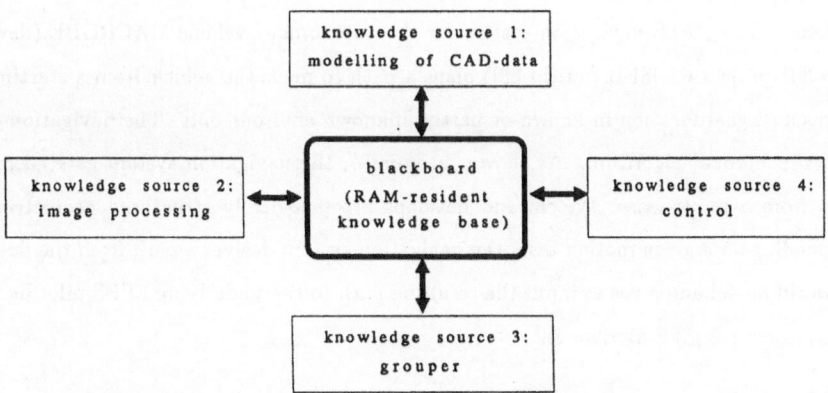

Figure 3: Knowledge base as blackbord in an image interpretation system

The blackboard is built as a single knowledge base kernel process. Every knowledge source corresponds to one knowledge base application process. The knowledge base does not hold whole images but only pointers to the images used. Beside the CAD models and the different partial structures, the agenda including the possible actions of the knowledge sources is kept in the knowledge base. The actions currently possible are automatically calculated by demons while writing the results of the actions of a knowledge source into the knowledge base. Moreover, the knowledge base manages all the communication between the knowledge sources.

In this application the knowledge base has proved to be very suitable for the implementation of a blackboard. Demons are predestined for an automatic calculation of the agenda. Communication between the knowledge sources can easily be implemented by means of the active component of the knowledge base. But, because demons need a lot of additional data for their calculations, the knowledge base kernel has to execute many database access calls. It was therefore necessary to use the RAM-resident knowledge base program products for that application, although there is no permanent storage of the dynamic data. The blackboard was implemented by [Wurstbauer 91] on a DEC-Station 2100 with Ultrix.

4.2 Navigation

The second application using the knowledge base was done in the field of autonomous, mobile loco-motion systems. The GPES navigation system for the autonomous vehicle MACROBE (developed by the SFB 331 project C1 [SFB-Bericht 88]) plans a path to move the vehicle from a starting loca-tion to a specified goal location in known or partly unknown environments. The navigation system is based on the Voronoi algorithm. As shown in figure 4, the navigation system gets geometrical information from a multi-sensor system and develops a topologically structured geometric world model. Depending on a given motion task, the navigation system derives a path from the developed geometric world model and gives as input the resulting path to the underlying LPES piloting system of SFB 331 project C2 [SFB-Bericht 88].

In former applications of the navigation system, the topologically structured geometric world model was stored as "pointered" MODULA-2 records. [Lainer 90] analysed these data according to their content and their internal relations, and modelled them as knowledge base objects. [Lainer 90] implemented the resulting objects with the VMS program product of the knowledge base and adapted the navigation system to the resulting comfortable knowledge base interface (see figure 4). To

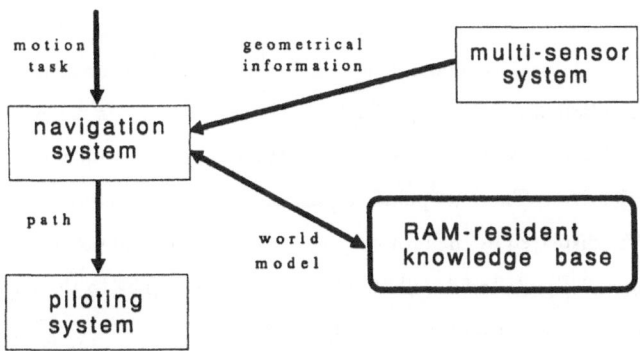

Figure 4: Navigation system connected to the knowledge base

reduce the time-consuming remote knowledge base service calls, [Lainer 90] embedded in a second inplementation the whole navigation algorithm in the form of demons into the knowledge base.

The implementation showed that the concepts of the knowledge base are well suited to model these topologically structured geometric data as knowledge base objects. Additionally, the complex algorithm of the navigation system can be implemented completely in the form of demons. As the record structure is implemented without any optimization, the run-time of the knowledge base version of the navigation system is longer than the former implementation. But the main advantage of the knowledge base version is the structured storage of the geometric data with comfortable access also for other applications.

4.3   Shop Floor Control System

As the third application, the knowledge base is integrated into the shop floor control system of SFB 331 project A1 [SFB-Bericht 88]. The shop floor control system consists of a scheduling component and an execution control component. The inputs of the scheduling component are jobs coming from a production planning system. First, the scheduling component expands the jobs with tasks for the flow of material as well as tasks for the manipulators. Second, the scheduling component gives the tasks to the execution control component. The execution control component then distributes the tasks to the autonomous systems, where they are executed.

The shop floor control system uses a large amount of data about the given manufacturing environment. For instance, it needs information about available robot systems and existing depots together with the status of tasks currently being executed. In former implementations of the shopfloor control system this information was stored as PROLOG facts in a file. As a new approach, [Kogler 90] designed a knowledge base which is able to keep the facts for the manufacturing environment of the SFB 331 project A1. The manipulation part of the class structure of this knowledge base is shown in figure 5 as displayed by the browsing facility of the program product of the knowledge base. In addition, [Kogler 90] built the distribution of the scheduled jobs to the autonomous systems by means of the active component of the knowledge base.

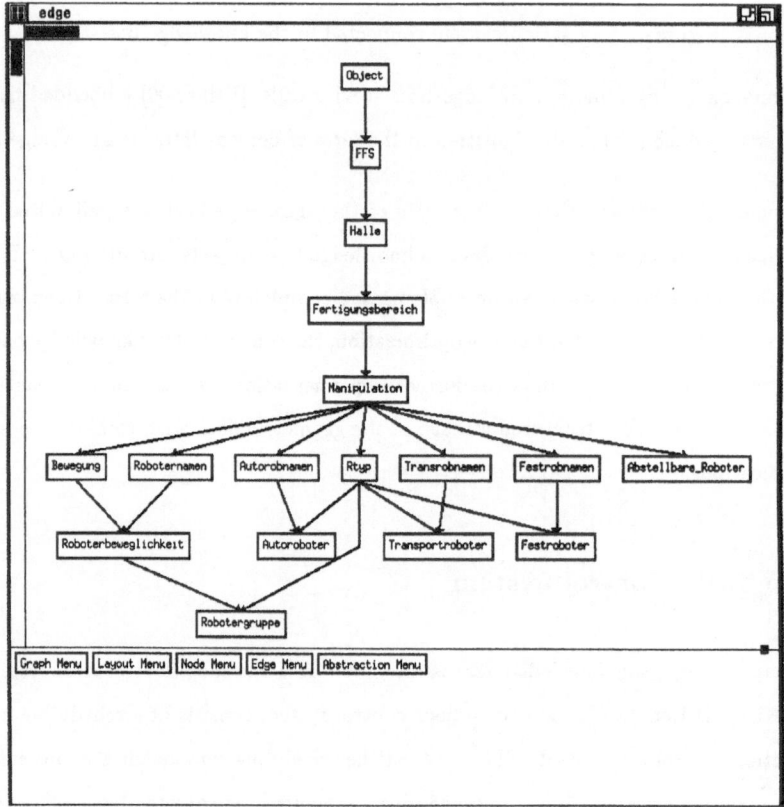

Figure 5: Manipulation part of the A1 knowledge base as displayed by the browser

The modelling of a manufacturing environment in a knowledge base provides a structured, redundancy-free description with comfortable access for the shop floor control system and also many other application processes. The knowledge base design is not fixed and can easily be modified according to other manufacturing environments. Finally, the whole communication between the execution control component and the autonomous systems is managed by the knowledge base, so that special communication facilities are not needed.

4.4 Task-Oriented Programming

The last application with a knowledge base is done in cooperation with SFB 331 project A2 [SFB-Bericht 88]. This project investigates the decomposition of high-level tasks of production control systems down to the level of code generation for manipulators, vehicles and machines. Furthermore, the execution of the decomposed task, the synchronization of all necessary units, and the flexible reaction on errors have to be organized independently from the production control layers. The concept developed so far is based on a system of distributed agents using a knowledge base. The agents can be created dynamically and may be connected by communication links to exchange messages. By this means the agents perform services and make use of services of other agents. The agents deal with so-called behaviour patterns, i. e. sets of rules and facts describing a possible solution for a given problem. To provide comfortable access for all agents, all environmental data and the behaviour patterns are kept in the knowledge base.

To evaluate the concept, a prototype application is currently being implemented for a model of a manufacturing plant built by Fischer-Technik. In this model two mobile robots, each with four axes, serve machine tools when performing pick-and-place tasks. The working space of the two robots overlap and the movements of the robots have to be coordinated with each other and with the activities of the machines. The machines of the model are a heating cell, a milling machine and a revolver lathe. The model also includes two conveyor belts, two depots and a measuring device. The presence of a workpiece can be checked through inductive sensors. The controllers of the machines and the robots are connected to the knowledge base in order to get their tasks and return their status. The application software for this model is currently being implemented.

5 Distributed Knowlege Base

The applications showed that one central demand on a knowledge base for CIM environments is that it must be able to store huge amounts of data with short access times to them. Until now the program products of the knowledge base have fulfilled this demand only parcially. The TRANSBASE version can store huge amounts of data but needs large access times. The RAM-resident knowledge base provides short access times for a limited amount of data. To fulfil this demand as a whole, it is useful to combine both knowledge base versions to one distributed knowledge base. The distributed knowledge base then consists of several local knowledge bases, which are TRANSBASE or RAM-resident program products of the knowledge base, extended by a distribution mechanism for the data.

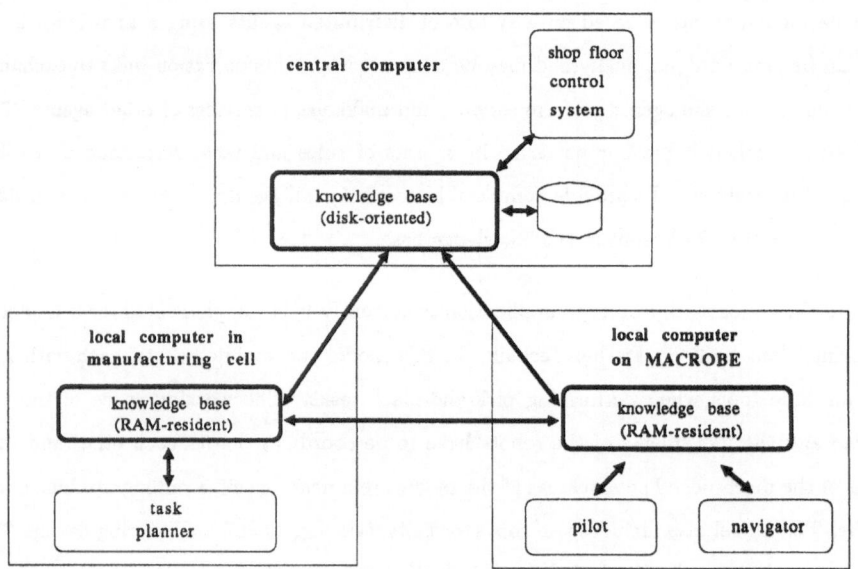

Figure 6: Principle of the distributed knowledge base

The principle of the distributed knowledge base is illustrated by the example in figure 6. The distributed knowledge base consists of three connected local knowledge bases. One local RAM-resident knowledge base is situated directly on board of the autonomous, mobile system MACROBE and provides short access times for the time-critical application processes *navigator* and *pilot*. The distribution mechanism automatically initializes the local knowledge bases and updates changing data of one local knowledge base in the other local knowledge bases if necessary. The development and integration of such a distribution mechanism is a central point in our present and future research.

6   Future Work

1. We are planning a closer integration of the knowledge base into other projects of the SFB 331 in order to validate the knowledge base and integrate the different projects into one unit. This integration will also lead to the refinement of the program products of the knowledge base. To support the knowledge engineer in designing the knowledge base for a special application, it is necessary to develop knowledge engineering tools. Additionally, the results of future research have to be included into the program product.

2. We are also progressing in improving the real-time features of the knowledge base. We are thus currently working on an exception handling concept for the program products of the knowledge base to eliminate needless error checking in correct service calls. Additionally, we want to improve the mapping between knowledge base objects and database tuples to speed up the access to the objects.

3. We will develop concepts for a distributed knowledge base as mentioned above and implement them. For this pupose it is necessary to find concepts for initializing the RAM-resident local knowledge bases with data of TRANSBASE local knowledge bases. Furthermore, we need mechanisms for the cooperation of the local knowledge bases in exchanging data between them. We also have to develop distribution strategies which provide local transparency as much as possible and which use replication of data depending on the actual applications.

4. As the whole operation in CIM environments takes place under time constraints, the integration of time-constrained relations into our knowledge base is necessary. For instance, the data of sensor systems refer to a certain time which corresponds with a certain status of the real world. Applications using the data of several sensor systems in general cannot mix data of different status. They have to take into account about the moment the data occurred.

   A first concept for versions of instances has been developed by [Wirth 90]. The structure of the knowledge base has not yet been versioned. This concept for versioning of data in the knowledge base is a useful facility for the user to take into account time-constrainted relations. Developing conceps for the representation of time-constrained relation in a distributed knowledge base and implementing them is one of our goals in the coming years.

References

[Bocionek 87]   Bocionek S.: Dynamic Flavors. Technischer Bericht. Technische Universität München, TUM I8708, June 1987

[Bocionek/Meyfarth 88] Bocionek S., Meyfarth R.: Aktive Wissensbasen und Dämonen-konzepte. Technischer Bericht. Technische Universität München, TUM I8811, September 1988

[Bocionek et al. 90]   Bocionek S., Meyfarth R., Schweiger J.: Handbuch zur A4-Wissensbasis-Shell. Benutzeranleitung Version 2.1. Technische Universität München, February 1990

[Dadam et al. 87]   P. Dadam, R. Dillmann, A. Kemper, P. C. Lockemann: Objekt-orientierte Datenhaltung für die Roboterprogrammierung. Informatik Forschung und Entwicklung, 2/87, 151-170, 1987

[Fischer 89]   Fischer M.: Untersuchung des Einsatzes von Regelprogrammen zur sensorgesteuerten Umweltmodellierung. TU München, Institut für Informatik 6, Diplomarbeit, November 1989

[Forster 89]   Forster J.: Entwurf und Implementie rung eines echtzeitfähigen (Roboter-)Umwelt modells unter Verwendung vorgegebener Wissensbasis-Tools. TU München, Institut für Informatik 6, Diplomarbeit, February 1989

[Kern 90]   Kern W.: Visuelle Unterstützung eines Roboters beim Anfahren einer geeigneten Greifstellung zum Greifen eines Quaders. TU München, Institut für Informatik 6, Diplomarbeit, November 1990

[Kogler 90]   Kogler M.: Modellierung der IWB- Fabrikumgebung in einer objekt-orientierten Wissensbasis. TU München, Institut für Informatik 6, Diplomarbeit, October 1990

[Lainer 90]   Lainer W.: Einsatz einer aktiven, objektorientierten Wissensbasis zum Speichern von topologisch strukturierten Geometriedaten. TU München, Institut für Informatik 6, Diplomarbeit, January 1990

[Meyfarth 88]   Meyfarth R.: Objektorientierte Datenhaltung für Wissensbasen unter Verwendung von B- Bäumen. Technischer Bericht. Technische Universität München, TUM I8815, November 1988

[Meyfarth 90]   R. Meyfarth: ACTROB: An Active Robotic Knowledge Base. Proceedings of the 2nd Int. Symposium on Database Systems for Advanced Applications (DASFAA 91). Tokyo, to be published in April 1991

[SFB-Bericht 88]   Zwischenbericht für den SFB 331 'Informationsverarbeitung in autonomen, mobilen Handhabungssystemen'. TU München, Juni 1988

[Schuster 90]   Schuster H.-D.: Entwurf und Realisierung einer Hauptspeicher-Version des MERKUR Tuple- Layer. TU München, Lehrstuhl für Prozeßrechentechnik, Diplomarbeit, June 1989

[TransAction 89]   TransAction Sorftware GmbH : Transbase Relational Database System - Manuals. Munich, 1989

[Wirth 90]   Wirth Michaela: Analytischer Vergleich wesentlicher Aspekte von Versionierungsverfahren und Systementwurf zur Integration von Versionen in eine objektorientierte Wissensbasis. Institut für Informatik 6, Diplomarbeit, November 1990

[Wurstbauer 91]   Wurstbauer T.: Realisierung einer Blackboard-Architektur mit einer objektorientierten Wissensbasis. TU München, Institut für Informatik 9, Diplomarbeit, Februar 1991

Architectural Features of Computer Systems for Autonomous Mobile Robot Applications

F. Färber, S. Helling, and A. Ruß

ABSTRACT

The design of a computer system operating in an autonomous mobile robot is dominated by the broad spectrum of requirements resulting from its applications such as planning, sensor data processing, motion control, etc. Increasable performance, high bandwidth of sensor data input and short response times to real world events are obtained best by a multi–microcomputer system, which allows to integrate different kinds of computing nodes and runs a distributed real-time operating system kernel.

In this paper we show appropriate methods of interfacing peripheral systems, integration of specialized processor nodes and focus on special architectural support for internode communication, which is crucial for system performance. An interface to a transparent interprocess communication is introduced, which allows various forms of communication semantics to be expressed and a high degree of parallelism to be utilized.

Main parts of the communication related software in each node are delegated to a dedicated communication processor, thus offloading the main processor and providing concurrent synchronization and communication with its counterparts in the communication subsystem.

1   REQUIREMENTS

Information processing in autonomous mobile systems has many different aspects and requirements. The table (figure 1) shows the very heterogeneous tasks to be performed as well as some quantitative data.

TASK	Period of Service	I / O [kB/s]	Memory [kB]	Processing Power
Motion Control				
Sensor Input	5 .. 20 msec	0.5 – 2.0	< 2	low
Control of Speed / Heading	5 .. 20 msec	–	< 10	low
Processing of Motion Commands	20 .. 100 msec	–	< 10	low
Modification of Motion Commands	> 100 msec	–	< 10	low
Actuator Output	5 .. 20 msec	0.2 – 1.0	< 1	low
Motion Planning				
Path Planning	1 .. 10 min	–	< 100	moderate
Trajectory Generation	1 .. 10 min	–	< 100	moderate
Local Planning				
Optimizing	ca. 5 min	–	< 100	moderate
Error Handling	ca. 5 min	–	< 100	moderate
Knowledge Base				
Rule Based Systems	n.a.	–	> 500	high
Global Knowledge	10 .. 100 msec	>10	>10.000	high
Perception System				
Preprocessing, Correlation, Interpretation, Environment Model				
in case of range sensors	100 msec	ca. 100	ca. 300	moderate
in case of visual sensors	20 .. 500 msec	> 5.000	> 5.000	high/very high
Communication				
to Factory Host	5 .. 10 min	< 10	< 200	moderate

Figure 1 : Quantitative Requirements for Information Processing in Autonomous Mobile Systems

Time constants vary from milliseconds to minutes, the required I/O bandwidth is between a few bytes to megabytes per second and also the processing power as well as the memory size depends strongly on the individual tasks.

Therefore a computer architecture must provide the following features:
- Performance scalable to the requirement,
- Enough memory also for complex sensor data evaluation and knowledge representation,
- Modularity and expansibility, especially for a research project, where the requirements are not yet well known,
- Ease of use, especially in the software development phase,
- Ease of interfacing to all sensors and actuators as well as to an external control- and development system.

2   SYSTEM ARCHITECTURE

The requirements of the applications for the computer system in an autonomous mobile system led to a distributed multicomputer architecture, which is composed of conceptionally heterogeneous nodes.

High processing power and short reaction latency to external events is provided by universal node processors or specialized processing modules dedicated to specific tasks by their attached peripherals. A special communication subsystem provides high bandwidth for internode communication, while the Ethernet link of the nodes is used for downloading and debugging purposes only.

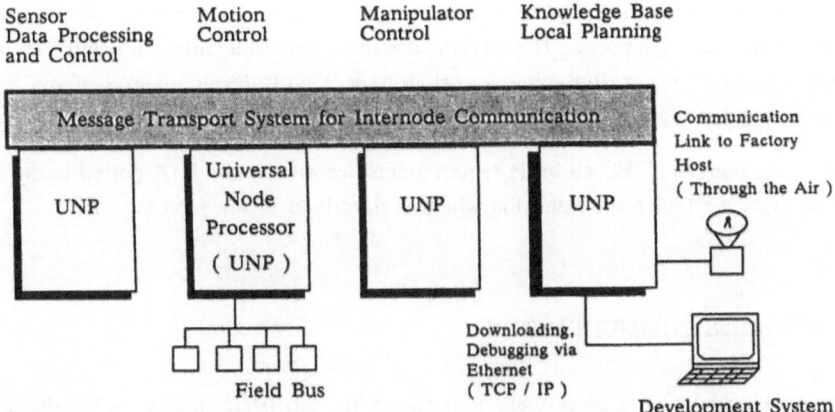

Figure 2 : Architecture of the Multicomputer System

2.1   REAL TIME KERNEL

Application programs for the multimicrocomputer system are decomposed into concurrently executing, communicating tasks. External events are intregated into this abstraction by appropriate drivers producing software events (e.g. sending a message, releasing a semaphore).

In order to combine the single nodes into an intregated system with a single system image a distributed operating system (kernel) is required [1]. We focused on the kernel functions, because in the real–time control area strict distinction between system and application code is not required.

The main issues for a distributed kernel are task– and address–space–management, memory– and device–management and interprocess communication [2]. We didn't intend to address all those topics, our primary interest was to provide good performance for non–local interprocess communication (IPC) to minimize the time penalty compared to the local case. One of the basic ideas behind our system is to locate sensor– or actuator–bound tasks on the node where the I/O is done and to locate non-bound tasks arbitrarily. A bound task for instance is one that does sensor-data preprocessing.

For the reasons mentioned above a small, fast real-time kernel (RTK [3]) was chosen as software basis on each node of the system. Among its services are routines for task-, interrupt- and memory-management and task-synchronization and -communication (local IPC). Very attractive and important are its task switching times, which are below 30 μs on a M68030 with 20MHz and short interrupt latency – the longest interrupt disabling interval in the kernel is 12 instructions long. There is no provision for virtual address space and thus all tasks on the same node share a common address space, so there are "light-weight" tasks only.

An additional feature that primarily addresses comfort of use, is a series of UNIX-style interfaces. Most important for porting C programs are the compatible stdio- ("printf" etc.) and heap-routines ("malloc" etc.). Run-time access to our UNIX-development-systems – e.g. remote file access – is supported through Ethernet and a low-level protocol.

Beside facilities for non-local IPC all basic requirements are met by the RTK ported to our node processor. For these we added an extension which is described in the next section.

2.2 INTERPROCESS COMMUNICATION

Facilities for non-local IPC are of primary importance for the performance of the distributed systems, as they glue together the single components [4]. We have looked at some design considerations for the user interface and the semantics. In systems without common memory the use of *explicit* message passing routines is the easiest way to implement IPC [5]. They represent a wide-spread and well-understood way to express communication in real-time control, so we have chosen them for our system.

For the design mainly four criteria are to be taken into account:

1) Type of the message:
   We implemented untyped messages of variable length (byte stream with preservation of message boundaries) to allow for arbitrary records or arrays to be transported. Variable message size was an application requirement, an upper limit of 16KByte has been settled for our applications, although fixed message sizes would have implementation and performance benefits.

2) Type of the connection between senders and receivers:
   There are no types of connections between a sender and a receiver, but we provided for an open-call in order to initiate communication between two tasks and to specify its parameters. This can be compared to a connect-statement in Internet's UDP datagram service. We will come back to this point later.

3) Addressing:

Addressing is done indirect by means of *ports*, which are named communication objects of the kernel and can be accessed system-wide. The idea behind ports is that they represent a way to a certain service independent of the implementor of the service. Thus they support location- and network- transparency. Furthermore indirect addressing permits the local kernel instances to assign task identifications autonomously.

4) Operations for the message exchange:

We decided to implement an asynchronous send-routine to give the possibility to exploit optimal parallelism to the user. In the real-time area synchronous kernel interfaces to IPC like rendezvous oriented approaches are not often used [6]. It requires further subsplitting of the task and installing of shared memory regions between the resulting (sub-)tasks altogether with the necessity for their synchronization. Nevertheless it must be mentioned that there is a respectable implementation complexity in non-synchronous communication resulting from the need for message buffering and flow control (which can be omitted in synchronous one). Receiving messages is blocking in our approach, but there is a routine which allows to specify ports from which messages may be awaited and accepted.

In our opinion this is a sufficiently expressive power in an environment where tasks and their communication relationships are rather static and tasks perform computing patterns like data-processing, sending information as long as available or awaiting information from other tasks. Therefore a task should not – within certain limitations – be prevented from spreading its information and thus setting others into the ready state.

Additionally we allow the calling task to specify the role it wants to take on referring to its communication with a specific port by supplying the previously mentioned open-call with appropriate parameters. The caller may be *producer* or *consumer* of information, or it may be *client* or *server* at a port which means a request/response style of information exchange. Choosing the latter variant a RPC (Remote Procedure Call)-style facility may be constructed, which varies from the classic one in nowait-send making it possible to have more than one request "on the run". For request and reply only one port is necessary independent of the count of clients attached. Thus *different semantics* can be expressed using the same routines for message exchange.

Additional setup-parameters are estimates for buffer requirements (of a sender). As a result of this call a logical connection between a task and a port is established and maintained (this may be compared to a process opening a file e.g. in an UNIX system).

To give an insight into some of the implementation details together with an idea of the purpose of the message transport system the essentials of a remote message exchange will be explained (see figure 3). During an open-call the kernel will allocate a control structure ("c_node") to manage the task/port- relationsships further. If herein a specific port is found to be remote, the remote kernel copy is ordered to allocate an 'alien' control structure also. Additionally buffer credit is negotiated.

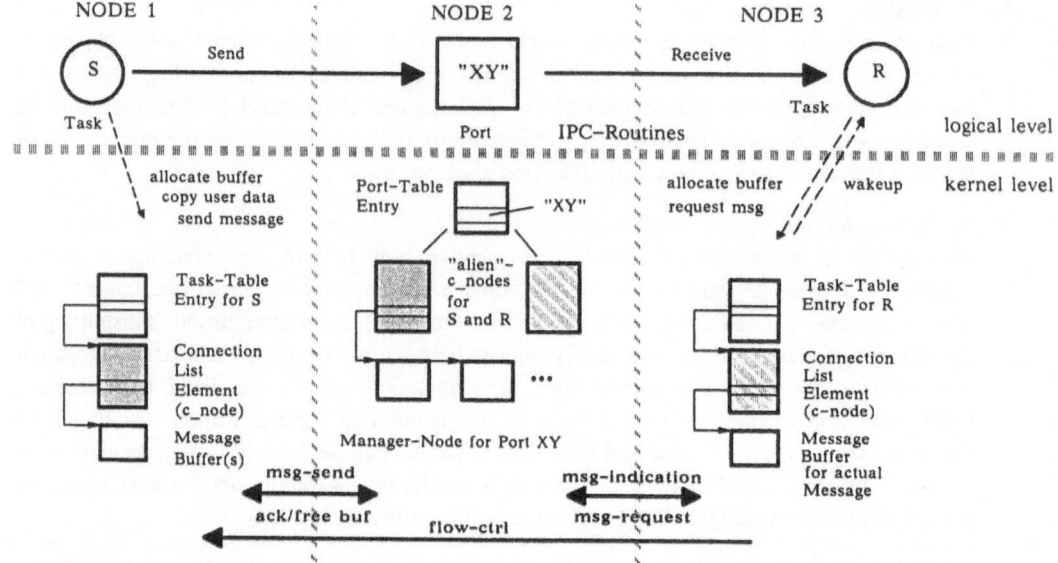

Figure 3 : Non-local Communication using Ports

Attached to each port structure managed on one node exclusively is a queue which may contain either messages or message requests. From a message request a c_node-structure can be addressed which leads to the destination task. Analogously from a c_node structure a destination port can be addressed. Each send or receive operation of an user task results in the kernel performing the corresponding operation on behalf of a c_node control structure and possibly modifying the state of a queue. The IPC management routines cope with performing the same operation upon matching control structures and resolving messages resp. message requests referring to the same queue. The kernel copy of the final destination of a message performs flow control operations relative to the original source of the message. This proceedings are illustrated in figure 3.

So far for the principal control flow involved in message exchange in the distributed system. We have done several studies to design the IPC management in a way that minimizes communication and synchronization demands of the multiple kernel instances possibly involved in one operation [7]. To dedicate the UNP to application code processing to a high degree, we decided to provide a message transport controller for each node as part of the message transport system, which releases the main processor from the burden of processing kernel code for handling communication requests and handles low level events autonomously [8], [9]. Thus the message transport controller is a key component of the distributed system and will be discussed in section 3.

2.3   DEVELOPMENT ENVIRONMENT

Initially we were building the UNP that it could play the role of an application processing element for multiple purposes to be covered inside an autonomous mobile robot. Thus we designed the processing node as a dedicated unit, which is the run–time environment for the applications only. Application development requirements are kept out of the design considerations. Instead we have chosen a UNIX host system as development environment [10]. First UNIX systems are a wide-spread and accepted programming environment, that will access even more application areas in future, and second UNIX systems offer a broadly accepted set of tools, which may be used in a more generally fashion, especially for constructing and maintenance of source code.

We have compilers (for Modula–2 and C) and linkers running on our UNIX system that produce executable code for the target in COF–Format. The operating system kernel for the target node is made available by means of libraries in UNIX archive format. In order to achieve independence of the UNIX host for users which have none available, we implemented a run–time system on our UNP that supports - within restrictions - the execution of a single UNIX-process. Thus our compilers and linkers can be run on the UNP. The main components of the run–time system are a small file system based on RAM, Internet's FTP (to load/store the RAM–disk from/to an arbitrary hosts secondary storage), UNIX system call–handling and a small Shell–oriented user interface.

For each target node the user decided to use for his application there will be an executable image. These images will be downloaded for execution over a V.24 tty–line or the Ethernet. In the latter case the protocol is based on Internet's UDP, to be independent from type of host or protocols specific to its manufacturer. Only one node of the multicomputer system needs to have access to Ethernet, the rest may be booted over the message transport system. The necessary download code is contained in boot prom of the target nodes, the corresponding host part is available for different machines.

In order to provide for remote debugging a target debugger part has to be linked together with the application and the kernel, which acts transparently to the kernel on behalf of a host debugger part which is connected by TCP/IP and Ethernet. The single target debugger instances work together using the MTS. The debugger functions are distributed between host and target in that low level functions like register and memory access and breakpoint–handling are performed by the target, whereas the higher level functions related to symbolic debugging are left to the host part. The host part has access to the symbolic information spent by the compilers via the file system of the host [11], [12].

Beside the conventional debugging functions known from available symbolic debuggers the debugger offers functions which take into account that the application under test consists of several tasks. For example the state of all active tasks can be displayed, watch points can be made sensitive to a specific task only, selected IPC–events may be traced and displayed and so on. Currently we are working on a window based user interface for the host part, because to study the run–time behaviour of a multi–machine program requires a clear and meaningful representation of the selected events and ease of use improves acceptance.

For building up the system the remote debugger of the RTK called RDB was used and extended by symbolic (COFF–based) functions. It uses a V.24 line and a polling driver as host–target connections and its target component has only few functions (read/write registers, memory cells). Because of the small functionality required on the target system it is a very useful tool in the early stage of system development.

To support mixed applications between host and multicomputer target system we implemented an emulation library of the target IPC–functions on UNIX. This library is linked to an UNIX process, which wants to communicate with other processes on UNIX or tasks on the target. In combination with specialized network server–processes on target and host, which use a TCP/IP connection, every remote communication between the two systems, that takes place on behalf of processes using the IPC–functions described above, can be accomplished. Thus easier shift of an application process existing on the host down to the target can be achieved.

3.   NODE ARCHITECTURE

3.1.   BASIC HARDWARE STRUCTURE

The multicomputer system is composed of possibly heterogeneous nodes. Most of the requirements for the individual nodes are realized by an universal node processor supported by a communication processor and a generalized interface to its dedicated peripherals.

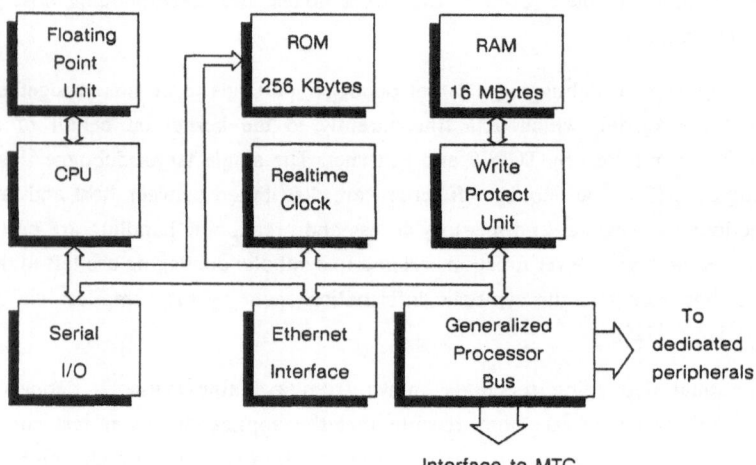

Figure 4 : Structure of the Universal Node Processor

The universal node processors are implemented in standard technology based on 32-bit CISC-processor and also in a prototype version based on a RISC-processor. Figure 4 shows the structure of the CISC-version. The design of the CPU core and the memory system is almost standard design for a single board computer except the write protect unit and the generalized processor bus:

The write protect unit protects code and data of a task. Only write cycles are protected; read cycles are not watched. The on-chip memory management unit of the microprocessor is left unused because it slows performance of the real time kernel and requires a lot of memory space for the translation tables in the case of small pages and many processes. The write protect unit covers less than one square inch board space, making its implementation reasonable.

The generalized processor bus is used to connect all peripherals (e.g. the communication processor discussed below). The generalized processor bus provides full address space (4 GB), full data transfer capability (26.67 MBytes/sec) and direct memory access for high speed peripherals. The bus is enhanced by signals for asynchronous events and on-board decoded select and handshake signals, thus reducing the implementation overhead at the peripheral modules.

The RISC-module is compatible to the CISC-module at the hardware level at the generalized processor bus and at the software level at the system call interface of the distributed real time kernel. If the processing power of 4 MIPS provided by the CISC-module is not sufficient for certain applications, it can be substituted by a RISC-module without changes or reimplementation in the peripheral module or in the application software. Main application areas for the RISC-module will be the field of image processing and administration of representations of the environment of the mobile robot.

The system architecture is open to integrate special processors for applications like artificial intelligence and image processing. Studies have shown that the speed up of special LISP, Prolog or SmallTalk processors is comparatively small compared to the speed up that can be obtained by RISC processors or sophisticated compiler technology [13]. Digital signal processors for fast image processing are suggested to be connected at the generalized processor bus. One kind of special processor is the communication processor of a node, which will be discussed next.

3.2. MESSAGE TRANSPORT SYSTEM

The communication coprocessor – called message transport controller (MTC) – offloads the universal node processor from communication related tasks. The MTC sites on a separate board which is attached to the UNP via the generalized processor bus. The two processors interact through shared memory and asynchronous event lines.

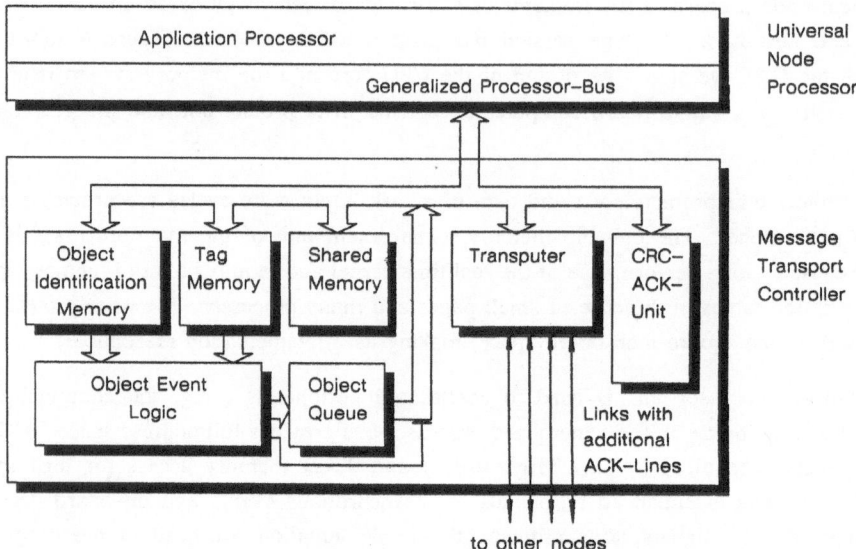

Figure 5 : Hardware Structure of the Message Transport Controller

The message transport controller is based on a 32–bit transputer (see figure 5) which combines high processing power with excellent communication support by high speed serial links, which in effect build up the medium for internode communication. Several alternative architectures have also been studied [14]. The performance of this conventional coprocessor structure [9] is enhanced by a special memory system for interprocessor communication and additional hardware support for internode communication.

The basic idea for minimizing the interprocessor communication is to install hardware supported demons which generate exceptions to the corresponding processor, if certain data structures in the shared memory are accessed by its counterpart [15]. This principle has been realized as follows:

Data structures in the shared memory – called objects – used to pass service requests, status indications and other information are specified by unique identifiers and their memory locations. The shared memory is enhanced by an object identification memory and a tag memory, which extend the 32–bit memory word to following format: /32–bit data/12–bit identifier/5–bit tags/. During each memory cycle to the shared memory the object identification memory and the tag memory are accessed simultaneously. If the object event logic detects a match between the information coded in the functional bits of the tag memory and the type of the actual memory cycle (read/write form UNP/MTC), it passes the object identifier to the object queue (realized as hardware fifo) and triggers the exception to the appropriate processor. One of the tag bits is used to guarantee exclusive access for an object during update from one processor.

The object oriented trigger mechanism allows the application processor to request services from the communication processor by merely accessing objects in the shared memory like message descriptors, etc without dealing with software fifos or rendezvous. The realization provides a fine granularity of 4 Byte and up to 4096 objects, so that it can be tailored to all applications of interprocessor communication.

Internode communication is done via the four serial links of the transputer at a data rate of 20 MBit/sec. The intention of the native transputer links as on-board communication lines results in several restrictions if used for a spatially distributed multicomputer system. The short comings in this case are: limited distance to 0.5 m, high noise sensitivity and most important insufficient support for checksum or CRC-information to detect corrupted data packets. The use of differential line drivers together with shielded cable enlarged the internode distance to 10 meters and at the same time, the influence of noisy environment inside and outside the autonomous system became negligible.

The CRC/ACK-unit, which is implemented as a gate array, supplies a 16-bit CCITT standardized CRC-polynom for each data packet. The receiving node uses the CRC/ACK-unit to check for transmission errors. The CRC/ACK-unit also drives separate ACK-lines which are fed to the transmitting node, thus allowing an indication of the success of a transmission within one microsecond. In the case of a failure a retransmission is started within 20 microseconds. In our implementation the CRC/ACK-unit gave a five times speed up for the low level protocol handling for short packets [16].

3.3. FUNCTIONAL STRUCTURE AND SUPPORT OF INTERPROCESS COMMUNICATION

From the kernel point of view the fundamental functionality offered by the MTC can be summarized as

- the interfacing mechanisms to the main CPU, which are shared memory with a programmable, impressed structure for signalling in each direction (interrupts are released by writing/reading memory cells and the interrupted part gets an index to the corresponding structure). We call this mechanism an object oriented interface,

- the communication processor, which provides for the processing power needed to perform operations upon the kernel structures demanded locally or remotely and to process the communication protocols for shipping of messages,

- the communication facilities, which are characterized by bidirectional, arbitration-free, high-capacity transmission links to other nodes, variable arbitrary packet sizes and hardware support for reliable transmission and acknowledging.

These mechanisms support the IPC concept described in chapter 2.2 . It should be outlined which special issues of IPC management are addressed by the features of the MTC.

First of all, the possibility to locate the kernel structures (which are used in any case) in common memory and to release interrupts mutually in the quite natural way of data access reduces the complexity of interface handling. Each interrupt is accompanied by an index to the corresponding data structure. This measurement resolves the need for searching through lists or tables and results in logical addressing being an easy business. The only operation maybe needed is to determine the alien structure to a triggered kernel structure and its location, which is cheap.

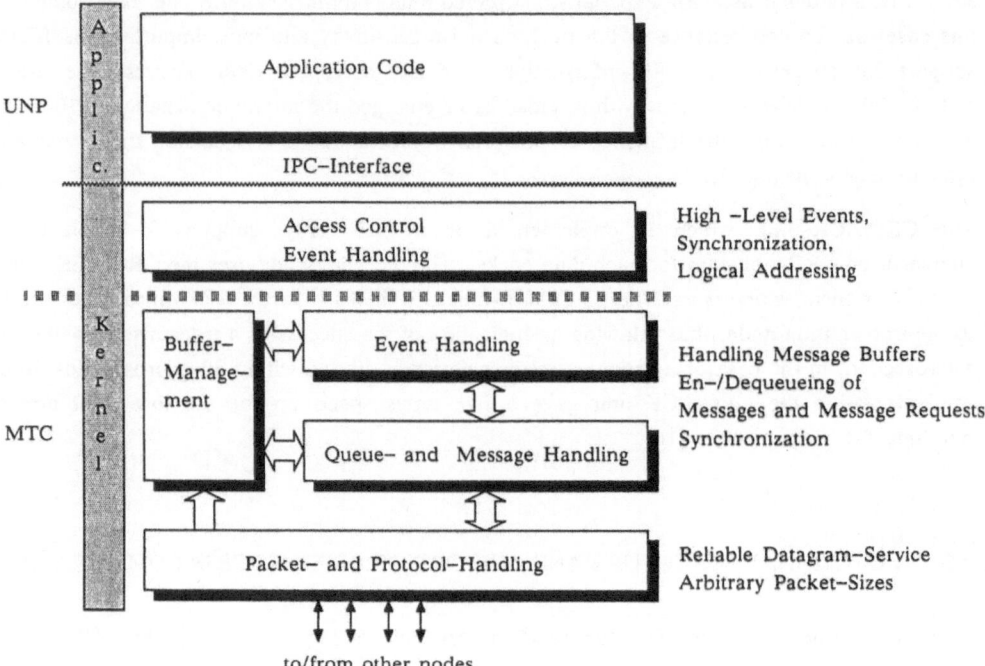

Figure 6 : Functional Structure of the Message Transport Controller

The processing capacity provided by the communication processor is used for managing the kernel structures involved in processing a certain communication request. This not only eases the load of the application processor, but also results in a high responsiveness if the local kernel is involved in an operation initiated remotely without disturbing the application processor whenever possible. If, for instance a message arrives from remote and there is no local task willing to accept it yet, the application processor is not interrupted. In an analogous way system messages for transactions on kernel structures are handled autonomously by the MTC's.

We have done several studies concerning the strategies for managing the kernel structures and message buffers. We have taken into account several alternatives and tried to determine the need for system messages to exchange between the kernel instances in each case. The different variants are valued on the resulting interkernel synchronization effort, which we wanted to keep as low as possible.

For non–local communication there are messages to be transported across node boundaries. To achieve a respectable performance for that case it is necessary to reduce protocol processing overhead and number of times a message has to be copied between address spaces or interface buffers.

Protocol processing requirements are very small in our approach,

- there is no kind of arbitration in accessing the transmission links, immediate access is guaranteed,

- arbitrary packet sizes may be transported, which makes message assembly or disassembly unnecessary (this is rather important, since receivers expect messages to be located in a contiguous address region as a whole)

- a kind of reliable datagram service is required to transport messages, therefore performing CRC in hardware is crucial for performance. Moreover the hardware acknowledge lines release the need of extra packets in the opposite direction and impose another performance gain.

Main part of protocol handling remains buffer handling. Preallocated buffers or fast deterministic algorithms influence the degree the high transmission capacity may be exploited.

In order to emphasize the design issues, a comparison to Ethernet, which is quite a standard solution may be given:

- There is nearly the same net transmission rate on the Ethernet and on the Transputer links,

- Our protocol introduces slightly less overhead for small packets, and substantially less overhead for large packets, because no packet assembly/disassembly is required,

- The Ethernet provides only one shared communication link for all nodes. Therefore the available communication bandwidth is split up between the participating nodes. In our solution pairs of nodes can communicate simultaneously without influencing one another,

- The acknowledging response packets can be spared, which saves a high portion of available transmission rates.

Some measurement results are given in comparison. We have taken a simple interface to raw Ethernet to send packets between two of our UNP's (20MHz). The other values are for two MTC's with 20MBit/s links (with T414 or T425 Transputers, 20MHz) and our transmission protocol evaluating acks. For large packets the Ethernet is at its limits, while the MTS would exploit the high potential of parallelism, when more than one link is active. The values in the table show, that to send acknowledge packets for making transmission reliable substantially reduces the net messages per second rate.

Messages/second

Bytes per packet	Ethernet (acknowledged)	Ethernet (unacknowledged)	MTC with T414 (T425) (acknowledged / only one link)
4	1400	2900	5300 (**5400**)
32	1400	2900	4600 (**5000**)
64	1400	2800	3900 (**4500**)
128	1300	2300	3100 (**3900**)
1024	770	1060	690 (**1070**)

The execution of the protocol code on our MTC can be optimized by extended use of the on–chip memory of the Transputer, which is three times faster than external memory. This will increase the throughput for short packets. Regarding the bare transmission capacity provided by the MTS, we think there is no bottleneck for user communication any longer. We will have to measure and evaluate to which extent we have addressed the overhead incurred in the higher levels [17], [8].

3.4.   INTERFACING OF PERIPHERALS

Each node in the autonomous mobile system is dedicated to a special task by its peripheral system which covers a broad spectrum of peripherals ranging from low data rate peripherals to high data rate peripherals. Figure 7 shows three methods of interfacing, that will cover all applications in the mobile robot:

	No. of Modules	Transfer Rate	Distance covered	Synchronization with CPU	Effort
Shared Memory (incl. DMA)	2 .. 4	.. 20 MB/sec	0.1 m	interrupt	moderate
Memory Mapped Input / Output	.. 16	.. 0.5 MB/sec	1 m	interrupt / polling	low
Field Bus	.. 32	.. 20 KB/sec	.. 100 m	interrupt / polling	moderate

Figure 7 : Interfacing Methods for Peripherals

Conventional interfacing via shared memory or memory mapped I/O are well supported by the generalized processor bus. It is very simple to add interfaces to other equipment – One chip of interface logic was sufficient in all cases until now.

Field buses are widely used in the field of factory automation to connect a large number of low speed peripherals to a processing node via serial buses that cover distances up to 100 meters. Worldwide standardization (IEEE, DIN) leads to a growing number of actuators and sensors supplied with integrated field bus interface. Our implementation of the field bus system is very close to bitbus and is characterized by a data rate of 2.4 MBit/sec and a capability to handle more than 5000 packets/sec with a deterministic real time protocol [18]. The field bus system consist of a master module, which is connected by shared memory to the universal node processor, and a variety of slave modules for analog and digital input / output. The field bus master serves the slave modules by polling at an user definable rate and provides an uniform interface to many different, distributed, low speed peripherals. Because of its open architecture additional peripheral modules can be easily integrated.

4   APPLICATION

The UNP- module is used for several dedicated control - and sensor data processing application within the SFB 331 (Sonderforschungsbereich) [19],[20]. Figure 8 shows two of them:

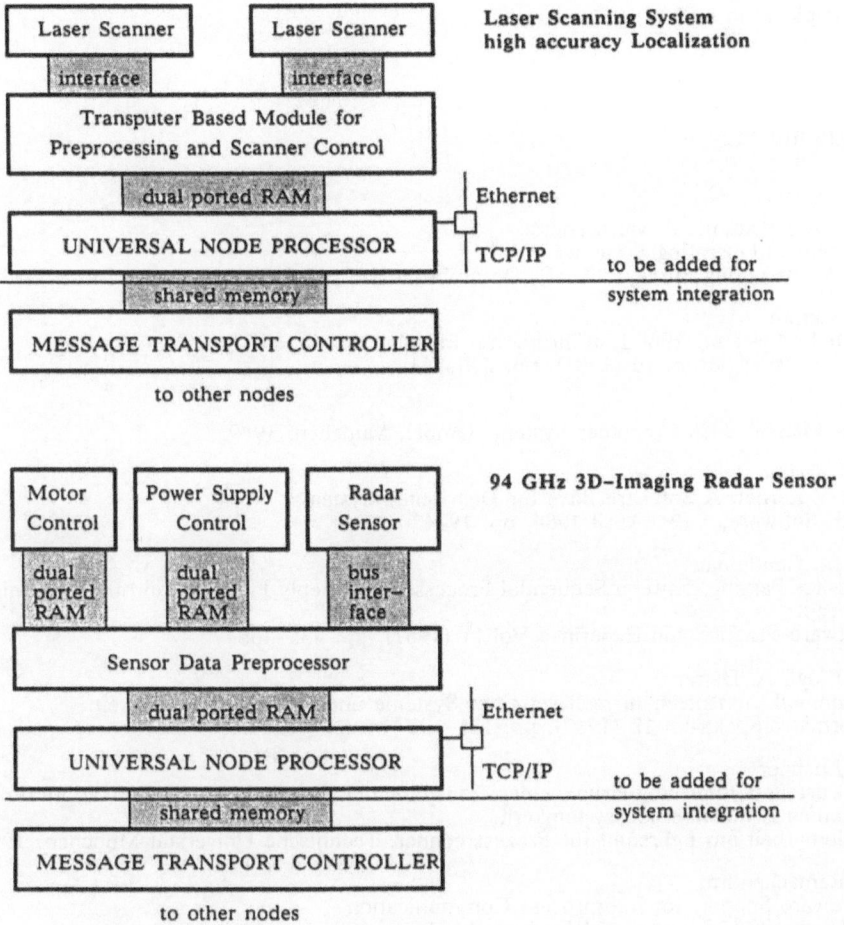

Figure 8 : Application of the UNP- Module

– Laser Scanning System for high accuracy localization: Here a Transputer Module is used for preprocessing and control, the UNP measures the position and identifies the artificial landmarks.

– 3D- Imaging Radar Sensor. The radar protocol control has been ported to the UNP, the distance measurement is done, data rates in the 100 kBytes/sec are processed.

Additionally other applications like a main-memory version of an object-oriented knowledge base for CIM-environments have successfully been ported to demonstrate the universal capabilities of the UNP-architecture.

The distributed multicomputer system as a whole is not yet installed for reasons of system complexity and variety of tools and packages to be supplied for the multicomputer system. Also there are still final tests and optimizations concerning the message transport system. The system will be completed in mid of 1991.

5   REFERENCES

[1]   A.S. Tanenbaum, R. van Renesse
       Distributed Operating Systems.
       ACM Computing Surveys 17 94), December 1985, pp. 417–470

[2]   M. Stumm
       Verteilte Systeme: Eine Einführung am Beispiel V.
       Informatik Spektrum 10 (1987), pp. 246–261

[3]   N.N.
       RTK Manual. PCS Computer Systems GmbH, München, 1989.

[4]   D. Cheriton
       The V Kernel: A Software Base for Distributed Systems.
       IEEE Software, 1 (2), April 1984, pp. 19–42

[5]   W. M. Gentleman
       Message Passing Between Sequential Processes: the Reply Primitive and the Administrator Concept.
       Software–Practice and Experince Vol.11 (1981), pp. 435–466

[6]   M. Pfügl, A. Damm
       Kommunikationsmechanismen verteilter Systeme und ihre Echtzeitfähigkeit.
       Informatik–Spektrum 12 (1989), pp. 121–132

[7]   W. Lochner
       Entwurf und Implementierung einer transparenten Interprozesskommunikation für einen verteilten Echtzeitbetriebssystemkern.
       Diplomarbeit am Lehrstuhl für Prozessrechner, Technische Universtät München, 1990.

[8]   U. Ramachandran
       Hardware Support for Interprocess Communication.
       Ph.D. thesis, Universtity of Wisconsin Madison, 1986

[9] J.D. Northcutt
     Mechanisms for Reliable Distributed Real-Time Operating Systems, The Alpha Kernel.
     Perspectives in computing; vol 16. Academic Press, 1987

[10] F. Tuynman, L.O. Hertzberger
     A Distributed Real-Time Operating System.
     Software-Practice and Experience Vol.16 (1986) 5, pp. 425-441

[11] T. Böhnke
     Entwurf eines remote Debuggers für unterschiedliche Host-/Target-Konstelationen und
     Implementierung der Basisfunktionen.
     Diplombeit am Lehrstuhl für Prozessrechner, Technische Universtät München, 1989.

[12] T. Bemmerl
     Realtime Highlevel Debugging in Host-/Target-Environments.
     Proceedings of Euromicro Symposium on Micro-Architectures, Development and Applica-
     tions, Venice 1987, pp. 387-400

[13] P. Steenkiste, J. Hennessy
     LISP on a Reduced-Instruction-Set-Processor: Characterisation and Optimization
     Computer, July 1988, p. 34

[14] J. Stegmaier
     Vergleich von Alternativen zum Nachrichtentransportsystem des SFB 331
     Multi-Mikrorechners
     Diplomarbeit am Lehrstuhl für Prozeßrechner, Technische Universtät München, 1989,

[15] P. Fischbacher, K. Gresser, G. Koller, A. Stein, M. Triller
     Basisfunktionen für die UNIX- Implementierung in einem fehlertoleranten
     Multimikrorechnersystem
     Abschlußbericht Deutsche Forschungsgemeinschaft, 1989, DFG Fa 109/8

[16] T. Shergowski, A. Ruß, J. Lorenz
     Cyclic Redundancy Check in einem FPGA
     Elektronik, May 1990, p. 118

[17] M.L. Scott, A.L. Cox
     An Empirical Study of Message Passing Overhead.
     Proceedings of the 7th International Conference on Distributed Computing Systems, Berlin,
     Sept. 21-25, 1987, IEEE

[18] T. Lingenthal
     Digitalisierung und Aufbereitung von Sensorsignalen, basierend auf marktgängigen
     digitalen Bausteinen
     Bundesministerium für Forschung und Technologie, 1990, Förderungskennz. 13AS0060

[19] W. Frank
     Hardwareoptimierung und Software-Entwicklung für einen Laserscanner zur
     Positionsvermessung eines autonomen mobilen Roboters
     Diplomarbeit am Lehrstuhl für Prozeßrechner, Technische Universtät München, 1990

[20] M. Hermann
     Integration des universellen Verarbeitungsmoduls in das 94 GHz Radar
     Diplomarbeit am Lehrstuhl für Prozeßrechner, Technische Universtät München, 1990

Learning

Location Recognition in a Mobile Robot Using Self-Organising Feature Maps

U. Nehmzow, T. Smithers and J. Hallam

Abstract

Self-organising structures are a dominant feature of the experimental mobile robots built in our "Really Useful Robots" project. This paper continues where [Nehmzow & Smithers 90] finished. It explains some initial experiments using self-organising feature maps, and how those maps can be used by a mobile robot to recognise locations in its environment. This location recognition capability is achieved without using sensory information. Instead information derived from the motor actions of the robot is used, and shown to be sufficient.

1   Introduction

Self-organising feature maps (SOFMs), [Kohonen 88], can be used for simple navigation tasks. In earlier work, [Nehmzow & Smithers 90], we have shown that "Alder", the first of the "Really Useful Robots" (RUR) (see figure 1), is able to use a SOFM to recognize particular locations in a simple enclosure after it has had sufficient time to explore and to "learn" about this enclosure. In these earlier experiments the input that was presented to the SOFM was derived from sensor signals and contained explicit information about "landmarks" that the robot encountered, and the distance travelled between them. These landmarks were the concave and convex corners in the robot's enclosure.

In the RUR project, we are investigating ways to make mobile robots more flexible, and in particular, to make them more able to cope with *unforeseen situations*. We believe that in order to achieve this as many decisions as possible should be left to the robot; in particular decisions about what and how information derived from sensors or other sources internal to the robot should

Figure 1: Alder, the first of the "Really Useful Robots"

be interpreted for the purposes of control in task achieving behaviour. For the same reason we aim to equip the robot with as little predefined knowledge about its environment as possible [Nehmzow *et al.* 89]. Information concerning landmarks is specialised information. Therefore, we decided to try to achieve similar results to those reported earlier, but this time using less specific input information. Whereas before we used processed sensor information (denoting corner types) to achieve location recognition, we now use *no sensor information* at all! Instead we use information about the history of the *motor action commands* of the robot controller. In this paper we describe how we achieved this, again using self-organising networks.

2 The Duality of Sensing and Acting

Sensing and acting are typically treated as separate functions in robotics. We believe, however, that sensing and acting are two aspects of the same function, and that they therefore cannot be successfully analysed in isolation. The actions of a robot, just like those of a person, determine to a large extent the sensory signals it will receive, which will in turn influence its actions. Breaking this tight interaction into two separate functions leads, we believe, to an incorrect decomposition of the robot control problem. While particular features of a robot's sensors and actuators do play an important role in determining its performance, these effects cannot be determined by their separate analysis. Acting and sensing have to be seen together; neither acting nor sensing alone will make the agent succeed.

The input vector we have chosen for the following experiments demonstrates this point: it contains no direct information about sensory input. The information it does contain is derived from the motor action commands of the robot controller, but these, as we have said, are themselves influenced by the sensory signals received by the robot as a result of its actions. We chose to use information derived from the motor action commands of the robot controller, rather than from sensor signals, because they form a smaller set of signal types, they are much less subject to noise, but they still adequately characterise the interactions between the robot and its environment as it seeks to achieve its task—wall follow, in our case.

For the location recognition experiments the robot is placed in an enclosure as shown in figure 2; it then follows the wall using a preprogrammed wall-following and obstacle-avoidance behaviour (alternatively these skills could be acquired by learning, see [Nehmzow et al. 89]). In other words, the robot is governed by its preprogrammed wall following behaviour which, of course, does use sensory information. The process of constructing the SOFM is, however, independent of the wall following behaviour, it simply "looks" at the motor action commands issued by the controller as the robot performs this wall following task.

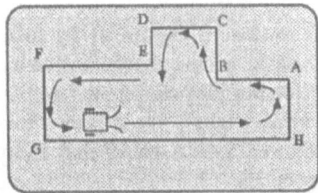

Figure 2: A typical environment for Alder

Every time a new motor action command is issued, that is, every time the wall-following or the obstacle-avoidance behaviour forces the robot to change direction a motor action vector is generated. This is a nine bit vector which contains information about the state the motors were in until this change, and thus the direction the robot has been travelling up to this moment (forward, left or right)[1], as well as information about how long it was in this state (see figure 3). This motor action vector forms the input to the SOFM.

Motor Action			Duration	
Forward	01	01	00000	less than 0.9 s
Left	01	10	00001	0.9 - 1.3 s
Right	10	01	00011	1.3 - 1.7 s
			00111	1.7 - 2.1 s
	Right	Left	01111	2.1 - 2.6 s
	Motor	Motor	11111	over 2.6 s

Figure 3: The motor action vector

Thus, from figure 3 we can see that no information concerning sensor signals is directly presented to the SOFM. The only information available to the network concerns motor action commands.

3 Early Experiments

3.1 Self-Organising Feature Maps

The mathematics for the kind of self-organising feature maps (SOFMs) that we use can be found in [Kohonen 88] or in [Nehmzow & Smithers 90]. Basically, they conform themselves to the structure of the space of input signals, clustering input signals in such a way that related inputs excite neighbouring areas of the network (they are *topology preserving*), and so that the relative sizes

---

[1]Strictly speaking, they contain information about the last command from the robot controller. Whether this command was actually obeyed by the robot or not is not sensed.

of these areas reflect the relative probabilities of the different inputs in the data presented. By making use of these two properties of SOFMs they can be used to perform a kind of unsupervised learning.

To investigate how this might be done using only data derived from motor commands we first carried out the following experiment in which we used a two-dimensional SOFM of ten by ten cells.

3.2 Experiments with a 10 x 10 SOFM

The robot was left to explore its enclosure (figure 2) by following the walls, generating input vectors as shown in figure 3 each time a new motor action command was issued by the robot controller for any reason. The following pictures show the response of a ten by ten cell SOFM to different input stimuli; the terms "right", "left" and "forward" describe the motor action command that was performed, followed by a number that indicates the length of time for which this action was performed (see figure 3).

```
 -----------        -----------        -----------        -----------
|.       ..|      |  .... ...|      |  .....////|      |  ..//////|
|..    .....|     |  /..///..|      |  ../////////|    |  ../+++//|
|//.   .....|     |++/..///..|      |///////////|      |/////+**//|
|//.   ...  |     |**/../// |       |////////...|      |/////+++..|
|///.....  |      |+++/////  |      |//+++///..|       |///++++++..|
|++///..  |       |/////////. |     |//+**///..|       |.//++++/..|
|*+//.    |       |/////...  |      |//+++//...|       |.//+++////|
|++/..    |       |/////.....|      |./+++/////|       |..//////////|
|..      |        |  ........|       |  ..././///|      |  ...///..|
|..      |        |  ........|       |  ..../////|      |  ..../...|
 -----------        -----------        -----------        -----------

Forward, zero      Forward, one       Forward, two       Forward, three



 -----------        -----------        -----------        -----------
|   ..//+++|      |   .//+++|        |         |        |         |
|   ...+++++|     |   ...//+++|      |         |        |         |
|//../+++++|      |....///++**|      |         |        |..    .. |
|/////+++//|      |....////////|     |    ..|          |..    ....|
|/////+*+//|      |..///+++//|       |    ..|          |.....   ...|
|..///+++//|      |..///+++//|       |    ..|          |  ........|
|..////////|      |..////////|       |.    .....|      |  ...////.|
|..////////|      |..///./...|       |.........|        |..//////////|
|   //...|        |   .....|        |++////..//|       |//+**//////|
|   .....|        |   .....|        |*+////.////|      |//+++++////|
 -----------        -----------        -----------        -----------

Forward, four      Forward, five      Left, zero         Left, one
```

```
------------      ------------      ------------      ------------
|    ...|         |    .....|       |   ..///\        |   ..///\
|    ....|        |   ///..|        |   /////\        |   .///\
|..  .....|       |.. .///..\       |   .//////\      |   ..///\
|...  ...//\      |.....//////\     |   .///++|        |   .///++|
|.///...//\       |....//////\      |   ...///++|      |   ...///+*|
|  ///..///\      |  ..///////\     |   ...//+++|      |   ...///++|
|  /////////\     |  ..//////|      |   .../+++*|      |   ...///++|
|  ///+++++|      |  ..//+++++|     |   .../++*+|       |   ...///|
|.///+++*+|        |  ///+*+++|     |  /.//++++|       |.....//////\
|/////+++*+|      |..///+++//\      |..////+///\       |.....//////\
|/////+++++|      |..////++//\      |..////////.|      |.....//////\
------------      ------------      ------------      ------------

  Left, two          Left, three        Left, four         Left, five


------------      ------------      ------------      ------------
|*++//..  |       |//+*+//.  |      |///++//...|       |..///++/..|
|++///    |       |//+++...  |      |//+++//...|       |../+++//..|
|  //.     |       |..+//... |      |..+*+/....|       |../+*///..|
|  ...     |       |..///... |      |..+++/.. |        |../+///..|
|          |       |.....     |     |..///.   |        |...////...|
|          |       |  ...     |     | ///...  |        |  ...//...|
|          |       |  ...     |     | ///..   |        |........|
|          |       |  ...     |     | ...     |        |........|
|          |       |          |    |          |        |   ...  |
|          |       |          |    |          |        |     .  |
------------      ------------      ------------      ------------

  Right, zero        Right, one         Right, two         Right, three


------------      ------------
|..///+*///\       |...//++*//\
|..////////\       |..////////\
|  //+/////\       |  /////////\
|  //////..|       |  /////////\
|  ...///..|       |  ...////|
|  ...///..|       |  ...////\
|........|         |........|
|........|         |  .. .. |
|    .   |         |     .  |
|        |         |        |
------------      ------------

  Right, four        Right, five
```

Figure 4: Excitation patterns of the SOFM for different types of motor action vector inputs. The cell with the largest excitation is indicated by "*", "+" means that the excitation of a cell is higher than 0.9, "/" denotes excitations between 0.9 and 0.7 and "." excitations between 0.7 and 0.5.

Two observations can be made from these pictures of the response of the SOFM to different types of input vector:

- The size of the excited area is roughly proportional to the frequency of occurence of the input signal that caused the excitation.

- Related inputs excite neighbouring areas. In this example, we can see that "forward" movements stimulate the central region of the network, "left" movements stimulate the lower region and "right" movements the top region of the net. Within these basic regions there are variations, depending on the duration of the movement.

Topographic mappings such as these are common in biological nervous systems. The striate cortex, for example (one part of the visual system of primates) is organized in topographic fashion, each half of the visual field (of a macaque monkey) is projected onto the striate cortex in a systematic map [Hubel 79]. [Churchland 86] gives a good survey of topographic mappings in neural systems (for topographic mappings of the visual cortex see, for example, [Allman 77]).

Similarly, somatotopic[2] maps can be found in the cortex. Very similar observations to those stated above can be made about such biological systems, ([Churchland 86]):

- Map distortion is based on the population of neurons representing a particular body area. Areas with a higher density of receptors are bigger. [A higher density of receptors will result in a greater number of sensory signals].

- A precise mapping of the body surface onto neurons of the somatosensory cortex is found. Neighbourhood relations of body parts are preserved.

If the nature of input stimuli is altered, for example by joining two fingers of a hand together, the mapping in the somatosensory cortex changes accordingly, see ([Clark *et al.* 88]).

The question that arose for us from these observations was: "Can we use this sort of SOFM to achieve location recognition?" In other words, can the mapping of different motor action types onto different regions of the SOFM network, as shown in figure 4, be used in a system for recognising locations? Our tentative answer is that it seems they can be. In the next sections we explain experiments we devised to do this, the system we finally used and the experimental results of the tests we have carried out so far.

3.3 Location Recognition, using one SOFM

Having done these initial experiments with a ten by ten cell SOFM, using the motor action vectors directly as an input to the network, we then tried to combine several of these vectors as input to a network of twelve by twelve cells. The idea was that the response of the network would be correlated with Alder's arrival at a physical location, thus allowing the robot to recognize particular locations. As it turned out, the number of motor action vectors within an input vector that gives best recognition results varies from location to location. Some locations are easy to identify using just a short history (i.e. small numbers of combined motor action vectors), others need a longer history (i.e. a large number of combined motor action vectors). There was no one length of input vector which gives equally good recognition performance with all the corners in the robot's enclosure.

In retrospect, this is not a particularly surprising result. The means by which locations are identified — their "signatures" — are patterns of actions extended over time. As the rate at which

---

[2]From σῶμα, body. What is meant here is that touch, pressure, vibration, temperature and pain sensors are mapped onto the cortex in an orderly fashion.

motor action commands are issued goes up at corners (obstacle avoidance), the temporal scale of these signatures is dependent upon the distances between corners. In order to be able to identify arbitrary locations it is therefore necessary to use input vectors of different lengths. A fixed size of input vector, whatever it is, will either include extraneous information (complicating the task of the SOFM) or miss possibly crucial signature components (eliminating necessary distinctions).

In some respects, the problem is similar to that encountered in the analysis of natural images in vision: the data is intrinsically multiscalar (though here it is temporal scale that is critical) and so must be subjected to a multiscalar analysis (for this analysis in images see [Marr & Hildreth 80] and [Canny 86]). This motivated our next experiments, in which different sized histories of actions were presented to separate SOFMs in parallel.

4   A SOFM Location Recognition System

As in our earlier work on location recognition (see [Nehmzow & Smithers 90]), the method used is one in which location recognition is based upon the robot recognising its arrival at some location. In other words, location recognition is not based upon recognizing some structural feature of the particular physical location, but on the sequence of actions that have taken place prior to arriving at the particular location. In our earlier work it was the sequence of previous features (corners) that had been detected immediately prior to the current feature (corner) that was used to determine recognition. In this work it is the sequence of motor action commands that have been issued prior to the arrival at the particular physical location that enable Alder to recognize the location. To achieve this we used a system of *seven* independent, two-dimensional SOFMs, working in parallel. Each SOFM consists, as in section 3.3, of a network of twelve by twelve cells. The input vectors to each of these networks are different, but all are built from motor action vectors as shown in figure 3. By combining 2, 4, 6, 8, 12, 16, and 24 of these basic motor action vectors we formed seven SOFM input vectors which correspond to increasingly longer histories of the robots motor action changes, (see figure 5).

The lengths of histories are chosen to cover adequately the expected spectrum of action periodicity. If we think of the sequence of actions generated as the robot circles its enclosure as a pseudo-periodic series, with period roughly equal to the average number of action-vectors generated in a single circuit, then the use of SOFMs tuned to different "frequency bands" allows us to sample the temporal structure of the series across its spectrum and associate these samples with the physical locations whose signatures they are.

In other words, the set of excitation patterns of the seven SOFMs produced as the robot arrives at a particular location (a corner, say) in its enclosure can be used to distinguish this location from all other locations the robot passes through as it wall-follows its way around the enclosure.

4.1   The Experiment

The robot was set to wall-follow its way around the enclosure. Every time a new motor action command was issued as a result of the built in wall-following or obstacle avoidance behaviour of the robot, a motor action vector as described in section 2 was generated. This vector, together with the respective number of previous motor action vectors was presented to each of the seven SOFMs. After a sufficient time, about five times round the enclosure, these feature maps had organized themselves into stable structures corresponding to the topological interrelationships and proportional densities of the input vectors.

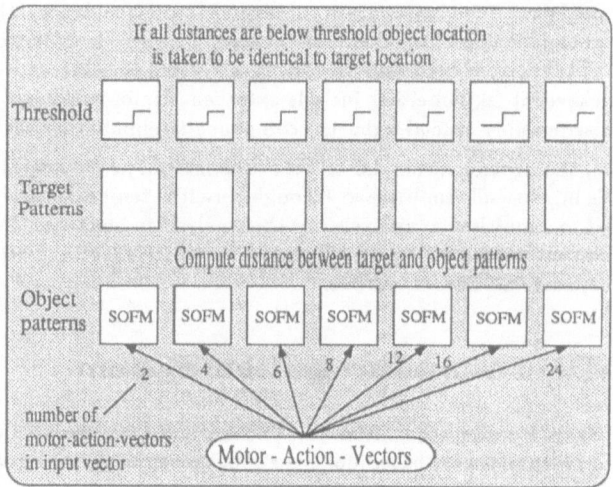

Figure 5: The System used for Location Recognition

After this "learning" period the excitation patterns of all seven networks at a particular location (the *target patterns*) were stored. All subsequent sets of seven excitation patterns generated by new input vectors (*object patterns*) were then compared to the set of seven target patterns. This was done by computing the Euclidean distance (or, alternatively, the city-block distance) between pairs of target and object patterns, see equation 2 in section 4.2. If the distance values between each of the seven pairs of object and target patterns are less than a threshold defined for each pair, the robot is taken to have arrived back at the target location and thus to have "recognized" the previously stored location.

4.2   The Mathematics

In mathematical terms, the system works as follows:

1. Compute the output o_{xyj} of each cell at position (x, y) of each network j:

$$o_{xyj} = \vec{w}_{xy} \cdot \vec{i}_j \quad j = 1, ..., 7, \tag{1}$$

   where $\vec{w}_{xy}$ is the individual weight vector of cell(x, y), and $\vec{i}_j$ is the input vector to network j.

2. Compute the distance between target pattern and corresponding object pattern:

   (a) Either the Euclidean distance is chosen,

$$d_j^e = \sum_{x=1}^{12} \sum_{y=1}^{12} (o_{xyj} - o_{xyj}^T)^2, \tag{2}$$

   where o_{xyj} is the output value of cell(x, y) of the object pattern j, and o_{xyj}^T is the output value of cell$_{xy}$ in target pattern j.

(b) Or, the city-block distance is computed

$$d_j^{cb} = \sum_{x=1}^{12} \sum_{y=1}^{12} |o_{xyj} - o_{xyj}^T|. \tag{3}$$

3. Determine whether object location and target location are identical:

   If $d_j < \Theta_j$ for all j=1 to 7 then object location and target location are taken to be identical. Θ_j is the threshold for network j. For the experiments reported here the Θ_j were chosen by the designer so that patterns obtained at target location and patterns obtained at other locations could be separated. Note that the actual values will, of course, depend on the distance metric chosen. Schemes can be devised that determine these thresholds automatically, but not without supervision.

4.3 Results

Data recorded from the robot was used to compute these results, however the actual computation was done off-line on a workstation.

In the experiments conducted so far the robot's task was to identify three particular corners in the enclosure shown in figure 2: corners H, E and F. To do this the set of seven excitation patterns of the seven SOFMs at the target corners were stored; all subsequent sets of excitation patterns were then compared to these by computing the Euclidean distances between respective pairs of patterns in the set of seven.

Provided a suitable set of thresholds was used (see paragraph 3 in section 4.2), the robot recognised corner H four times in the subsequent five rounds, and corners E and F in five out of the five times. At no time was a non-target corner erroneously "identified" as a target corner.

4.4 Changes in Parameters

Different Metrics

Instead of using the Euclidean distance to estimate similarity between excitation patterns, the "city-block" distance can be used with identical results (see equation 3 in section 4.2). City-block distance is computationally cheaper than Euclidean distance.

One-dimensional Networks

Instead of using two-dimensional networks, *one-dimensional networks* can also be used. This considerably reduces the computational cost of the scheme. We used one-dimensional networks of twenty-five cells to test this. In the case of corners H and E the results were the same as the ones presented above. In the case of corner F, one erroneous identification took place, as well as the five correct ones as before. This seems to indicate that one-dimensional networks can be used instead of two-dimensional ones in this task, except with slightly less reliability. A plausible conjecture for this is that the neighbourhood relationship plays an important role: each cell in a two-dimensional network has eight neighbours, through which excitation can spread, whereas a cell in a one-dimensional network has but two neighbours. The actual number of cells (twelve by twelve as opposed to one by twenty-five) will have some influence, too; but in our experience this is not as important as the neighbourhood relationships.

5 Summary and Conclusion

5.1 Summary

The robot's task was to recognize particular locations (for example corners) in its world--a simple enclosure. In an earlier publication ([Nehmzow & Smithers 90]) we showed that this task can be accomplished using self-organising feature maps. The input vector used then contained explicit information about landmarks encountered: that the robot was at a corner, what sort of corner (convex or concave) it was, and information about previous corners encountered.

In the subsequent experiments reported here we tried to reduce the explicit information content in the input vector. We also tried to generate input vectors to the network(s) that contain no direct information about sensor signals. The motivation for the first was to avoid using predefined knowledge, because we believe that predefinition of knowledge limits the robot's flexibility; the reason for the latter was to prove our claim that sensing and acting are in fact closely related.

The input vector we chose thus contained information only about the motor commands of the robot controller, and their duration. Vectors put together from varying numbers of these motor action vectors (2,4,6,8,12,16 and 24) were presented to seven separate self organising feature maps, each two-dimensional of size twelve by twelve cells. In order to identify a particular corner, all seven excitation patterns had to be close enough to a stored set of target patterns. If this was the case, the robot was said to have identified the target location.

The location recognition system performed well in this experiment, recognizing corner H (see figure 2) in four out of five times, corners E and F in five out of five times, with no erroneous identifications in either case. Using self-organising feature maps to recognize locations adds a high degree of freedom to the robot controller. The robot is able to build its own representations of the environment, independently of the designer.

5.2 Conclusion

There are two main conclusions we can draw from the work presented. First, our claim that sensing and acting are closely coupled is confirmed by the success of the robot in recognizing locations based upon the sequence of motor activity which leads to arrival there. The "sensor" being used to provide features on which the recognition is based is actually the behaviour of the robot.

Choosing an input vector that contains no direct information about sensory signals makes the system independent of the actual sensors used. Whether tactile, ultrasonic, infrared or other sensors are used: the location recognition system stays the same.

Second, with this approach the features to be identified by the SOFMs are spread out over time. This contrasts with our early work [Nehmzow et al. 89] in which the robot controller learns instantaneous reactions to sensor stimuli. The idea, successfully demonstrated here, of using multiple channels to avoid an assumption of a single natural timescale, merits considerable further investigation since, at first sight, it should be capable of wide application to problems of eliciting temporal structure.

5.3 Open Questions and Future Work

We intend to implement a whole system as described above on a mobile robot. This will show us how feasible it is to use this method in applications where real-time processing is important, and where signal noise can bring problems. Implementing such a system on a mobile robot will also enable us to investigate further the questions concerning network size and structure of input vector.

The question of identifying relevant structure over time merits further study and we intend to investigate the limitations of the method used here as well as test alternative approaches.

Acknowledgements

We thank Peter Forster, our colleague on RUR, for his constructive and helpful contributions to this work, and for his comments on earlier versions of this paper.

The work reported here is supported by a grant from the UK Science and Engineering Research Council (grant number GR/F/5852.3). Other facilities were provided by the University of Edinburgh.

References

[Allman 77] John Allman, *Evolution of the visual system in early primates*, Progress in Psychobiology and Physiological Psychology **7** pp.1–53, 1977

[Canny 86] John F. Canny, *A Computational Approach to Edge Detection*, IEEE PAMI **8(6)** pp. 679–698, 1986

[Churchland 86] Patricia Smith Churchland, *Neurophilosophy*, MIT Press Cambridge Mass. and London, England, 1986

[Clark et al. 88] Sharon A. Clark, Terry Allard, William M. Jenkins and Michael M. Merzenich, *Receptive fields in the body surface map in adult cortex defined by temporally correlated inputs*, Nature **332** 31 pp. 444–445, March 1988

[Hubel 79] David H. Hubel, *The visual cortex of normal and deprived monkeys*, American Scientist, **67** No 5 pp. 532–543, 1979

[Kohonen 88] Teuvo Kohonen, *Self-Organization and Associative Memory*, Springer Verlag Berlin, Heidelberg, New York, 1988

[Marr & Hildreth 80] David C. Marr and Ellen C. Hildreth, *Theory of Edge Detection*, Proc. R. Soc. London **207**, pp. 187–217, 1980

[Nehmzow & Smithers 90] Ulrich Nehmzow and Tim Smithers, *Mapbuilding using Self-Organising Networks*, in: Jean Arcady Meyer and Stewart Wilson (eds.), *From Animals to Animats*, Proceedings of SAB90 Paris, pp. 152–159, MIT Press Cambridge Mass. and London, England, 1991

[Nehmzow et al. 89] Ulrich Nehmzow, John Hallam and Tim Smithers, *Really Useful Robots*, in: T. Kanade, F.C.A. Groen and L.O. Hertzberger (eds.), *Intelligent Autonomous Systems*, Proceedings of IAS 2, pp. 284–293, ISBN 90-800410-1-7, Amsterdam 1989

Strategies for Learning Elementary Mobile Robot Operations

R. Dillmann

Abstract: Efficient robot application programming implies the use of task oriented programming techniques. A task description is mapped with the help of a symbolic planner onto a sequence of fault tolerant robot operations which are called elementary operations. Elementary operations represent dedicated classes of basic robot operations. However, geometric constraints, kinematic restrictions or physical process laws have to be considered for each individual task. Todays robot action planners are specialized to generate sequences of elementary operations fullfilling these constraints. Instead of generating and coding each elementary operation again according to the actual task, it can be adapted to various exceptions and generalized using machine learning techniques. Symbolic learning techniques allow to match memorized elementary operations with real tasks and support its generalization. Generalized plans and elementary operations can be applied to a class of similar tasks. Memorization and similarity search is based on a formal representation. Two learning methods based on an analog reasoning strategy, on explanantion based generalization and on induction are presented. They are applied to learn elementary operations and plans for travelling tasks and classes of assembly operations. The work reported is part of the SFB 314 (Artificial Intelligence)

1. Introduction

An autonomous mobile robot system which operates in a real environment implies various machine learning approaches. Machine learning strategies may enable a robot system to better planning of action, robot skill acquisition, use of sensors for cognition, reactive behaviour, world modelling, the acquisition of facts and process knowledge. Learning can be related to symbolic robot control levels or to subsymbolic layers of a robot system. On the symbolic control level autonomous mobile robot systems require various types of knowledge. According to the type of control architecture used, knowledge about planning, about plan execution and plan execution monitoring can be distinguished. Furthermore knowledge about handling of exceptions and knowledge about sensory perception is required to enable the system to decide how to behave and to react within an partly uncertain real world environment, see Fig.1. Modelling of all possible exceptions which may occur caused by an inaccurate world model, unforeseen events and uncertainties is cumbersome because of the extreme model complexity expansion. Because of this fact most of todays existing autonomous robot systems operate on simplified world models. Planning is performed with the help of heuristics and with geometric algorithms applied to abstract world models which are simplified approximations of the real world (Taylor 1976, Lozano-Perez 1980, Brooks 1982, Laugier 1988, Kampmann 1989 and others). If a plan is once generated it is to be interpreted and executed in real time and to be matched with the real environment of the robot. In case of unforeseen exceptions and failing of error recovery strategies replanning is necessary. Modelling of error recovery and robot behaviour is cumbersome if planning, cognition and interaction does not meet the complexity of the real world.This means that all possible process states and state transitions have to be modelled. This yields to models which are difficultly to handle, to adapt and to modify.

One alternative to overcome this problem is to memorize experiance from the past and to apply it to future robot tasks with the intension to do the task better. In other words machine learning mechanisms can be applied which allow training and consequent refinement of the knowledge of the robot system. The task of a robot machine learning component is to refine robot skills, to refine behaviour, to evaluate and to optimize the robot system performance, to refine the control laws, to modify the robot control system and to acquire world model data. With the help of machine learning algorithms knowledge about the task to be executed, facts, measurements, observations, symptoms, situation and goal related experience may be derived. For robot performance evaluation predictive robot states, expected sensory data and data about interaction with the environment have to be matched with real robot states and world states. This implies learning strategies which integrate task knowledge with real robot data and sensory data. For robot skill acquisition, machine learning strategies are of interest, which are able to match internal symbolic information about the world with information from the physical system and its real environment. Learning robot systems must be able to interprete success and failures of actions and to convert this information into more efficient control strategies. Thus, experience from the past can be used to reach a given task goal and to improve the robot performance. Memorization of experiance and its conversion into operational knowledge is a basic feature of machine learning systems.

An overview on general learning methods is given in the bibliography in Michalski, Carbonell and Mitchell, 1983, 1986. In the area of robot machine learning the principles of self organizing and self optimizing systems (Saridis,1977) are to be applied. It is assumed, that a self organizing system system has enough degrees of freedom to modify its own structure. For supporting the own optimal organisation of a robot control system artificial stimuli (training cycles) can be used generating system responses which are necessary for this type of feedback control. Self modifying systems based on neuronal networks have been introduced by Rosenblatt, 1957, with the definition of the perceptron. Connectionist control system are mainly applied to subsymbolic robot control problems. Other learning mechanisms are based on models of finite automata with stochastic approximation and estimation procedures (Fu,1970, Wee and Fu,1969, Zadeh,1965,1973). Inductive learning is the classical approach of symbolic machine learning where general descriptions are induced from given examples. The technique of learning from examples has been applied mainly in the area of concept learning for image description (Winston, 1975, Vere, 1975, Mitchell, 1982, Michalski, 1980, 1983), but also to planning of robot action sequences and sensory perception. According to the learning control mechanism learning algorithms supporting incremental learning, active learning (experimenting), data driven or background knowledge driven learning, empirical similarity based learning (SBL) or analytic explanation based learning (EBL) can be distinguished.

Some symbolic machine learning strategies have allready been applied to robot control problems, but with limited success. A learning strategy for robot program generation under uncertainties has been proposed by Dufay and Latombe, 1983. A two phase approach for constructing robot assembly programs has been realized. A training phase produces traces of task execution including acquired sensor data. In the second phase an induction algorithm based on graph manipulation mechanisms transforms these traces into an executable general robot program.

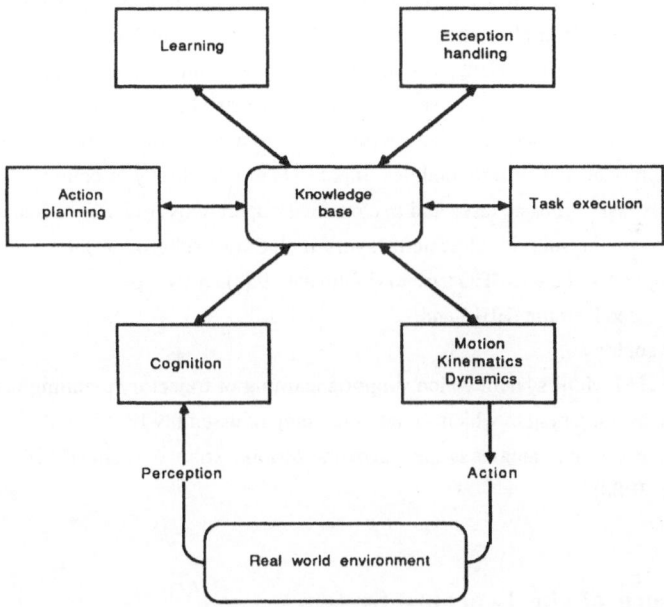

Figure1: Functional schema of an autonomous robot system

Sussman´s classical robot learning system HACKER (Sussman, 1975) is based on a symbolic robot simulator and suports a learning system with robot plan execution traces. This program consists of a symbolic action planner, a plan critisizer, a simulator, an execution tracer and a debugger which has to identify and locate recognized errors. Induction algorithms are applied to generalize detected errors with the purpose of learning generalized error classes. The generalized error classes are memorized and used by the planner and the critisizer to avoid in subsequent planning phases allready learned error types. Task level learning learning is proposed by Aboaf, 1988. Instead of executing a large number of trial motions and of extensive model calibration like adjusting all parameters of the kinematics, dynamics, actuation and sensing the task level command is refined and modified. The learning procedure compensates for structural modelling errors on the lower level component models of the robot. Learning from experiance at the task level is done by trial motions that actually attempt the task and which provide a concise method of achieving the task goal.

The task of robot plan generalization or elementary operation (EO) generalization is to match instances of examples to an initial plan or EO description and to generalize or to specialize the concept that it includes or excludes the examples within the concept. The achieved EO should be consistent and complete with respect to all examples. Inductive inference operators can manipulate EO representations by means of

- generalizing conditional expressions in EO descriptions
- turning constants into variables
- matching several EOs into one EO
- adding alternatives.

For the Karlsruhe Autonomous Robot System (KAMRO) a set of learning strategies is under developement. One strategy is implemented in the HALMOR system which supports planning of EOs for mobile platforms. It allows knowledge acquisition on the symbolic planning level with the help of explanation-based learning techniques. The second system is designed to generalize EO sequences with the help of analog reasoning techniques. The KAMRO system architecture (Rembold, 1988) is based on hierarchically interconnected system modules. It consists of a hierarchy of control modules which are related to complex tasks, to basic tasks and to elementary operations (EOs). The learning elements refer to the structure of the individual system layers including background knowledge.The basic learning strategies applied to the KAMRO system (Dillmann,1988) are of type

- Explanation based learning (EBL) and
- Learning by analogy.

In the following the HALMOR system which supports learning of trajectory planning is discribed. In addition the analog reasoning system which supports learning of assembly EOs is outlined. The testbed for evaluation of the learning strategies is the Karlsruhe Mobile Robot System (KAMRO, Rembold, 1988, Hörmann et.al 1988).

2. Basic Structure of the Learning System

A suitable model for a learning system is the simplified model proposed by Dietterich, 1982, which consists of a knowledge source, a learning element, a knowledge base and a performance element. The knowledge source may be a teacher, the robot target system itself or the system environment from where the required knowledge and experience will be derived. The learning element is an inference engine, which manipulates the knowledge base to extend or to refine existing knowledge, see Fig. 2. The performance engine evaluates prior knowledge and activates the learning element to continue knowledge acquisition from the environment till the learning goal (control law, behaviour, skills, task-knowledge or world-knowledge, concept) is reached. A task performance evaluator controls the learning cycle. In case of failure the learning element receives stimuli to refine, to modify, to enforce or to extend the robot control knowledge. Learning capabilities required for the KAMRO system are related to the EOs derived from the task sequence plan. The plan is assigned to agents and decomposed into terms of EOs and assigned to appropriate elementary operation modules (EOMs). The goal of the learning system is the generalization of EOs and the acquisition of control knowledge for the planners. The HALMOR system is associated to the KAMRO system with its hierarchical top down control system architecture (Hörmann, 1988). It supports knowledge acquisition for trajectory planning and obstacle mapping. Global path planning with HALMOR is done at the top driven by an abstract world model. The plan is decomposed through various levels down to the bottom of the system. The mobile

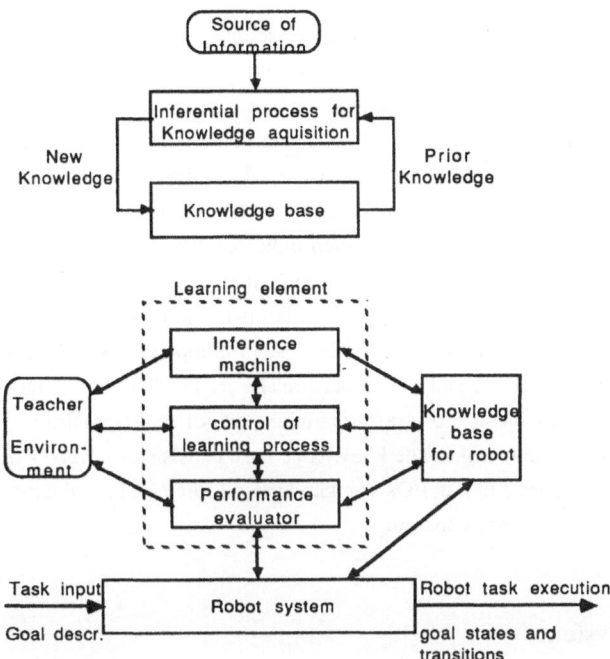

Figure 2: Basic structure of a robot machine learning system

platform is assigned to a specific travelling task. The path planner for the mobile platform consists of two major subparts: A floor-level planner and a detailed room - level planner. The workareas the robot has to serve are located in different rooms of the overall area. The central component of the path planner is the map of the environment. The path planner computes collision-free trajectories within each room based on a polygonal decomposition of the freespace. In addition it generates reference landmarks needed for navigation and obstacle information needed for sensor guided trajectory control.The real-time mobile platform control system needs this information to move along the computed trajectory. In case of non critical error situations it recovers the error locally in real-time. In general, local obstacles cause the activation of an obstacle avoidance algorithm which circumvent the obstacle and which enforces the continuation of the trajectory. If the obstacle can not be passed, the plan execution will be interrupted and the high level planner starts to generate a new alternative trajectory. The task of the HALMOR system is to learn macro trajectories to support path planning and the automatic acquisition of knowledge about current obstacle configurations while executing a computed plan. Macro trajectories may be task related. They depend on various classes of start and goal configurations of the mobile platform. Geometry related macro trajectories depend on geometric obstacle configurations inside the workspace.

The second learning methodology used for learning robot EOs is based on the learning by analogy paradigm as introduced by Carbonell /Carbonell 83/. It is applied to simple assembly tasks to be performed by KAMRO. Learning by analogy is related to the acquisition of new facts, concepts or rules by referencing and modifying existing knowledge. New knowledge is learned by searching similarities and differences between allready existing knowledge beeing proved to be useful to solve a given task or an actual problem which seems to be similar. Analog reasoning involves a basic concept of problem solving, a target concept and a premise. Analog mapping defines relations between the initial concept and the target concept. It is characterized by matching the base concept and the premise. This involves induction. Analogic mapping of unknown features of the target concept involves deduction. By this point of view, analog learning can be seen as inductive and deductive learning combined. The basic concept may be represented by a robot action plan in terms of EO sequences which perform a given task. The premise is given by a distance metric which indicates, that a second task EO sequence is similar to the initial one. The target concept is represented by the modified initial action plan (analogic plan) performing the second task. The next step of learning is to infer from both concepts a general schema representing a generalized plan. The basic system component is a learning element, a knowledge base, a similarity analyzer and a task state analyzer. The learning element supports the generalization of assembly task classes, of grasp operations and of transformation operators which adapt former task solutions to new tasks. The knowledge base consists of a library of task solution plans, of T-operators and of grasp and join EOs.The knowledge source is either the target system or a human tutor, who teaches the system by showing.

3. The HALMOR System

The HALMOR system developement follows two important goals. One is related to learning of macro-trajectories to support path planning (Spandl, 1991). The other is related to the acquisition of statistic knowledge about a-priori unknown obstacles while travelling along a computed path.
The learning goal for the geometric path planner is the generation of macro-trajectories to decrease the overall planning time. Especially often required standard travelling tasks can be learned and memorized. Macro-trajectories are part of a collision-free path which may be reused in the context of a future path planning task. Macro-trajectories may be either task-related or geometry related. A task related macro-trajectory depends on various sets of start and goal conditions. These conditions represent geometric similar configurations of the robot in its environment. Similar configurations are for example an often repeated transport task of a mobile platform between multiple workstations with slight deviations of its exact start and/or goal configuration. A geometry - related macro-trajectory depends on the various classes of obstacle configurations within a room. This may happen if two obstacles form a narrow critical area within a room.
The second learning goal is the acquisition of information about a-priori unknown obstacles detected during path execution. Such obstacles have to be classified and have to be entered into the world model for updating. This obstacle information has to be considered during path planning for a new travelling task. However no knowledge about previously computed macro-trajectories should get lost. The HALMOR system operates on three obstacle classes:
- Stationary obstacles
- Quasi stationary obstacles
- Dynamically moving obstacles
Dynamically moving objects and quasi stationary obstacles cannot be included immediately into the world model because they may change their position with time. The map has to be extended with

statitical coloured obstacle information. Thus, statistical information about the probability of obstacles may be usefull. The geometric path planner considers hypotheses on possible configurations of quasi stationary obstacles or dynamically moving obstacles with the help of acumulated statistical data (coloured map).

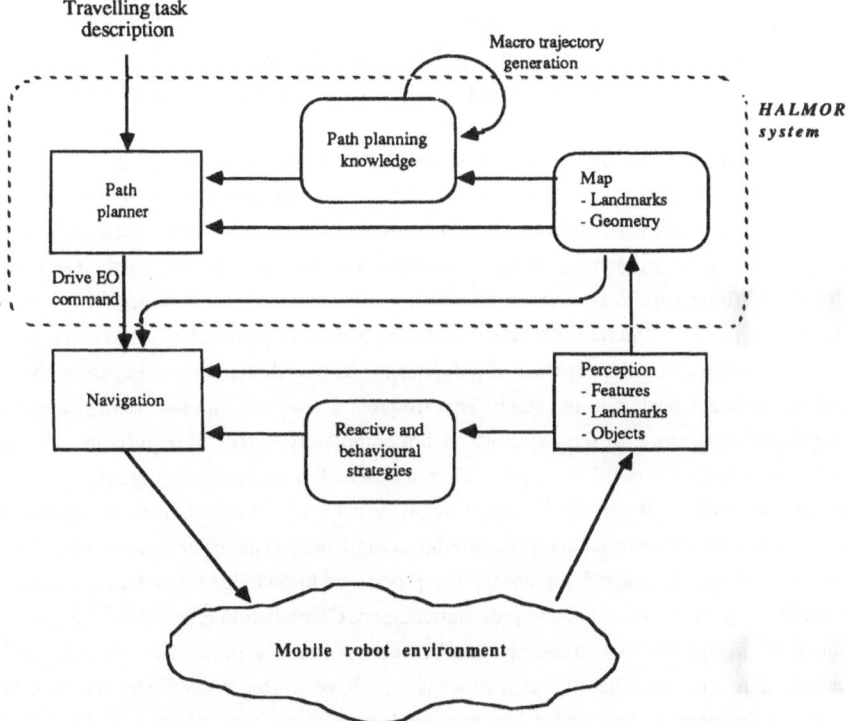

Figure 3: Basic structure of the HALMOR system. Path planning and navigation is based on the geometric map augmented by landmarks and macro trajectories. The map is updated by the sensor perception system of the vehicle.

The macro trajectory learning strategy of the HALMOR system follows systems which apply chunking to operators in state-based problem spaces like STRIPS (Fikes et al. 1972), SOAR (Laird et al. 1986) and PRODIGY (Minton et al.1989). Instead of an explicit operator based planning method developement, planning skill aquisition is performed in multiple steps with the help of a production system (Anderson 1983). The system is initialized with knowledge in declarative form.This knowledge can be interpreted by a set of general problem solving rules for a particular domain.Repeated use of this knowledge triggers a compilation of the declarative knowledge into a procedural form.New production rules are generated which already contain the declarative knowledge and yield to a faster application. Further training leads to a refinement of the acquired knowledge using generalization, discrimination and strengthening mechanisms.

For macro-trajectory learning HALMOR is incrementally constructing a virtual problem space with the help of available date from the path planning module. The learning component intercepts gross motion and fine motion planning. The declarative knowledge is represented by the acquired macro-trajectories. The procedural component consists of special control rules which solve a path planning task in the virtual problem space.The problem solver operates on the data from the gross motion planner (start and goal configuration, sequence of polygons to be traversed) and generates subtasks for fine motion planning. Learning is performed in two subsequent steps. The system memorizes efficiently polygon sequences and the corresponding computed trajectories. In addition, often used trajectories are compiled into the operational part of new rules. The convex polygons are represented by a connectivity matrix, which consists of rows and columns representing the polygon´s neighbours. Thus parts of a complete trajectory traversing freespace polygons can easily retrieved.

The internal structure of the basic path planner is not visible to the learning component. The virtual problem space generator for fine motion planning only access path planner data via an interface.Each fine motion planning cycle causes an extension or actualization of the virtual problem space. If problem solving in the learned virtual problem space fails, the original planner is to be activated. The central component of the learning system is a forward chaining rule interpreter. It receives all inputs and outputs of the motion planner. The rule interpreter maintains a stack of goals it has yet to solve. At the beginning of a new planning task the sequence of polygons together with the information about the start and goal configuration are pushedon the stack. This initiates a search in the knowledge base for a suitable usefull macro-trajectory. If the trajectory cannot completely fullfill all requirements the fine motion planner will be activated to fill in the gaps. Thus, a new macro-trajectory may result.

If one trajectory or parts of it are used frequently, it will be transformed from its declarative representation into procedural representation (knowledge compilation). The resulting new rules shorten the effort for search in the virtual problem space. The process of knowledge compilation consists of two distinct steps which are combination and proceduralization. Combination is possible whenever two or more productions are applied on a fixed sequence. The new production implements the effect of the sequence in one step. Proceduralization is a process which generates specialized versions of a production encoding declarative knowledge into new production rules. This allows to make repeated search for declarative knowledge unnecessary because the declarative information is directly available in the operational part of the new rule. The robot commands for the computed path are directly implanted into the right hand side of a production. The left hand side of the rule will match to the same goal and the same conditions the rule-interpreter accessed before in the declarative data base. With this type of knowledge compilation the speed of trajectory planning could be considerably shortened. Future work with the HALMOR system is related to learn sensory controlled trajectories with the help of obstacle features and landmarks which are part of the geometric map.

4 Learning Plans and Elementary Operations by Analogy

In the domain of robot based assembly a lot of similar assembly tasks can be found, which may be solved by direct application or modification of an allready existing similar plan generated in the past by a planner. We say that such tasks correspond to a common task class. Typical similarities in robot assembly problems are:

- Differences in parameters of involved objects (e.g. diameter of peg/hole, size proportions objects etc.)

- Differences in schema type of object classes (e.g. instead of a cylindrical peg a polyedra type peg,..)
- Differences concerning the world state; objects with different relations among themselves
- Differences concerning the goal state within the same assembly class
- Differences of the assembly class

If these differences are calculated with the help of a distance metric algorithm, a decision can be made if similarities are given. Similarities are are existing if the distance value is smaller than a given threshold value. In case of similarities between two assembly tasks the exact differences between them can be analyzed. With the result of the analysis the existing solution plan for the allready solved task can be adapted and modified to the new task. Carbonell introduced so called T-operators which transform the solution plan to the new problem. The following type of transformations can be performed:

- Parametric changes: Change of approach frame, gripper width, join position, compliant force
- Object specific changes: The task solution must be transformed to other object characteristics
- Strategic modifications: Insertion of additional operations like repositioning of overlapping workpieces, turning of objects into stable position.

If the system is not able to find a solution for a new task, the system tutor can demonstrate one possible solution. Thus, the system can generalize and/or specialize its knowledge by using learning from examples. There is no use of a large conventional planning system which is able to construct a solution. This would arise typical problems of planning systems like modelling completely of all exceptions, modelling of all object classes and its combination as well as the integration of multiple specialized planners like fine motion, gross motion, grasp and join operations planners. An automatic approach based on a small system kernel with basic problem solving capabilities is a more promising approach. The system we propose is shown in Fig.4. The system is based on a general frame based representation. The task description consists of an object to be joined with another object, a basis object, some environment objects, the robot and a goal description consisting of a set of relations which have to be fullfilled. Each object is an instance of a generic object class which is composed of elementary objects. Each object may be positioned and oriented individually by a frame.

The given task description is analyzed by the geometric analyzer, which adds current spatial relations between object features like parallel surfaces or contact of surfaces. The similarity analyzer matches the extended task description with the task classes in the plan library. In case of a positive match the result is a memorized solution sequence consisting of a number of states and a list of corresponding object features between the new task and the task class from the library. The solution is not a parametrized EO sequence such as GRASP, TRANSFER, JOIN, etc. but a sequence of state relations that have to be reached. This is usefull because it is easier to transform and adapt state sequences to a new task than to transform values of specific positions and orientations. The solution sequence found or one of them is to be transformed. The strategic transformer has access to three smaller transformation and planning modules which compute legal grasp positions, collision-free paths and possible joining motions. Thus, transformation can be hierarchically subdivided. A simulating and critisizing module observes the generation of the new solution and checks it for possible formal errors. If no errors are detected, the resulting sequence of EOs are transferred to the target robot system. In addition the task description and the resulting solution sequence are memorized and used as positive example for generalization by induction. If the EO sequence is not applicable another T-operator is searched which may transform the

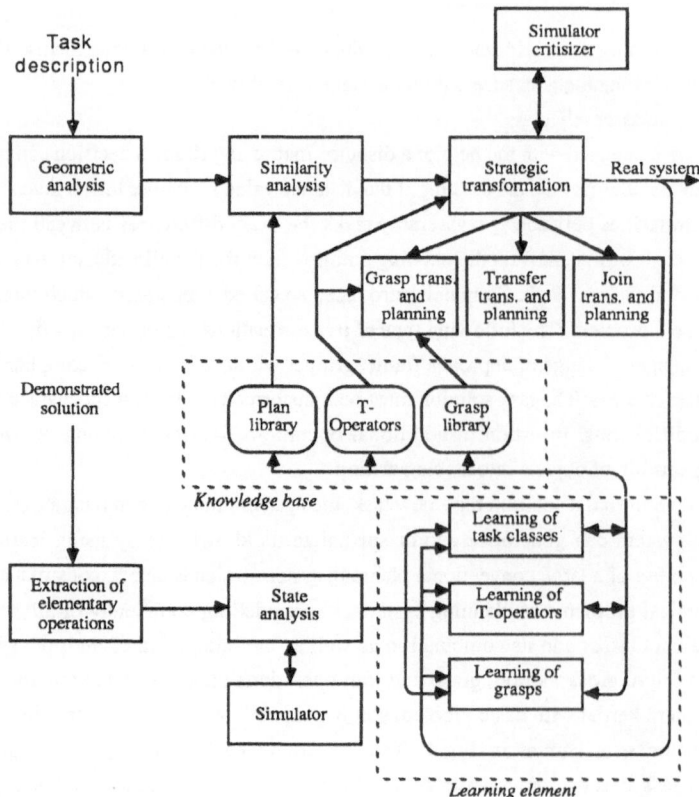

Figure 4: Structure of the analog problem solver with learning capabilities

failing plan into a correct one. If no T-operator can be found, the next possible analog plan solution is checked if there exists one. Otherwise the user can teach one solution by showing one correct example.The example is to be traced, analyzed and transformed into a sequence of elementary operations with parameters (reconstruction). The state analysis is supported by a simulation system to compute the states reached by the user´s training example. The given solution will then be compared with the task solution sequence that could not be transformed successfully. This comparison may result in a new T-operator that can be used in future problems of a similar type. Finally the solution will be transferred to the learning component. If the new task description can be parametrically or structurally distinguished from the found class description, this class will be specialized. The new task represents a new task class. Otherwise the two solution sequences are combined to form one solution sequence with conditional branches.

The type of knowledge to be learned is as indicated above of five types: 1. Task classes with solutions, 2. T-operators, 3. Grasp classes, 4.Transfer motions, 5. Joining strategies. The most important type is the first type. A task class is described by a set of generalized task descriptions. The problems of matching two tasks or a task class with a class are as follows:

- Determination of the correspondance of different environmental objects of the robot
- Generalization of the objects included in the task problem
- Determination of corresponding object features (surfaces, vertices, etc.)
- Generalization of the relations in initial and final states.

According to this type of problems it is not possible to simply substitute variables for constants. Alternatively similarities of attribute values have to be found and generalized in an attribute type way. Object types, type of subobjects and primitives have to be generalized according to the object and primitive hierarchies. Attribute values such as diameter, height, weight, etc. are seen as point intervals. These are then generalyzed according to a set of eighteen rules for generalizing sets of real intervals. Sets of relations used for describing a world state have to be generalized.

For simple blocks-world assemblies the system has been partly implemented and evaluated. Future work is related to the problem of memorizing sensor based EO sequences (no linear sequences) and the problem of calculating analogies.

5. Conclusion

To support autonomous system with learning capabilities additional learning components have to be added to the system. The task of these learning components is to acquire planning and control knowledge for the EOMs. The HALMOR system and an analog learning component for KAMRO have been discussed. Each method has its own strenghts. HALMOR is suitable for learning of efficient path planning and navigation. The second learning system approach follows the strategy of learning by analogy. This is a combined learning strategy which is based on both deduction and induction. The modelled learning algorithms are complex and difficultly to handle. Further work is required to make the algorithms more operational and to integrate them into the KAMRO system. The goal is to equip the KAMRO system wih self organizing capabilities which enables the system to refine its capabilities with time via subsequent training phases with the help of real world examples.

Acknowledgement:

This work is part of the KAMRO project sponsored by the Deutsche Forschungsgemeinschaft. This project represents AI-Robotics research in the framework of the Sonderforschungsbereich No. 314 "Künstliche Intelligenz". Part of the work is sponsored by the EEC ESPRIT - BRA program, Project No. BR 3275, B - LEARN.

References

Aboaf,E.W.: "Task-Level Robot Learning", MIT Technical Report no. 1079, 1988

Amari, S.: " Neural Theory of Association and Concept Formation", Biol. Cybern. 26, pp 175 - 185, 1977

Anderson,J.R.: "The Architecture of Cognition", Harvard University Press, 1983

Brooks,R.A.: "Solving the Find-Path Problem by Good Representation of Free Space", 2nd American Association for Artificiel Intelligence Conf. August 1982

Brooks,R.A.: "A Robust Layered Control System for a Mobile Robot", IEEE Journal of Robotics and Automation, RA-2, April 1986, 14-23

Brooks,R.A.: "A Robot That Walks; Emergent Behaviours from a Carefully Evolved Network", MIT A.I. Memo 1091, February 1989

Carbonell, J., G.: "Learning by Analogy: Formulating and Generalizing Plans from Past Experience", in Michalski, Carbonell, Mitchell (eds.): Machine Learning: An Artificial Intelligence Approach", Tioga Press, 1983

Dietterich,T.G. et. al.: "Learning and Inductive Inference", in Cohen,Feigenbaum (eds.): "Handbook of Artificial Intelligence, section 14", Stanford University, Stanford 1982

Dillmann,R.: "Machine Learning Strategies for Knowledge Aquisition in Autonomous Robot Systems", Proc. of the SYROCO 88, October 1988, Karlsruhe

Dufay, B., Latombe, J.,C.: " An Approach to Automatic Robot Programming Based on Inductive Learning", Int. Symp. on Robotics Research, Bretton Woods, 1983

Fikes,R.E., Hart,P.E.,Nilson,N.J.:"Learning and Execution of Generalized Robot Plans",AI,Vol.3 1972, p.251-288

Fu,K.S.: "Stochastic Automata as Models of Learning Systems", in Mendel,Fu (eds.), "Adaptive, Learning and Pattern Recognition Systems", Academic Press, 1970

Giralt,G., Chatilla,R.,Vaisset,M.: "An Integrated Navigation and Motion Control System for Autonomous Multisensory Mobile Robot", Proc. of Robotics Research, the first international symposium, 1984, p.191-214

Hinton, G. E., Anderson,J.A.: " Parallel Models of Associative Memory", Hillsdale, N.Y. Erlbaum, 1981

Hörmann,A.,Hugel,T.,Meier,W.: "A Concept for an Intelligent and Fault-Tolerant Robot System", Journal of Intelligent and Robotics Systems, Vol.1, 1988, p.259-286

Kampmann,P., Schmidt,G.: "Multilevel Motion Planning for Mobile Robots Based on a Topologically Structured World Model",Proc. of the IAS 2, Amsterdam, Dec.,1989. p. 241 - 252

Laird,J.E., Rosenbloom,P.S., Newell,A.: "Chunking in SOAR:The Anatomy of a General Learning Mechanism",Machine Learning,Vol.1,p.11 - 46,1986

Laugier, C.: "Planning orf Robot Motions in the SHARP System", Ravani,B., Eds.:" CAD Based Programming for Sensory Robots", Springer Verlag Berlin, Heidelberg 1988

Lozano-Perez,T.: "Automatic Planning of Manipulator Transfer Movements´, Memo 606, AI Lab. MIT, Dec. 1980

Michalski,R.S.: "Pattern Recognition as Rule-Guided Inductive Inference", IEEE Trans. on Pattern Analysis and Machine Intelligence, vol.2, pp.349-361, 1980

Michalski,R.S.: "A Theory and Methodology of Inductive Learning", Artificial Intelligence, vol.20, pp.111-161, 1983

Michalski,R.S., Carbonell,J.G., Mitchell,T.M.: "Machine Learning: An Artificial Intelligence Approach, Vol.1", Morgan Kaufmann Publishers, 1983

Michalski,R.S., Carbonell,J.G., Mitchell,T.M.: "Machine Learning: An Artificial Intelligence Approach, Vol.2", Morgan Kaufmann Publishers, 1986

Minton,S.,Carbonell,J.G.,Knoblock,C.A.,Kuoka,D.R.,Etzioni,O.,Gil,Y.: "Explanation-Based-Learning - A Problem Solving Perspective", Artificial Intelligence, Vol.40., p.63 - 118, 1989

Orlando,N.: "An Intelligent Robotics Control Scheme", proc. of the Conference on Automation Control, San Diego, 1984, p.204-208

Rembold;R.: "The Karlsruhe Autonomous Mobile Assembly Robot", Proc. of the IEEE Conference on Robotics and Automation, April 24 - 29, 1988, Philadelphia

Rosenblatt,F.: "The Perceptron: A Perceiving and Recognizing Automateon",Project PARA, Rep.No. 85-460-1, Cornell Aeronautical Laboratory,1957

Rumelhart, D.,E., McClelland, J.,L.,Eds.: "A general framework for Parallel Distributed Processing", MIT Press, 1986

Saridis,G.N.: "Self-organizing Control of Stochastic Systems",Marcel Dekker, 1977

Spandl,H., Pitschke,K.: "Learning of Macro-Trajectories for an Autonomous Mobile Robot", EWSL` 91, Porto, March 1991

Sussman,G.J.: "A Computer Model of Skill Acquisition",American Elsevier, 1975

Taylor,R.H.: "Synthesis of Manipulator Control Programs from Task-Level Specifications", Memo AIM 228, AI-Lab., Stanford, July 1976

Vere,S.A.: "Induction of Concepts in the Predicate Calculus", proc of the 4th IJCAI, Tbilisi,1975

Wee,W.G., Fu,K.S.: "A Formulation of Fuzzy Automata and its Application as a Model of Learning Systems", IEEE Trans. on Syst. Sc. and Cybernetics, Vol.5, p. 215 - 223, 1969

Zadeh,L.A.: "Fuzzy Sets", Inform. Control, Vol.8, p.338-355, 1965

Zadeh,L.A.: "Outline of a New Approach to the Analysis of Complex Systems and Decision Processes", IEEE Trans. on Systems, Man and Cybernetics, Vol.SMC-3, No.1, p.28-44, 1973

Applications

Safety Aspects for Autonomous Robot Systems

E. Freund, R. Mayr, F. Dierks, U. Judaschke, U. Kernebeck and B. Lammen

Abstract

A high-level control system called MACS (Multi Agent Control System) is presented. It is aimed to enhance the flexibility, autonomy, safety and fault tolerance of an integrated manufacturing system including several robot work cells and autonomous mobile robots. Special attention is directed to the safety aspects. A description is given of the hierarchical control structure of the whole system (including planning and task-scheduling), algorithms are presented for cooperative and autonomous collision avoidance, some results are shown concernig advanced vehicel control and finally the sensor task is dealt with.

Introduction

This paper deals with the design of a high-level control system called MACS (Multi Agent Control System) aimed to enhance the flexibility, autonomy, safety and fault tolerance of an integrated manufacturing system containing several robot work cells and mobile autonomous robots. Special attention will be directed to the safety aspect which is closely related to the aspect of efficiency: if there is no high-level control and/or safety instance, the proper and safe operation of the whole system can only be guaranteed by a strict but inflexible separation of the working areas of men and robots and by slowing down autonomous robots to act safely but ineffectively. MACS will be designed to overcome some of these drawbacks.

A main task of the Institute for Robotics Research (IRF) is the integration of already existing and rather sophisticated partial solutions worked out in several different projects during the last years. IRF-work so far is concentrated in the fields of multi-robot-systems (CIROS)[1], autonomous vehicles (PROMETHEUS)[2], multisensor integration (ROSI)[3], security systems (CARE)[4] and diagnosis (XDiRo)[5].

The CIROS-project includes the development and implementation of algorithms for collision avoidance, coordinated control of robots, automated planning, task scheduling and automated error

---

[1] (Control of Intelligent RObots in Space)

[2] (PROgraM of an European Traffic with Highest Efficiency and Unprecedented Safety)

[3] (RObot Sensor Integration)

[4] (Carefull Activity of Robots in their Environement)

[5] (EXpert System for the Diagnosis of Robots)

recovery [2, 4, 11, 7, 10]. The PROMETHEUS-project focusses on autonomous and coordinated collision avoidance [6, 12] as well as on advanced vehicle control [12, 13, 15]. The ROSI-project concentrates on the design of high-dynamic distance/orientation control loops using laser distance sensors and the amalgamation of position/orientation and force/torque control loops [20, 21, 23, 24, 17, 18, 9, 8]. In the course of the CARE-project [1] sensors are employied to detect obstacles in machine environment, situation analysis based on the sensor information is done and strategies for intelligent collision avoidance between men and robots respectively mobile systems are worked out. Finally the IRF is with the XDiRo-project engaged in the field of diagnosis [19] where a knowledge-based real-time fault detection system for robots is implemented using limit and trend checking, analysis of the signals frequency spectrum and observer algorithms.

Future task will be to integrate these parts to form the MACS-szenario like this:

- several autonomous multi-robot work stations are linked togethter by use of autonomous mobile robots,
- there is a hierarchical structure of planners and task-schedulers which support task-oriented programming (MACS);   the multi-robot work stations and the mobile robots are to act as autonomous as possible,
- the autonomous multi-robot work stations use sensors to do their tasks (e.g. welding, grinding, assembly etc.) as do the mobile systems (e.g. environment identification, navigation, docking, machine loading and deloading, tool replacement etc.),
- the whole system is designed to tolerate the presence of men even during operation.

In this paper a description will be given of the hierarchical control structure of the whole system (including planning and task-scheduling) and of the cooperative and autonomous collision avoidance strategies. In addition some ideas concerning advanced vehicle control will be presented and finally the sensor task will be dealt with.

Relevant Categories of Agents in Modern Production Plants

The agents in modern production plants can be subdivided in three relevant categories with respect to their technical capabilities and their possibilities of intervention. Fig. 1 schematically describes how the agents in modern production plants can be subdivided in three categories and lists the marks of categorizing. Obstacles like low-level material-handling systems and human operators form a group of agents which cannot be controlled by the central control unit. In the next group stationary systems like manufacturing units, assembly stations and material storage systems are combined to a category.

On the contrary, autonomous mobile robots are able to change their operating position. This additional degree of freedom increases the flexibility of manufacturing but also requires a lot of necessary equipment for controlling the working spaces of each autonomous mobile robot. Therefore the sequences of motion are coordinated by the central control unit and each autonomous mobile robot is equipped with an sensor based on-line collision avoidance system. Furthermore stationary devices observe the execution of tasks.

Categories of Agents

Category	Marks of Categorization
1	autonomous mobile robots
2	stationary systems
3	obstacles which are not controlled by the central control unit

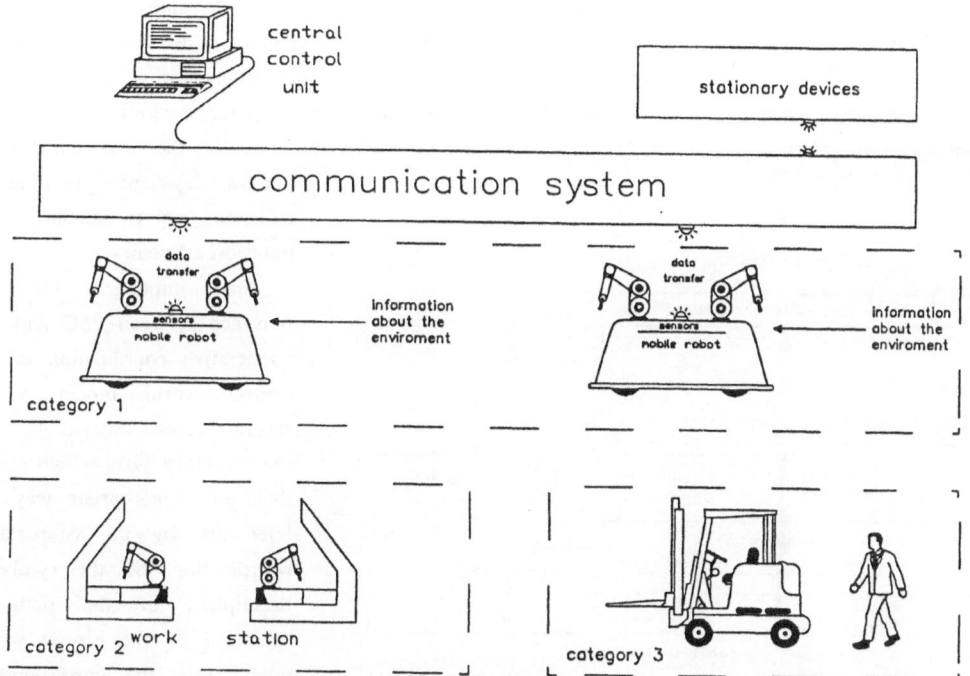

Fig.1: Categories of agents

Hierarchical Structure of the System

In the previous chapter different categories of agents, mobile and stationary subsystems, have been introduced which may work together in a large application. Therefore different agents must be coordinated und supervised so that a high degree of flexibility and safety can be achieved for the whole system. To keep track on the things that happen the user who has to control a multi agent application needs tools which provide the necessary information. But these information must be in an abstract form so that the user is not confronted with all details of the tasks and there execution. In this way he has the possiblity only to think about the tasks and not about their execution. The control system itself takes care of the correct task execution even if a failure occurs. In this case it is able to solve the problem by replanning and eventually by use of redundancy.

To meet the requirements mentioned above the control system for multi agent enviroments called MACS consists of a central control unit and several agents connected by the communication system, as shown in Fig. 2. On top of the control system the user describes the tasks of the system with the help of the Men Machine Interface (MMI). The planning and scheduling component (PSC) which is used for task orientated programming, execution supervision and automated error recovery gets the task description by the MMI and determines the agents for the task, the devices and the task dependent paths. Thereby elementary instructions for movements of the mobile robot, for movements of the robot arms or for endeffector handling are generated and tranfered to the cooperative coordination component including the information about the temporal succession. With these instructions the cooperative coordination gets the symbolic information about the trajectories of the agents. Its task is to coordinate the movements of the mobile systems to avoid collisions and to maximize the transport efficiency.

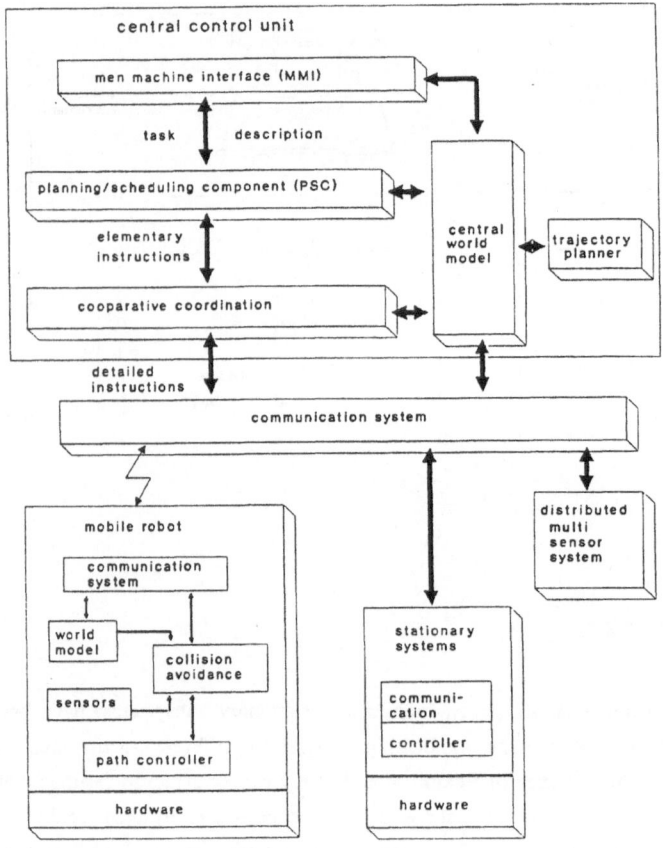

Very important for the interface between PSC and the cooperative coordination is the central world model which receives planed trajectories from the trajectory planner and stores them in an apropriate way for later use by the cooperative coordination. So the symbolic description of the path is compiled by the central world model into the corresponding numeric trajectory description which is used by the cooperative coordination. On the other hand the central world model also gathers and dispatches information according to the environment and to the tasks so that informations from the robots or sensors become available to the components of the central control unit. Moreover the central world model processes the data so that they become available in different kinds of

Fig. 2: Hierachical structure of the multi agent control system (MACS).

abstraction. For example the position of a mobile robot is not only represented as a numeric frame but also as a symbolic expression like "in front of the milling station 4". The trajectory planning component is divided into two parts. One part generates the trajectoriers for the robot arm considering collision avoidance. The other part is responsible for the path of the mobile subsystem concerning collision avoidance with other stationary systems. With the help of the central world model and the trajectory planner the cooperative coordination generates detailed instructions which are dispatched by the communiction system. When the instructions are executed the planning and scheduling component supervises the execution so that an automatic error recovery can be initiated.

A similar structure to MACS can be found in intelligent stationary systems. For example the same structure as described above is realized in the CIROS project [2], which concerns applications of multi robot systems in space.

Planning and Scheduling Component

The planning and scheduling component (PSC) in the central control unit is responsible for the task management and the resource allocation, so that the agents work with a maximum of productivity. Therefore the PSC should coordinate the agents according to the actual tasks and their availability. Especially the automatic error recorvery demands the capability of the PSC to handle interrupts. These interrupts may result from the user for solving an new and more important task or it may result from an agent which needs an error recovery. A task consists of a conjunction of goals each of them describing one property of an object. For example a goal could be the position to which a printed circuit board has to be moved. In this case the single goal describes the whole task, while the position is a property of the object represented by the printed circuit board. The PSC's task is to determine the agent which is to get the printed ciruit board, the gripper used for this task, the place where the printed circuit board actually is located and the path of the mobile robot and the robot arm. The path is build up according to the demands of the task by coordinating different symbolic path elements.

The output of the PSC consists of elementary instructions for the movement of the mobile robot and the robot arm coordinated with endeffector instructions. The PSC increases the fault tolerance of the whole system with the ablity of automated error recovery. To handle errors it is important to get information about the reason why the error occures which is realized by a stepwise classification of the error. Classes of errors are:

unforseen changes

Something happend which wasn't known when the task was planed but the system is still able to operate. For example gripper slip or closing of a part buffer which has to be open.

breakdown

The execution was not possible because a breakdown of an agent or an other object occured or may occure in the next time. For example breakdown of a robot because the motor current exceeds.

<u>exception</u>

The execution was principally possible but in this special case it doesn't work. For example jammed parts.

According to the different classes of errors which may occur, the automated error recovery is different. In the case of unforeseen changes a replanning will solve the problem if the actual state of the system is updated. If a breakdown is ascertained a repair may be possible but for a quick solution a redundant system could be brought into action by the PSC for executing the remaining part of the task. In contrast to this an object out of tolerances causes an exception. In this case the defective object has to be exchanged.

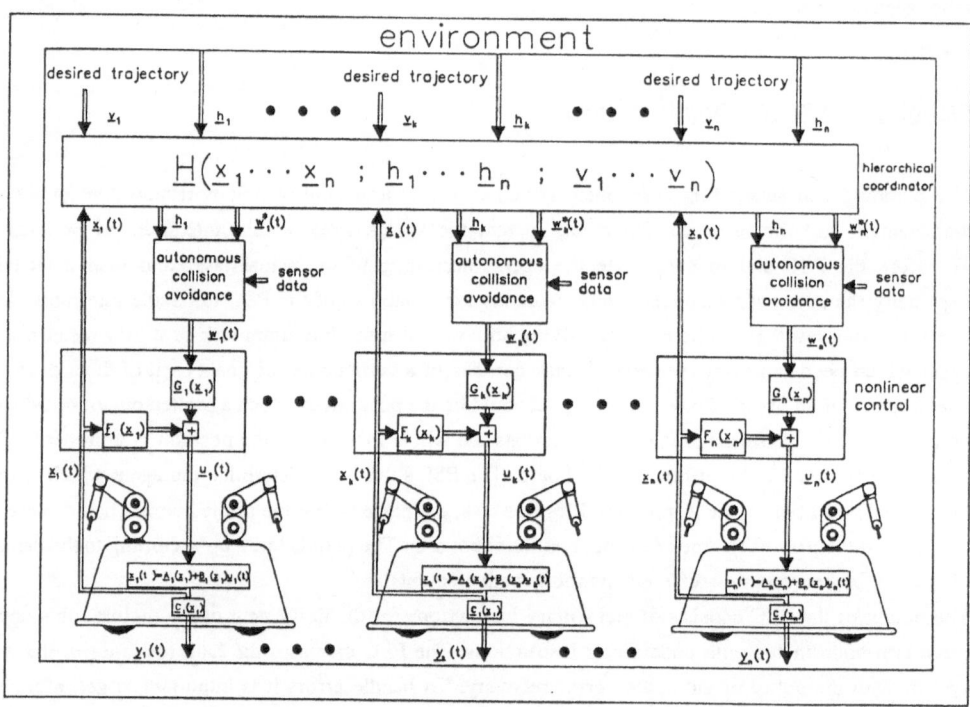

Fig. 3: Principal structure of the cooperative coordination

Strategies for Cooperative Coordination of Motion

Based on the central control's capabilities sequences of motions of each mobile robot can be improved by means of cooperatively coordinated manoeuveres between autonomous mobile robots. Therefore the central control system obtains a module for cooperative coordination of motions. This module analyses the specific situation with respect to the different tasks of transportation. Considering elementary instructions for movement desired trajectories of motion are calculated and compared. As a result characteristic parameters of the situation are generated and a danger of collision can be detected. In order to increase safety and efficiency of transportation the desired trajectories of autonomous mobile robots are cooperatively coordinated referencing technical, physical and environmental constrains. Therefore the reference inputs $\underline{w}_1^*(t), ..., \underline{w}_n^*(t)$ of the n autonomous mobile robots are coordinated by a hierarchial coordinator of the type

$$
\begin{bmatrix} \underline{w}_1^*(t) \\ \vdots \\ \underline{w}_n^*(t) \end{bmatrix} = \underline{H}(\underline{x}_1, ...,\underline{x}_n; \underline{h}_1, ..., \underline{h}_n; \underline{v}_1, ..., \underline{v}_n) \tag{1}
$$

where $\underline{v}_1, ..., \underline{v}_n$ represent the desired paths based on elementary instructions for movement of the autonomous mobile robots. Physical and environmental constraints such as lane markings are included in $\underline{h}_1, ..., \underline{h}_n$, while $\underline{x}_1, ...,\underline{x}_n$ represents the state vectors of each autonomous mobile robot. The principal structure of the hierarchial coordinator is shown in Fig. 3.

For further explanation the hierarchical coordinator can be subdivided as follows [6, 12]:

$$
\underline{H}(\underline{x}_1, ...,\underline{x}_n; \underline{h}_1, ..., \underline{h}_n; \underline{v}_1, ..., \underline{v}_n) =
$$

$$
\underline{H}^a \begin{bmatrix} \underline{x}_1(t) \\ \vdots \\ \underline{x}_n(t) \end{bmatrix} + \underline{H}^b(\underline{x}_1, ...,\underline{x}_n; \underline{h}_1, ..., \underline{h}_n; \underline{v}_1, ..., \underline{v}_n) + \underline{E} \begin{bmatrix} \underline{v}_1(t) \\ \vdots \\ \underline{v}_n(t) \end{bmatrix} \tag{2}
$$

Thereby the control dynamics of the autonomous mobile system can be adjusted by $\underline{H}^a$, while $\underline{H}^b$ admits the possibility of cooperatively coordinated sequences of motion between the autonomous mobile robots. The corresponding input gains of the desired trajectories based on the taks of transportation $\underline{v}_1,... ,\underline{v}_n$ are chosen by the diagonal matrix $\underline{E}$. The diagonal matrix $\underline{H}^a$ is set up as follows

$$
\underline{H}^a = \begin{bmatrix} \underline{H}_1^a & \cdots & \underline{0} \\ \vdots & \ddots & \vdots \\ \underline{0} & \cdots & \underline{H}_n^a \end{bmatrix} \tag{3}
$$

with the sub-matrix $\underline{H}_k^a$ for the autonomous robot k

$$
\underline{H}_k^a = \begin{bmatrix} \underline{H}_{k,1}^a & \underline{0} & \underline{0} \\ \underline{0} & \underline{H}_{k,2}^a & \underline{0} \\ \underline{0} & \underline{0} & \underline{H}_{k,3}^a \end{bmatrix} \quad \text{with } k = 1, 2,... , n \tag{4}
$$

and

$$H^a_{k,i} = \begin{bmatrix} \dfrac{\bar{\alpha}^0_{k,i}}{\lambda_{k,i}} & \dfrac{\bar{\alpha}^1_{k,i}}{\lambda_{k,i}} \end{bmatrix} \qquad \text{for } k = 1, 2,\dots, n \text{ and } i = 1, 2, 3, \tag{5}$$

where the parameters $\bar{\alpha}^0_{k,i}$ and $\bar{\alpha}^1_{k,i}$ can be choosen for each degree of freedom (x-y-position and yaw angle). The (n x 1)-matrix $\underline{H}_b$ has in general the nonlinear form

$$\underline{H}^b = \begin{bmatrix} \underline{H}^b_1 (\underline{x}_2,\dots,\underline{x}_n;\, \underline{h}_1,\dots,\underline{h}_n;\, \underline{v}_1,\dots,\underline{v}_n) \\ \vdots \\ \underline{H}^b_n (\underline{x}_1,\dots,\underline{x}_{n-1};\, \underline{h}_1,\dots,\underline{h}_n;\, \underline{v}_1,\dots,\underline{v}_n) \end{bmatrix} \tag{6}$$

with $\underline{H}^b_k(\underline{x}_1,\dots,\underline{x}_{k-1},\underline{x}_{k+1},\dots,\underline{x}_n;\, \underline{h}_1,\dots,\underline{h}_n,\underline{v}_1,\dots,\underline{v}_n)$ as (3 x 1)-matrices for $k = 1,\dots$ s. $\underline{H}^b_k$ does not contain the state vector $\underline{x}_k(t)$ of the autonomous mobile robot k as this would cause undesired feedback couplings in the dynamics of the considered autonomous mobile robot.

The input amplifying matrix $\underline{E}$ is correspondingly structured to

$$\underline{E} = \begin{bmatrix} \underline{E}_1 & \cdots & \underline{0} \\ \vdots & \ddots & \vdots \\ \underline{0} & \cdots & \underline{E}_n \end{bmatrix} \tag{7}$$

with the sub-matrix $\underline{E}_k$ for the autonomous mobile robot k

$$\underline{E}_k = \begin{bmatrix} \dfrac{\bar{\lambda}_{k,1}}{\lambda_{k,1}} & \underline{0} & \underline{0} \\ \underline{0} & \dfrac{\bar{\lambda}_{k,2}}{\lambda_{k,2}} & \underline{0} \\ \underline{0} & \underline{0} & \dfrac{\bar{\lambda}_{k,3}}{\lambda_{k,3}} \end{bmatrix} \qquad \text{with } k = 1, 2, \dots, n. \tag{8}$$

In (8) the new input gains for the x-y-position and the yaw angle are determined by $\bar{\lambda}_{k,i}$ with $i = 1, 2, 3$.

Autonomous Collision Avoidance

Planning and coordination of the courses of the autonomous mobile robots is accomplished by the modules of the higher hierarchical levels in the overall system. The individual course of each particular robot is determined for a complete task or at least for the part which is currently executed. All obstacles and operational constraints which are known in advance can be taken into account. The trajectories of the robots are calculated based on the information provided by the world model and by the instructions of the planning and scheduling component with respect to the courses of other autonomous mobile robots in the proximity. However, suddenly appearing unexpected obstacles cannot be ruled out in a real industrial environment. Therefore the execution of the task has to be supervised during the journey by sensor based on-line algorithms which are able to recognize imminent collisions and to initiate autonomous countermeasures if necessary. Several sensor technologies, which can be applied here to detect and locate obstacles, will be discussed later on in a separate chapter. For similar problems in the automation of vehicles in a road traffic environment autonomous collision avoidance strategies have been developed recently [13, 14]. The results of this work can be transmitted to autonomous mobile robots in a factory environment. The schematic structure of the collision avoidance module is shown in Fig. 4. The module receives its input data about the own desired trajectory $\underline{w}_k^*$ and the course of the obstacle $\underline{x}_{ob}$ from the cooperativ coordination module and from the on-board sensor system, respectively. Considering the current values of the obstacle course $\underline{x}_{ob}$ and of the own desired trajectory $\underline{w}_k^*$ the collision avoidance permanently anticipates the ensuing change of the distance between the obstacle and the vehicle. In the submodule "collision parameter anticipation" a characteristic set of parameters is calculated which enables the subsequent "collision detection" submodule to recognize imminent collisions. In case of a collision danger the original desired trajectory of the vehicle finally is modified to a collision free path. This can be achieved by a change of the desired speed of the vehicle or by a lateral shift of the desired position of the vehicle while it is approaching the obstacle. The modified trajectory $\underline{w}_k$ is calculated with respect to technical, environmental and physical constraints. Of course the collision avoidance algorithms have to take sudden changes in the motion of the obstacles into account. If the obstacle reacts in an unexpected way, e.g. by changing the direction or velocity of its motion, the collision avoidance has to modify its own manoeuvre in an adequate way. To illustrate this, the following simulations in Fig. 5 have been carried out. On the left (a, b) the execution of the first collision avoidance manoeuvre is shown. The vehicle is approaching from the left hand when a slowly moving obstacle

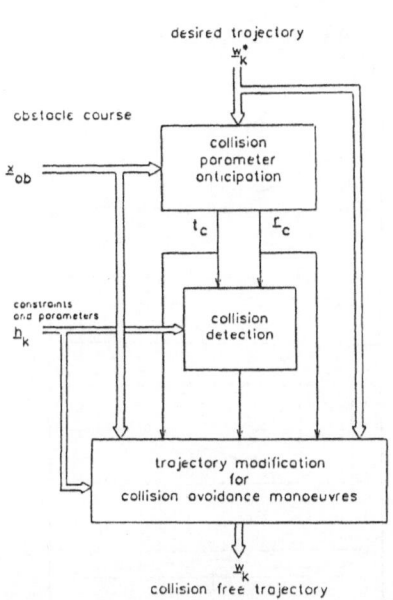

Fig. 4: Structure of the collision avoidance module

appears and crosses the road in front of the vehicle. The position of the obstacle is marked with a circle. In the upper half of each graph the velocity of the vehicle is given. The obstacle is moving ahead without paying attention to the vehicle. The collision avoidance recognizes the imminent collision and initiates a breaking manoeuvre. There's no need for a lateral deviation from the original course. As soon as the vehicle has passed the obstacle it is automatically accelerated again to its original desired velocity. On the right (c, d) a similar situation is shown. First the obstacle moves straight ahead again. Corresponding to the first manoeuvre the vehicle decelerates again to avoid the collision. The obstacle suddenly changes the direction of its movement. During the collision avoidance manoeuvre the countermeasures are permantly revised according to the current sensor data. In this simulation the worst case is assumed, i.e. the obstacle turns in direction to the vehicle. As soon as the new situation is realized, the collision avoidance adapts its manoeuvre. Its no more possible to avoide the collision by a plain braking maneouvre. Therefore the desired position of the vehicle is now shifted to the left to make way for the obstacle. After having passed the obstacle the vehicle returns to the original course.

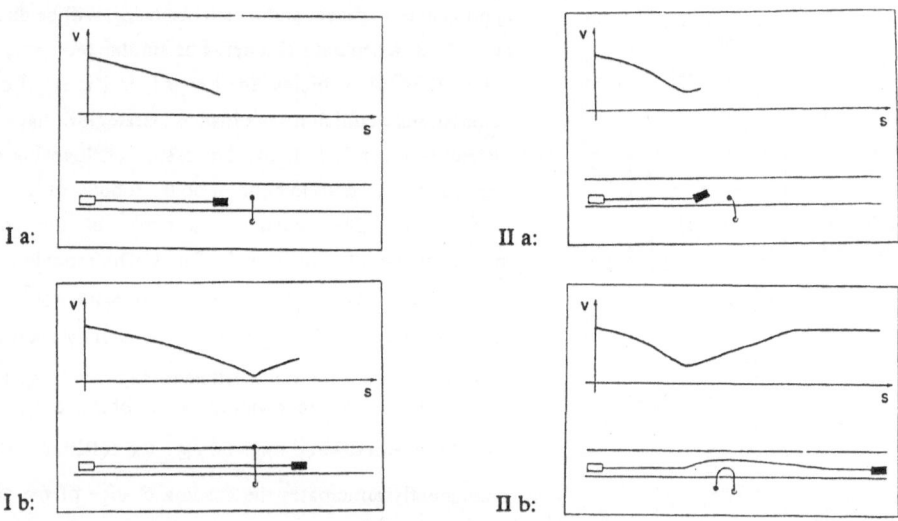

Fig. 5: Collision avoidance manoeuvres, I (a, b) and II (c,d)

The Level of Path Control

The task of the level of path control is to stabilize the mobile robot on the collision free desired course as well as to control the dynamical behavior of the yaw angle. So the signals for the drive and the steering system of every wheel have to be generated by the control system. These control signals depend on the collision free nominal trajectory, the desired yaw angle and the variables measured by sensors mounted on the mobile system. In order to get the position and other important state variables these signals must be processed by mathematical algorithms. In a second step of support the coordinates of the position of the centre of gravity and the yaw angle can be determined by an inertial navigation system. In this way

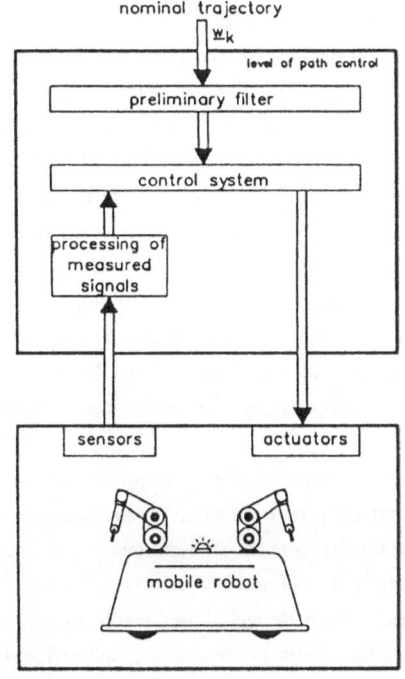

nominal trajectory
$\underline{w}_k$
level of path control

preliminary filter

control system

processing of measured signals

sensors   actuators

mobile robot

Fig. 6: The level of path control

the controllers influence the actuators for the steering system and the drive of every wheel so that the mobile robot will follow automatically its collision free desired path with respect to the desired yaw angle. As the equations of movement of the mobile robot contain extreme nonlinearities, the control strategies are developed by the method of nonlinear decoupling and control. Thereby all kinds of detrimental couplings between the lateral and the longitudinal dynamics are compensated. Based on this method which is described in [4, 15] the laws of nonlinear control are of the form

$$\underline{u}(t) = \underline{D}^*(\underline{x})^{-1} \{ -\underline{C}^*(\underline{x}) - \Delta \underline{w}(t) - \underline{M}^*(\underline{x}) \}$$

with the vectors $\underline{u}(t)$, $\underline{w}(t)$ and $\underline{x}(t)$ containing the control signals, the nominal trajectories and the state variables. By the matrices $\underline{D}^*(\underline{x})^{-1}$ and $\underline{C}^*(\underline{x})$ the detrimental couplings are compensated while the pole placement of the remaining linear system of the controlled mobile robot is implemented by the diagonal matrix Δ and by $\underline{M}^*(\underline{x})$. For compensating the deviation between the desired and the actual position of the centre of gravity of the mobile system, which represents the control error, a preliminary filter is added to the control system. A mobile robot equipped with the described system is insensitive to disturbances and follows its desired trajectory with high accuracy. Fig. 6 shows the structure of the level of path control.

The Sensing Task

This chapter deals with the problem of sensors and sensor data processing in autonomous mobile robots. Sensor systems in autonomous mobile robots are to do tasks like position estimation (i.e. navigation), obstacle detection and location as well as environment identification. First the sensor aspect of the navigation task will be discussed; second the sensor aspect of the security task will be described and at last the structure of a sensor based security system will be analysed [1].

a) Sensors for Navigation

There are a couple of sensor systems to find out an autonomous robots position (see fig. 7). One can distinguish between predictor techniques, tracking techniques, landmark techniques and path/landmark independant techniques.

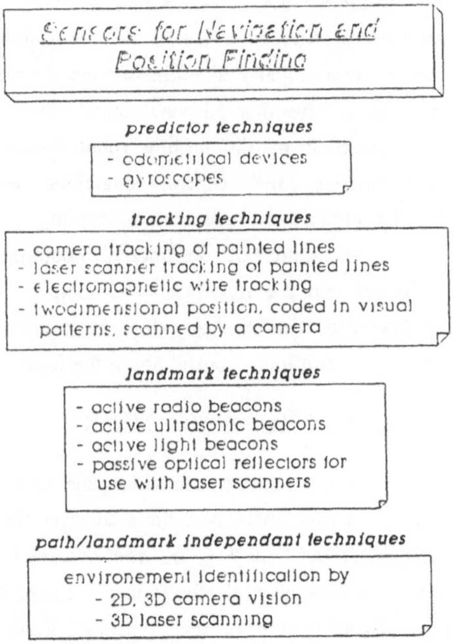

Sensors for Navigation and
Position Finding

predictor techniques

- odomelrical devices
- gyroscopes

tracking techniques

- camera tracking of painted lines
- laser scanner tracking of painted lines
- electromagnetic wire tracking
- twodimensional position, coded in visual
  patterns, scanned by a camera

landmark techniques

- active radio beacons
- active ultrasonic beacons
- active light beacons
- passive optical reflectors for
  use with laser scanners

path/landmark independant techniques

environement identification by
- 2D, 3D camera vision
- 3D laser scanning

Fig. 7: Sensors for navigation and position finding

Often a combination of predictor, line tracking and landmark techniques is used for the navigation task. With a tracking sensor (e.g. a laser scanner tracking a painted line on the floor) the autonomous system is guided along a floor fixed path containing only a certain number of switches. Incremental sensors (e.g. encoders for the wheel's positions) feed a position predictor to estimate the autonomous robot's position with respect to the guideline. Finaly a landmark sensor (e.g. a wall-fixed reflector with a certain barcode mark on it read by a laser scanner mounted on the robot) is used to intermittingly calibrate the position predictor.

Such a combination of sensors provides a rather robust and easy to implement navigation system exept for the fact that the autonomous mobile robot is of course path-bound.

For the landmark techniques active or passive beacons like radio-, ultrasonic-, active/passive light-beacons can be employied as landmarks. The autonomous mobile robot takes bearing to find out its position. Other interesting landmarks are pattern tapestries on the ceiling or the floor, absolutely or relatively coding the position and optically detected by a camera [25].

The landmark techniques are relative robust; their advantage over the line tracking techniques is the two-dimensional path-independant position information. Their disadvantage is the need of placing landmarks so that at least two of them can be seen from every point of the environment to enable cross bearing; this can cause prohobitive costs. All these problems are overcome by using the environment as "landmark" itself. Employing a laser scanner or an 3D-vision system together with an internal representation of the environment's shape makes it possible to find out an autonomous robots position.

b) Sensors and Data Processing for the Security Task

Fig. 8 shows the three essential elements of a sensor based security system: *perception, situation analysis* and *reaction*. The concept of "perception" has been extended to include a priori knowledge about the "percepting" robot itself (e.g. kinetics), a priori knowledge about the percepted environment (e.g. endangered surroundings) and knowledge about the desired task the mobile robot is supposed to do. The information coming from the perception modules has to be processed to perform a situation anaysis. This includes a risk prediction, an obstacle identification and at last a post-periculum analysis for the security system to learn from bygone emergencies. When the situation analysis modules have detected a risk, the reaction modules are braught into action. If an obstacle has been detected a warning is given; if it is possible, the security system will try to dodge the obstacle by intervening in the path planning modules; if no time is left to react smoothly an emergency stop has to be done immediately by intervening in the power modules. This methods include the previously described strategies of cooperative and autonomous collision avoidance.

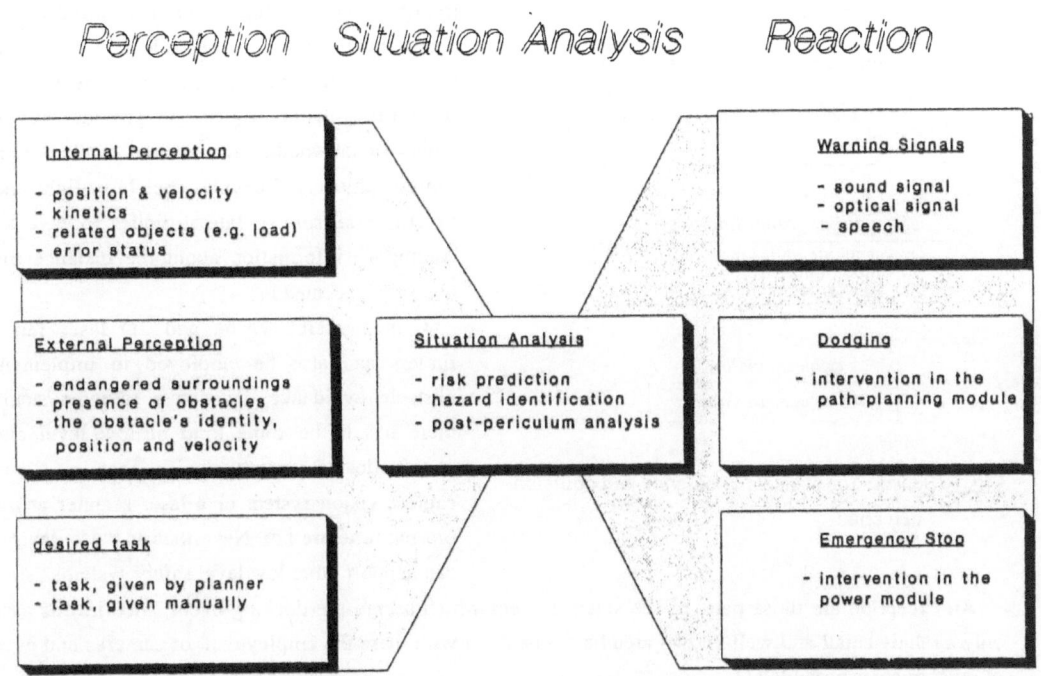

Fig. 8: Elemets of a sensor-based safety system

The better and the more reliable the sensory and the data processing algorithms are the faster an autonomous robot can be allowed to move and the more effectiv the whole system becomes. Special attention has to be drawn to the docking event when the danger of squeezing arises.

The main challenge is to find sensors, that have a suitable measuring range, resolution, respose time and recognition reliability to fit the high standards set by the legislative for safety systems [3, 22].

Some widely used sensor system are listed in fig. 9. Almost every autonomous robot has some kind of mechanical bumpers or whiskers. These sensors are cheap and reliable but can only detect obstacles in relative small surroundings. Since the stopping distance of the mobile robot has to be smaller than the obstacle detection range of the sensor, mobile robots equipped only with mechanical sensors can only travel at rather small speeds.

To overcome this, one can employ some sort of rangefinder, either acoustical (ultrasonics) or optical.

Ultrasonic rangefinders are rather cheap; their major drawbacks are their poor lateral resolution and the poor reflection characteristics of some potential obstacles (e.g. all kinds of coat). To overcome the

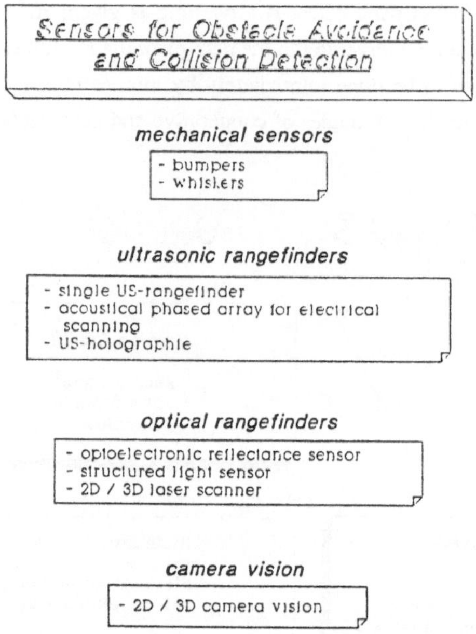

Fig. 9: Sensors for obstacle avoidance and collision detection

poor lateral resolution one can use an array of ultrasonic transducers to improve the directional pattern of the whole device or use holographic techniques.

Optoelectronic reflectance sensors use triangulation techniques to get information about the distance, an obstacle is located in. Simple reflectance sensors have mechanical fixed triangulation angles and provide only a binary signal whether there is an obstacle or not. More sophisticated devices user laser-light and CCD-line sensors or lateral-diodes to give an continuous information about the distance the obstacle is located in.

Camera stereo vision and 3D laser range finders can also be employed to implement obstacle avoidance; but for a safety systeme there has to be some kind of low level and uncomplicated sensor device that a stereo camera vision system or a laser scanner at the present time are not. Nevertheless these devices can support other low level safety systems.

An exception are those parts of the safety system which secure the docking event since in this case only a relativ small and well known area has to be dealt with; here the employment of cameras and even of sensormats is possible [1].

Conclusion

In this paper a description of an Multi Agent Control System (MACS) is given, aimed to enhance the flexibility, autonomy, safety and fault tolerance of an integrated manufacturing system containing several robot work cells and autonomous mobile robots.

MASC integrates several projects the Institute for Robotics Research (IRF) has performed during the last years, mainly CIROS (multi-robot-systems), PROMETHEUS (autonomous vehicles), ROSI (multisensor integration), CARE (security systems) and XDiRo (diagnosis).

The autonomous robots having their own world model, path planner, collision aviodance module and sensorium are lead by an automated task planner and scheduler employing a cooperative coordination module. The automated task planner also manages the cooperation between the mobile robots and the stationary multi robot systems supported by a distributed sensor system.

The relevant categories of agents in modern production plants are analysed and a detailed description is given of the overall system's hierarchical structure, the planning and scheduling component, the strategies for cooperative coordination of motion, the autonomous collision avoidance, the level of direct vehicle control and finally a survey of the sensor task with respect to the navigation and the security problem is presented.

References

[1] **Bühler Ch, Dierks F, Hein HW, Loevenich D** : Untersuchungen zum Arbeitschutz bei Mobilen Robotern und Mehrrobotersystemen, Projekt-Nr. irf-care-90-1, 1. Arbeitsbericht, Oktober 1990

[2] **Bühler Ch, Kaever P, Kernebeck U, Roßmann J** : Strukturierung intelligenter Robotersteuerungen für Weltraumanwendungen, 2. Arbeitsbericht zum Projekt CIROS, C-Z-289, IRF.8.89

[3] **DIN-Norm VDE 0837** : Strahlungssicherheit von Laser-Einrichtungen

[4] **Freund E** : Decoupling and Pole Assignment in Nonliear Systems, Electronic Letters, Vol. 9, No. 16, Aug. 1973, S. 373-374

[5] **Freund E** : Hierarchically Structured Control for Intelligent Robots in Space, Proceedings International Symposium on Europe in Space - The Manned SpaSystem , Strasbourg, France, 1988

[6] **Freund E** : On the Design of Multi-Robot Systems, Proceedings First International Conference on Robotics (IEEE Computer Society in Cooperation with other Member Societies), Atlanta, Georgia, 1984

[7] **Freund E, Bühler Ch** : Control of Intelligent Robots in Space, NASA Conference on Space Telerobotics, Pasadena, California 2.89

[8] **Freund E, Bühler Ch** : Entwurf nichtlinearer Regler für Industrieroboter aufgrund von Referenzbahnen und Sensorkorrekturen, Robotersysteme 5 (1989), S. 77-84

[9] **Freund E, Bühler Ch** : Robot Control in Manufacturing Combinig Reference Information with Online Sensor Correction, Preprints of the 6th IFAC Symposium on Information Control Problems in Manufacturing Technology (INCOM 89), Madrid, 1989, S. 515-521

[10] **Freund E, Bühler Ch, Roßmann J**: Teleoperated and Automatic Operation of Two Robtots in a Space Labratory Environment, 41st Congress of the International Astronautical Federation, Dresden, FRG,10.90

[11] **Freund E, Hoyer H** : Mobile Robots and Obstacle Avoidance in Integrated Manufacturing, CIM Europe Conference, Cheshire, England Mai 1987, S. 243-257

[12] **Freund E, Judaschke U** : Automated Vehicle Guidance as a Component of a Traffic Control System, Proceedings of the 2nd PROMETHEUS Workshop, Stockholm, 1989

[13] **Freund E, Lammen B** : Collision Avoidance in Vehicular Traffic, Proceedings 20th International Symposium on Automotive Technology and Automation (ISATA), Firenze, Italy, May 1989

[14] **Freund E, Lammen B** : On-Line Collision Avoidance for Automatically Guided Vehicles, Proceedings of the 2nd PROMETHEUS-Workshop, Stockholm, Sweden, Oct.1989

[15] **Freund E, Mayr R** : Verfahren und Vorrichtung zur automatischen Führung der Längs- und Querbewegung eines Fahrzeugs, Deutsche Patentanmeldung DE 3830747 AI, Internationale Patentanmeldung PCT/EP89/01043

[16] **Frohn H** : Robot-Vision-Systeme in der Fördertechnik, F+H Fördern und Heben, 40 (1990), Nr. 7, S. 462-468

[17] **Gerke M** : Nichtlineare Systementkopplung und Regelung in verschiedenen Koordinatensystemen, Studienarbeit am Institut für Roboterforschung, Dortmund, 1987

[18] **Gerke M** : Sensorkoordinierung und Regelung am Beispiel des Industrieroboters Manutec R15, Diplomarbeit am Institut für Roboterforschung, Dortmund, 1988

[19] **Hein HW, Siedler T** : Automatische Diagnose für Robtoer im Weltraum, BMFT-Projektabschlußbericht, Diagnose-EB-889, IRF 8.89

[20] **Mohr U** : Sensible Reaktion - Multisensorintegration für Roboter, Roboter, Oct. 1990, S. 22-24

[21] **Mohr U** : Sensorkoordinierung und Regelung in problemorientierten Koordinaten, Interner Projektbericht irf-sens-87-1, Institut für Roboterforschung, Dortmund, Juni 1987

[22] **Nicolaisen P, Link W** : Arbeitssicherheit beim Einsatz von Industrierobotern; in: Warnecke HJ, Schraft R : Handbuch Handhabungs-, Montage- und Industrierobotertechnik; Teil II, Kap. 13, Nachlieferung 1987

[23] **Pesara J** : Multisensorintegration und Reglersysthese bei einem Industrieroboter Manutec R15, Studienarbeit am Institut für Roboterforschung, Dortmund, April 1989

[24] **Pesara J** : Regelung in problemorientierten Koordinaten bei einem VW-G8 Industrieroboter, Diplomarbeit am Institut für Roboterforschung, Dortmund, Nov. 1989

[25] **Stevenson JT, Jordan JR** : Absolute position measurement using optical detection of coded patterns, J. Phys. E21; 1988, Nr. 12, S. 1140-1145

Seam-Welds Inspection of Underwater Structures Using Submersible Mobile Robot

G. Duelen, R. Bernhardt, V. Katschinski and D. Ozdes

Abstract

The paper reports on a project with the objective to inspect seam welds on underwater structures (e.g. oil platforms). A robot is mounted to a submersible vehicle which also carries the required equipment and tools for cleaning, measuring and inspection. The robot's tasks as well as the path of the submersible vehicle are preprogrammed using an off-line programming and simulation system developed at the IPK. In the first phase of the project, the overall concept was simulated by computer. Additionally, a laboratory model was constructed using a Manutec R 15 industrial robot and a tube intersection of a research platform in order to test the system. In further project phases, real underwater experiments are planned. Design work has begun concerning a waterproof robot and a mechanical interface between submersible vehicle and robot. This includes additional translational and rotational axes for robot positioning. The execution of robot tasks and the transport of the robot by the submersible vehicle to the underwater docking position have to be executed autonomously. For autonomous transport, a navigation system is required. Therefore, a feasibility study concerning analytical principles and performance evaluation was conducted in combination with a market survey of available components. Some of the study results are briefly presented in the paper.

1   Introduction

In the last few years many efforts have been made to develop highly flexible robots demanded in hostile environments such as in space and beneath ocean surfaces (subsea). In these environments, man should be replaced by remotely operated, automated and autonomous systems in order to reduce risks and costs.

Known concepts for such systems are based on an unstructured environment and use the capabilities of the human operator at the lowest level, namely, the execution of the task. The concept presented here is based on the approach to use human operator capabilities for planning and monitoring tasks and using automated systems for the task execution.

In order to work both on a theoretical and a practical level, a concrete scenario has been defined /1/: For the inspection of seam welds on underwater constructions like oil platforms, a mobile system should be used. It should consist of a submersible vehicle and a pressure resistant industrial 6 DOF robot. The robot should be linked with the vehicle via two additional axes (one rotational and one translational) for positioning the robot's base. The system should be remotely controlled and execute preprogrammed tasks autonomously. These global tasks can be decomposed into the following detailed subtasks:

1. Autonomous transport of the entire system from a known start position to the robot operation position along a preprogrammed path. When programming the path, a priori information about the subsea structure is taken into account. The fine motion of the submersible vehicle is planned autonomously, supported by suitable navigation systems.

2. Docking of the vehicle at a subsea structure element within approximately 0.5 meters and positioning the robot in its working area via two additional axes.

3. Identification of the docking configuration in relation to the structure (approximate accuracy of 0.1 meters) by using the robot, suitable sensors and sensor processing facilities.

4. Cleaning of the area to be inspected by using a high-pressure
   cleaning method.

5. Repetition of step 3 to improve the accuracy of the docking
   configuration by a factor of 5 to 10.

6. Inspection of the welding seam by a suitable method (e.g.
   hypersonic).

For the different tasks, a variety of tools are required which
are made available by a tool exchange system mounted on the
submersible vehicle.

The concepts and approaches as well as a partial realization of
such a system is presented in this paper. Also included is a discussion of analysis concerning navigation systems with a suitable
accuracy for guiding the submersible vehicle along its
preprogrammed path.

2. Remote Operation Concept

Human sensor-motoric faculties include all possible movements but
they are limited by speed and precision. Furthermore, the human
operator has a reaction time which cannot be infinitely
decreased. Highly repetitive routine tasks lead to a subjectively
sensed high degree of stress. The control of complex dynamics
requires high level training. However, in contrast to the mentioned facts, the cognitive capabilities of a human can be
characterized by /2/:

- An encompassing knowledge of the system behaviour.

- The collection, organization and storage of readily
  retrievable experience.

- The ability to enter into a complex exchange of information
  with his environment and his problem solving capabilities.

As a result, it can be stated that sensor-motoric areas as well
as their task-oriented optimization should be handled by a technical (automated) system which avoids the drawbacks of a human

operator /3/. The imitation of the cognitive capabilities of a human are at present still in a state of research. Present results allow practical solutions only for a limited number of specific tasks. Based on the capabilities of the human operator and an automated system, the tasks are distributed according to figure 1.

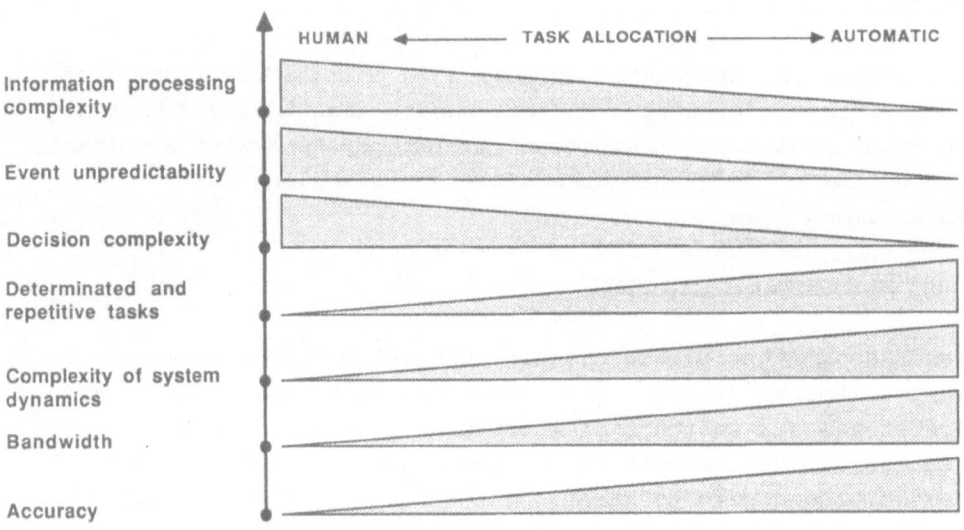

Fig. 1:   Man/Machine Task Distribution

As a basic rule, the design of such a system has to optimize the distribution of tasks between man and machine. This also means that the human operator will remain as an integral part of the overall system. For the remote operation concept, the basic "man in the loop" approach has to be assumed as shown in figure 2.

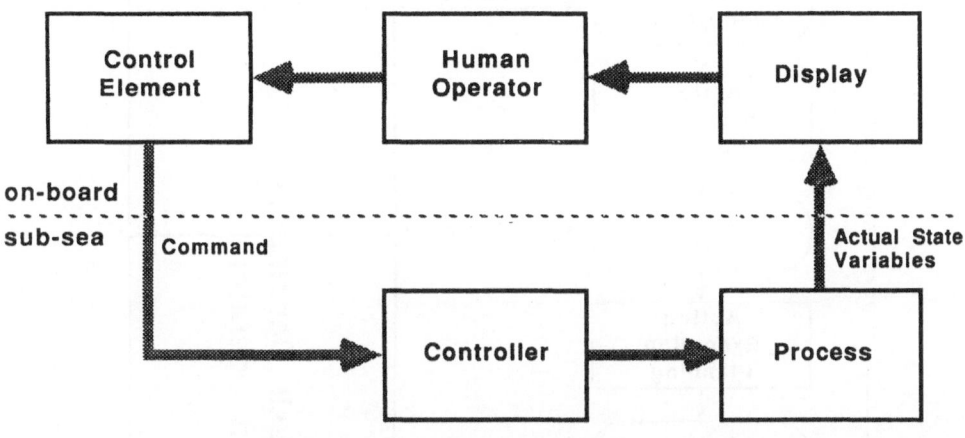

Fig. 2:   Man in the Loop Approach

The possible remote control concept is dependent on the task allocated to the human operator and the automated system. Task allocation also has strong influences on the overall system design. Assuming an immobile system as an example, fundamental statements for the task allocation are derived in chapter 2.1. These are based on a hierarchical approach to structuring the task, and a comparison of human operator and automatic system characteristics. This basic concept is then extended for mobile systems in chapter 2.2 /4/.

2.1   Hierarchical Control System Structure

Independent of the robot task, a sequence of activities is required which decomposes the task into a series of executable motions. This decomposition process can be represented in an execution sequence of subtasks in a hierarchical manner shown in figure 3.

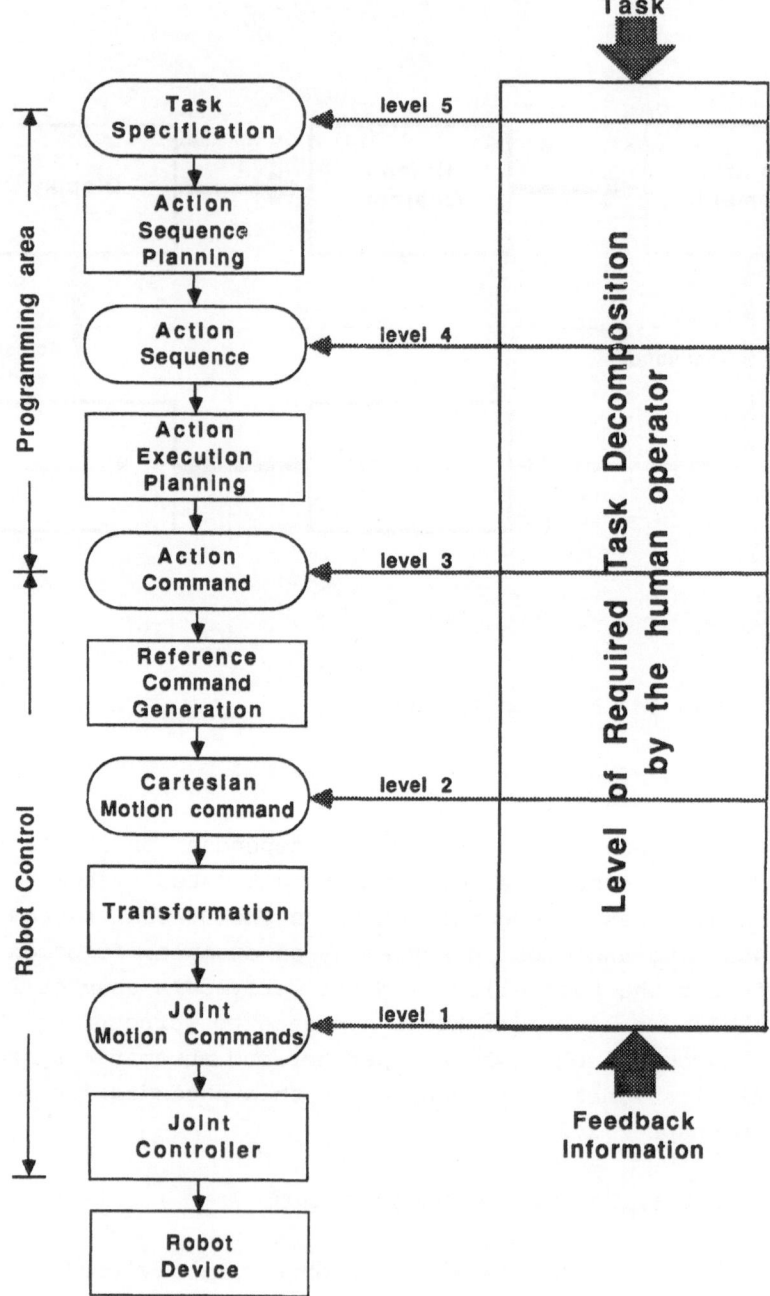

Fig. 3:   Task Decomposition

- At the action sequence planning level, the global task definition is decomposed into a sequence of required basic
  actions.

- At the action execution planning level, realtime action commands are converted into Cartesian motion commands.

- After transformation to joint level, the motion commands are
  executed by the joint controllers.

With respect to task allocation, it has to be decided which of
the subtasks should be done by the human operator and which
should be taken over by automatic systems. With this in mind,
different interaction configurations between human operator and
automatic system are possible.

In the area of industrial robots, the upper levels are related to
the programming process of the robot control systems. Due to the
repetitive character of the executed tasks of industrial robots,
the action commands are combined to form an application program
which defines the tasks for longer time periods.

2.2   Extension of the Remote Control Concept for Mobile Systems

In reference to figure 3, the basic structure of the robot system
is shown in figure 4. An identical system structure is possible
for the vehicle control system as shown in figure 5. Based on
this structural identity, the vehicle can also be remotely
controlled by a human operator with the same concept as explained
for the immobile system. This is true if the vehicle has a
control system similar to "Autopilot Systems" in airplanes.

From an engineering point of view, in this project the control of
the vehicle and the robot can be separated into control of /5/:

- Global motion of the vehicle to bring the robot system into a
  position such that task space and robot system operation space
  are identical.

- Local motion of the robot system for task execution assuming a
  fixed vehicle position.

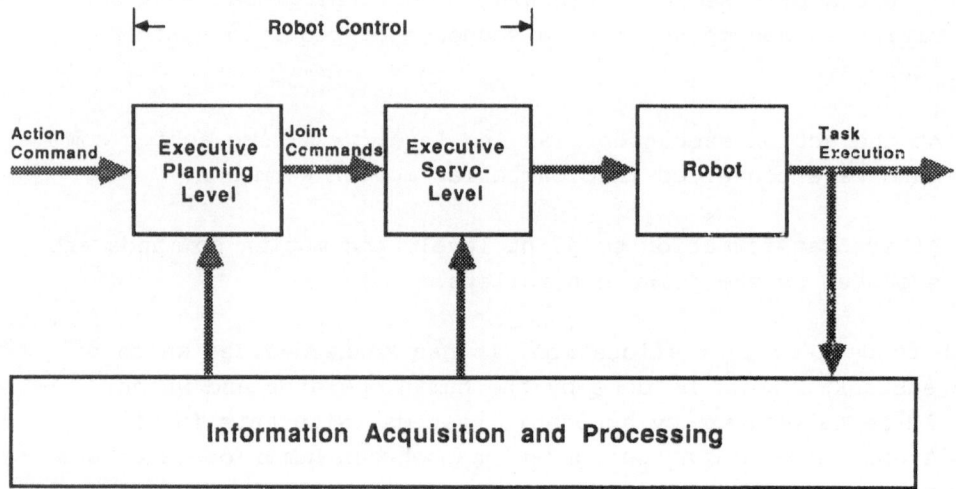

Fig. 4: Basic Robot System Structure

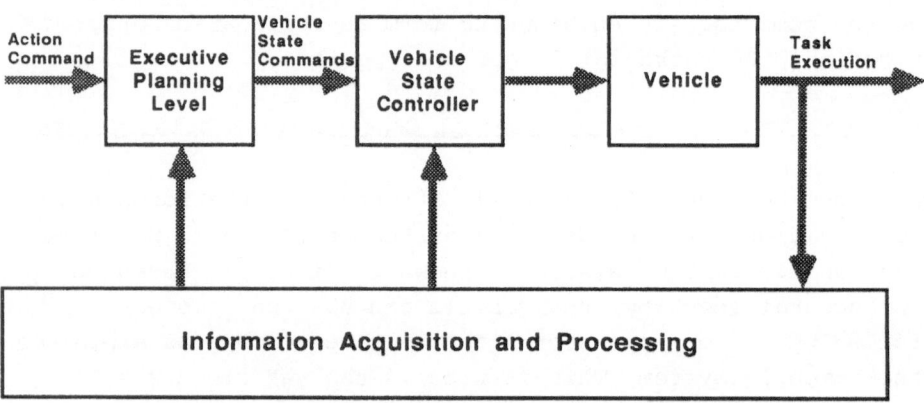

Fig. 5: Basic Vehicle System Structure

Concerning the first item, an analysis on navigation systems has been conducted. The main results are briefly presented in chapter 3 of this paper /6/.

Concerning the second item, a system concept has been elaborated and partially completed which is strongly based on approaches adopted in manufacturing technology. The results are presented in chapter 4.

3 Analysis of Navigation Systems

For very precise position location and steering problems, the navigation systems can be divided intc two groups, indirect and direct measurement systems.

Indirect Measurement Systems:
In this group, the required position coordinates are obtained either by double or by single steps of integration. This group is also known as the integrating or calculus navigation. Systems based on the solution of Newton's equations of motion and on the doppler principle are included in this group. These are exemplified by inertial (INS), Doppler radar, and doppler-sonar navigation systems. Also included in this group are those systems which measure the range rate or range difference rate. The error performance of this type of system degrads as a function of time.

Direct Measurement Systems:
In this group, the required position coordinates are obtained by direct measurement to and/or from the reference locations whose coordinates are known. The mathematical operations are based only on pure arithmetic, trigonemetry and/or algebra, (i.e. the inte-gration process is not used). Hence, systems are called the non-integration or algebraic navigation. Systems based on range, bearing, time or time differnce belong to this type of classification. It is exemplified by the Global Positioning System (GPS), Very High Frequency Ommidirectional Range (VOR), Distance Measurement Equipment (DME), Long Range Navigation (LORAN), and some acoustical navigation systems. The error per-formance of this type of system degrades as a function of distance and/or with Geometric Dilution Of Precision (GDOP).

Certain classes fall in between these two groups. A good example is the Navy Navigation Satellite System (NNSS). The NNSS (also known as the TRANSIT system) provides the geodetic position based on the doppler principle. Its radio line-of-sight, however, is limited to approximately 20 minutes per fix and its errors, due

to other sources, deal predominantly with integration of the
doppler measurements.

The above groups are then sub-classified as active or passive
systems. For example, the INS, GPS, VOR, LORAN and some acoustical systems (e.g passive SONAR and range/bearing measurements to
beacons) are typical passive systems. On the other hand, Doppler-
Radar, Doppler-SONAR and acoustical systems based on transponders
are typical active systems.

The many direct measurement systems based on electromagnetic
propagation have made precise navigation possible nearly all over
the world. Electromagnetic propagation in water, however, is
limited in that other forms of data transmission are required.
Currently, the most efficient mode of underwater signal propagation is acoustic sound pressure. Consequently, the candidate
navigation systems under consideration are the INS, Doppler-SONAR
and acoustic range/bearing measuring systems.

In order to select a navigation system which can fulfill the requirements of the project, one must quantitatively evaluate the
error performance of the systems being considered. This has been
done in the study already mentioned /6/. In this paper only a
summary of these results are presented.

3.1   Error Performance of INS and Doppler SONAR

The most accurate indirect measurement double-step integrating
system known is the INS. The INS short- and long-time error performances are determined by the accelerometer and gyro errors,
respectively. The position error due to gyro bias linearly
degrades as a function of time and is modulated with a Schuler
oscillation. The position error due to accelerometer bias is
oscillatory at the Schuler frequency.

Today's state-of-the- art Ring LASER Gyros (RLG) have a bias
residual of 0.005°/h. RLG's also have a random walk error, but
this has been excluded for the sake of simplicity.

As a result of the calculations, it can be concluded that even
the most advanced INS is not accurate enough for this

application, (i.e. an accuracy less than 1 meter cannot be reached).

Another candidate considered was the Doppler-SONAR system. The accuracy of this system which appears in the current technical literature is shown to be 0.01 knots. This corresponds to positioning error of 0.30866 meters/minute.

This seems to be an optimistic value. Nevertheless, this system cannot meet the requirements due to the much longer operation time required for inspection.

Due to unsatisfactory quantitative results mentioned above, the remaining candidate navigation systems are the acoustic range/bearing measuring systems.

The position accuracies of certain types of acoustical navigation system reported (even as early as 1969) are less than 15 feet (4.572 meters). Assuming a state-of-the art improvement of a factor of 10 in 19 years, one can expect a system accuracy of 1.5 feet (0.457 meters).

3.2 Acoustical Navigation Systems and Expected Performance

The basic operational theory of the operation of acoustical navigation systems is trivial. Three measurements are used to generate the vector position of an object in an a priori selected coordinate frame. Two possible configurations which make these measurements are described below:

• Short Base-Line System:
  The on-board equipment repertoire consists of a data pro-
  cessor-transceiver, three bottom-mounted hydrophones and a
  velocimeter. A transponder is anchored at a precisely known
  position near the ocean floor. At periodic intervals, the
  transceiver emits a command signal. Upon receiving this
  command, the bottom transponder returns a signal which is
  received by the hydrophones. The arrival of the return signal
  at geometrically separate hydrophones is processed by the
  transceiver and the data processor as three different measurements in order to generate the vector position of the boat

with respect to the chosen coordinate frame. If the center of
the coordinate frame is the bottom transponder, then the boat
position vector is defined with respect to that origin.

• Long Base-Line Systems:
  The on-board equipment consists of a data processor-
  transceiver, a single hydrophone mounted at the bottom of the
  boat and a velocimeter. Three transponders forming an equi-
  lateral triangle are anchored at precisely known positions
  near the ocean floor. The transceiver periodically emits three
  interrogation signals at different frequencies, each one addressing a particular transponder (some transceivers use a
  single frequency for interrogation). Upon receiving its specific command, each transponder emits a signal at a different
  frequency. The on-board hydrophone receives these signals on
  separate channels. The signals are passed to the processor-
  transceiver, and the vector position of the boat is obtained
  with respect to any known coordinate frame.

Each concept is based on an accurate measurement of the round-
trip propagation of the sound signal. The simple mechanization
methods are shown in figures 6 and 7. The on-board velocimeters
are used for accurate measurement of sound in the ocean. They can
either be separate instruments or they can be implemented ana-
lytically by measuring temperature and salinity. In most cases,
the seasonal salinity profile of the operating region is known
and stored in the computer.

The propagation of sound in an inhomogenous medium is subject to
consecutive bending effects. If it becomes necessary, an
analytical compensation is possible.

Critical to the calculation of measurement accuracy were the data
obtainable from different system manufactures. As it was
concluded that there are systems available which are based on
each of the two principles and meet the accuracy requirement of
less than 0.5 meters. However, this is dependent on a variety of
parameters and system configurations which have to be carefully
chosen.

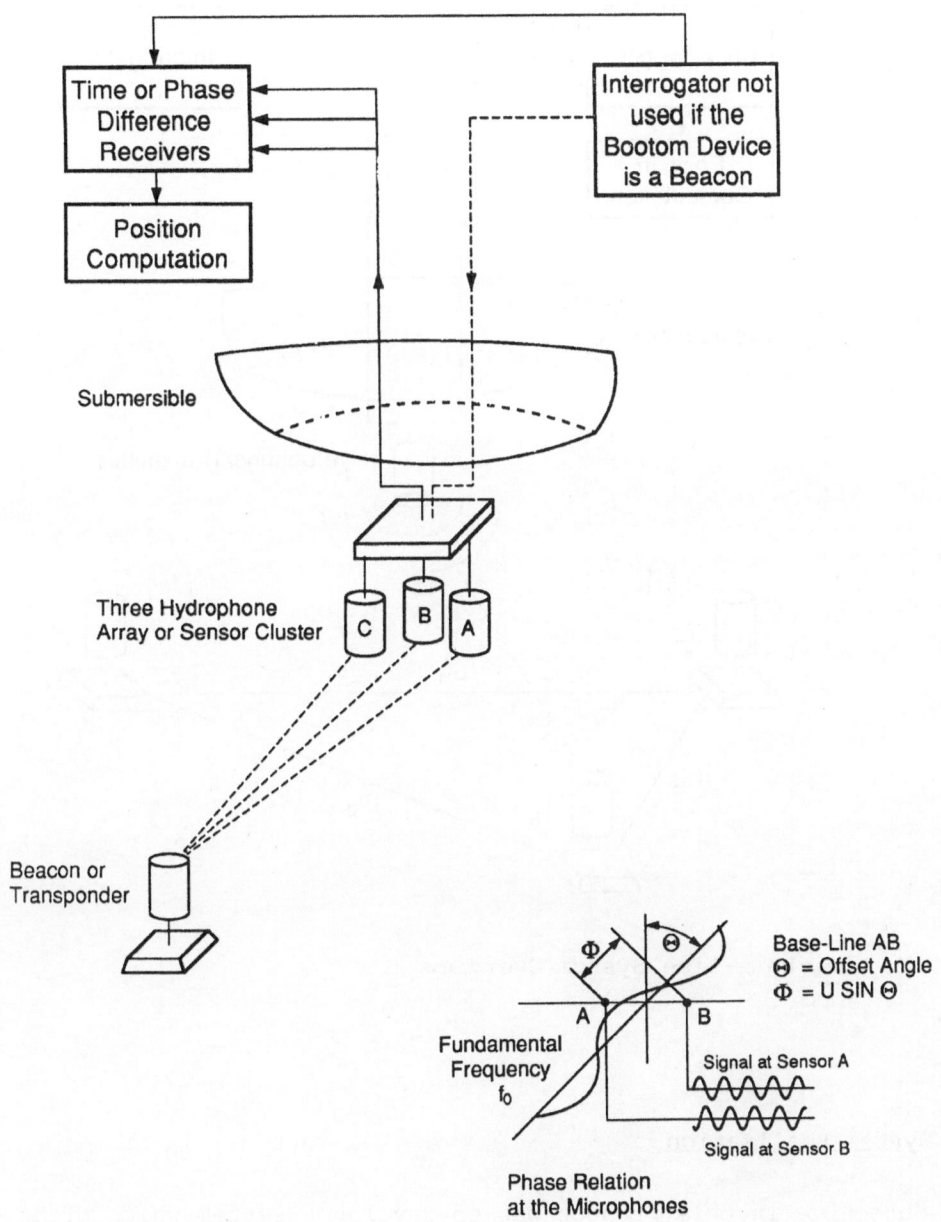

Fig. 6: Short Base-Line System Overview

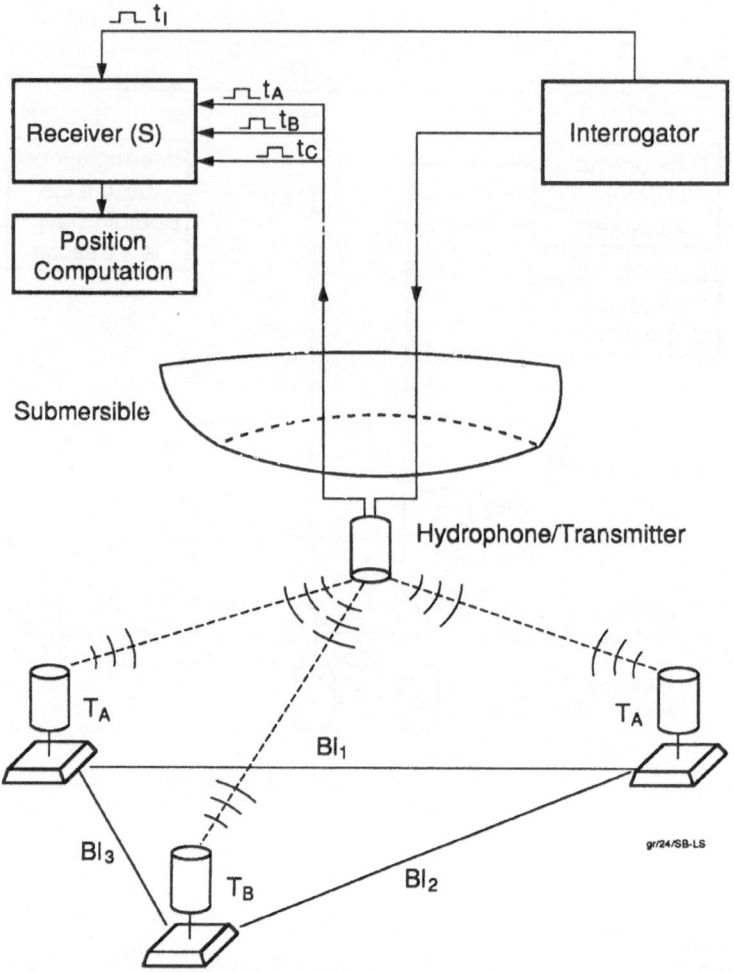

Fig. 7: Long Base-Line System Overview

4   System Realization

The objective in this project was to develop a system which plans
and programs the path of the submersible vehicle as well as the
task of the robot. Via the simulation system, the generated
application programs are tested and the execution of the robot
task is visualized.

The developed off-line Programming and Simulation System generates interactive and menu-oriented application programs for all
active components (figure 8).

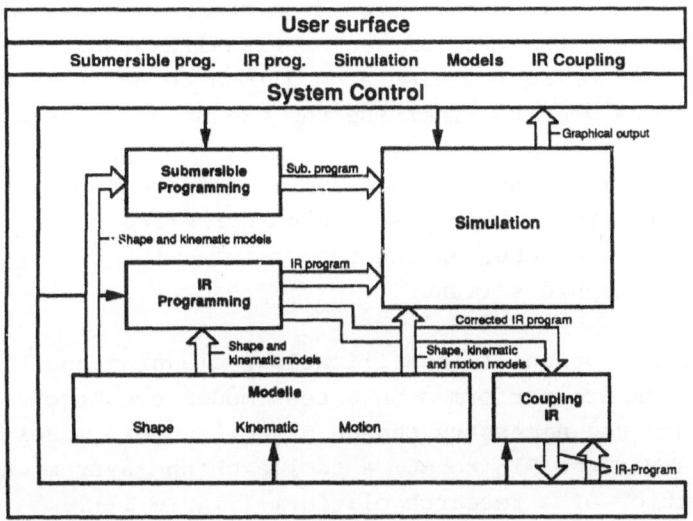

Fig. 8: Off-line Programming and Simulation System

The principal programming procedure is based on the application
of specific functions. These functions allow the user to describe
the task in an application-oriented way. All information is stored in a data structure called "Program Information". The data
structure includes all necessary geometrical, technological and
sequential information in a controller-independent representation. The code generator generates application programs in the
goal language of the specific controller. The programming concept
allows programs to be generated for a subtask which are then
executed after being verified by the simulation system. The
information (e.g. sensor data) gathered during this program
execution can be used for programming the next subtask.

The task of the simulation system is to detect programming errors
and to verify the executability of the programs. Therefore, the
same application programs which are downloaded to the real
controller are used within the simulation. The overall scenario
will be viewed on the graphic system. The simulation also

includes the visualization of the submersible's trajectory (figure 9).

Based on the joint values delivered from the real controller, the actual execution of the program can be visualized on the graphic system. This monitoring function also allows to define a specific fictitious camera view to supervise the execution (figure 10).

The fictitious camera can be fixed at any component within the overall scenario. It is thus possible to correlate a video image from a real camera mounted on the submersible with the synthetic image from the graphic screen.

The automated execution of identification, cleaning and inspection tasks are performed on a test model constructed in a lab and include the necessary change of tools. This model consists of a Manutec R15 robot, a tool exchange system, and the tube intersection of a research platform (figure 11).

Fig. 9: Simulation and visualization of the programmed submersible trajectory

Fig. 10: Camera View of the task execution simulation

Fig 11: Execution of the program at the real installation   (lab)

Conclusion

Based on the principle of remote control concept, a specific application was realized by a prototype. Experiences in the area of off-line programming of industrial robots can be consequently used for the system development. The user interface was designed to allow the end user an application-oriented formulation of the inspection tasks. A controller-specific knowledge is not necessary. The system offers a menu-oriented, graphical interaction interface to the end user.

The robot should have the ability to inspect a large section of the seam welds while in a fixed docking configuration. Therefore, additional axes are necessary for positioning the robot's base. Objectives of further developments at the IPK are the design of additional axes to guarantee a large work area for the robot.

Further reasearch activities have to be undertaken to prepare a waterproof industrial robot and to choose an adequate inspection method for detecting various welding failures.

5   References

1. G. Duelen, V. Katschinski, Programmierung und Simulation eines mobilen Inspektionssystems für Offshoreanlagen, .Autonome Mobile Systeme, 5. Fachgespräch München, November 1989.

2. F. Klix, Information und Verhalten, Verlag Hans Huber, Bern/Stuttgart/Wien, 1971.

3. U. Kirchhoff, Beitrag zur Identifizierbarkeit des BBN Modells und die Bedeutung des Modells als Beschreibung der Arbietsweise des Menschen im teiautomatischen Flugführungssystem, ILR-Bericht 35, Institut für Luft- und Raumfahrt, Technische Universität Berlin, 1978.

4. B. Schubert, G. Schultheiß, G. Duelen, U. Kirchhoff,F. L. Krause, R. Rieger, Advanced Development Of Automated Remote Handling, Proceedings of the 2nd Workshop on Manipulators, Sensors and Steps Toward Mobility, Manchester 24-26 October 1988.

5. G. Duelen, U. Kirchhoff, Trends of Automation in Space Application, International Advanced Robotics Programme, First Workshop on Manipulators Sensors and Steps Towards Mobility, Karlsruhe, May 11-13, 1987.

6. D. Ozdes, Navigationssystem für die Bewegungssteuerung eines
Unterwasserfahrzeuges zur Bohrturminspection, FO Bericht Nr.
89-1, April 1989.

6. Acknowledgement

The work reported on in this paper has been partially done in
ccoperation with the following institutions/companies:

GKSS-Forschungszentrum Geesthacht GmbH, Institut für
Anlagentechnik, Max-Planck-Str., D-2054 Geesthacht.

Fraunhofer Institut für Produktionsanlagen und
Konstruktionstechnik, Pascalstr. 8-9, D-1000 Berlin 10.

Fa. Ozdes, Alte Straße 1, D-7802 Merzhausen.

The basic developments for the off-line programming and
simulation system have been done by the IPK in the frame of the
European R&D project (ESPRIT 623), 'Operational Control of Robot
System Integration into CIM'.

Layers of Task Planning for Intelligent Autonomous Systems in a Flexible Manufacturing Environment

T. Kupec, R. Stetter, J. Milberg,
K. Fischer and H.-J. Siegert

Abstract

The handling, exchange and transportation of workpieces, tools and fixtures compose an important step towards a highly automated flexible manufacturing system. One step towards flexible automation is the use of autonomous mobile robots. This type of robot combines the functions of locomotion and manipulation in a single system. Several layers of information processing are nesessary for the integration of mobile robots in a production flow. This paper discusses three layers, the control layer, the cell layer and the component layer.

The control layer for a flexible manufacturing system is the highest. A knowledge-base contains the configuration of the flexible manufacturing system. Based on this, the manipulation and locomotion tasks for the autonomous mobile systems are added automatically to the working plans. These expanded plans are used for an exact scheduling of the manufacturing tasks. A constraint-based scheduling algorithm plans the tasks in real time. Finally, they can be dispatched to the machines and the mobile robots.

The planning tasks of the the cell layer are based on the jobs from the control layer. The execution of jobs between the mobile robots and the machines must be coordinated and synchronized. A rule-based planning system generates these jobs. It is able to process the tasks asynchronously, to react on new occurrences in real time.

At the third layer, the component layer, the interactive planning and simulation system USIS was developed. It generates robot programs free of collisions. USIS also has an integrated consideration of physical aspects, planning of gripping operations and the simulation of sensors.

1. Introduction

The handling, exchange and transportation of workpieces, tools and fixtures compose an important step towards a highly automated flexible manufacturing system. One step towards a flexible automation is the use of autonomous mobile robots. Furthermore, the increasing complexity of automated systems requires new solutions to reach a satisfying avalibility of the FMS and optimized control strategies to improve the return of investment [LUTZ 86]. A flexible manufacturing system is composed of flexible manufacturing cells. A cell consists of one machine and its peripheral systems such as handling, measuring and inspection. These cells connect the flow of material and the processing of information by way of the system [WEST 86].

At the center for production automation and robot technology of the "Institut für Werkzeugmaschinen und Betriebswissenschaften (iwb)", a flexible manufacturing system for the production of rotationally symmetric and prismatic parts was developed. The arrangement of a production line enables an optimized layout for accommodating the automated flow of workpieces and tools [KLIP 88].

This system is composed of the following flexible automated machines (see figure 1):
- NC-machine tools,
- coordinate measuring machines for an integrated quality control,
- tool measuring machine and
- store for parts.

For an automated flow of material, automated guided vehicles (AGV) and autonomous mobile robots (AMR) are used. Manipulation systems are able to load and unload the machines with tools and workpieces. At the iwb an autonomous mobile robot systems was developed, which combines the functions of locomotion and manipulation. A mobile robot is comprised of functions; the manipulation for the execution of manipulation tasks and the locomotion for the mobility in a specified area. An industrial robot produces the manipulation and the automated guided vehicle provides the locomotion. Service components are available for the exchange of data and for the supply of energy. The logical systems "transportable robot units" and "locomotion units" are comprised of these components and form a closed unit. The units are layed out so that they can be separated. Each unit can function as an independent system, taking on and executing orders.

The transportable robot either can be put into use from the AGV, or can be put on a support so that the AGV is free to accomplish other tasks.

In order to fulfill the spectrum of tasks that specifically apply to production, one must supply the transportable robot unit with the following additional components (see figure 2):
- a small tool magazine enables the maintenance tasks to be carried out for the machine tools,
- grippers for diverse tasks are stored in a magazine,
- a measuring system for the determination of the diameter of rotationally symmetric workpieces enables the realization of quality control tasks during the production process and

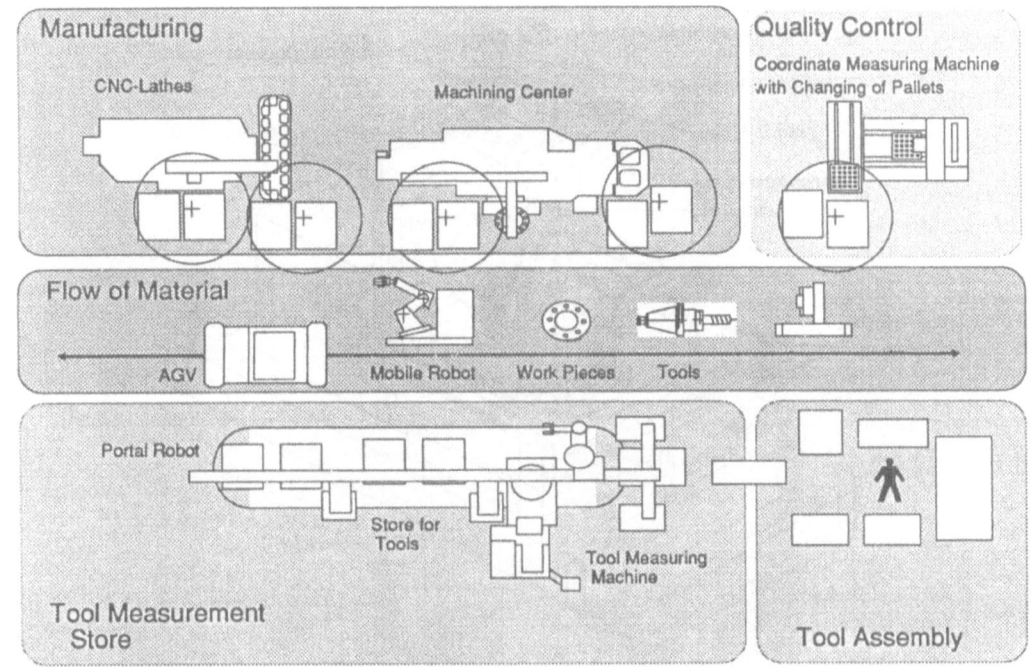

Figure 1:   Layout of the flexible manufacturing system at the iwb

- a laser measuring system, developed specially for the use on mobile robots enables the determination of positions without making contact.

The energy supply includes the battery recharger, the batteries and a direct-current-alternating-current-inverter for the production of the necessary alternating-current voltage. The operation box of the robot control is composed of logic construction groups, production units and operation fields. Additional equipment, the gripper station, the work piece measuring device and the machine tool store, is mounted on the robot control.

Flexible use of a mobile robot in a production surrounding occurs due to the autonomous feeding and preparataion of a machine. The automatic planning of handling operations, as well as sensors used for monitoring the enforcement of the handling jobs, help to accomplish the autonomous execution or accomplishment of these tasks. A machine generally occupies a robot for a short amount of time. The handling time is relatively short as compared to the time it takes for the processing to occur. As a result, more than one machine at a time is fed. A robot travels back and forth between several machines and exchanges work pieces and tools independently. The robot not only forms an autonomous unit, but also is temporarily synchronized with a machine during a handling procedure. Since the robot is able to make its own decisions, it is able to adapt itself and optimize its jobs according to the local aspects.

The SFB is concerned with the information processing structure for the integration of mobile robots into this manufacturing surrounding. Several institutes from different faculties have come together to research

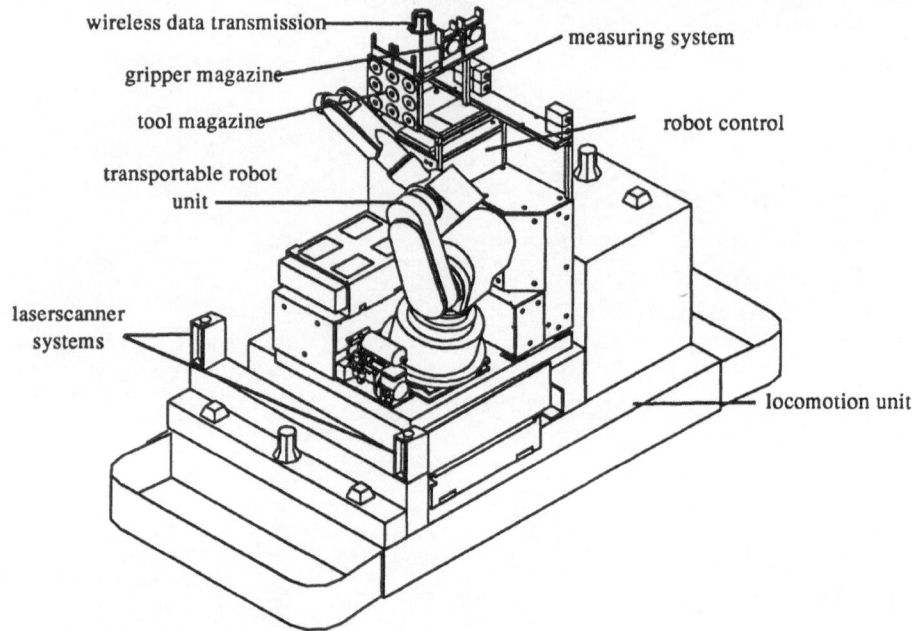

wireless data transmission

gripper magazine

tool magazine

transportable robot
unit

laserscanner
systems

measuring system

robot control

locomotion unit

Figure 2:   *Components of the mobile robot system*

this field. The job assignment is highly complex and variable. In order to control the flexible manufacturing system, a distribution of the job assignments into hierarchically classified layers are necessary. The restriction of relevant tasks in each layer, avoids an overloading of technical information. A classification of the following functional layer is logical [OTTA 86] (see figure 3):

- Planning layer,
- control layer,
- cell layer and
- component layer.

According to the definition of such a model composed of layers, each layer can only communicate with either the next higher or next lower layer. One can structure the flow of information in horizontal and vertical relationships [DUEL 86]. A vertical flow of information is necessary for the conveyance of assignments from one layer to the next. The horizontal flow of information enables the synchronization of the system as well as the conveyance of job assignments within a layer.

If these hierarchically classified layers are assigned to a flexible manufacturing system with autonomous mobile robots, the tasks of the individual layers as shown in figure 3 ensue.

The planning layer is responsible for tasks concerning goals for the whole manufacturing system, for example, production planning and control system (PPC-System). The major purpose of the planning layer is to establish general descriptions while taking certain restrictions into consideration, such as delivery

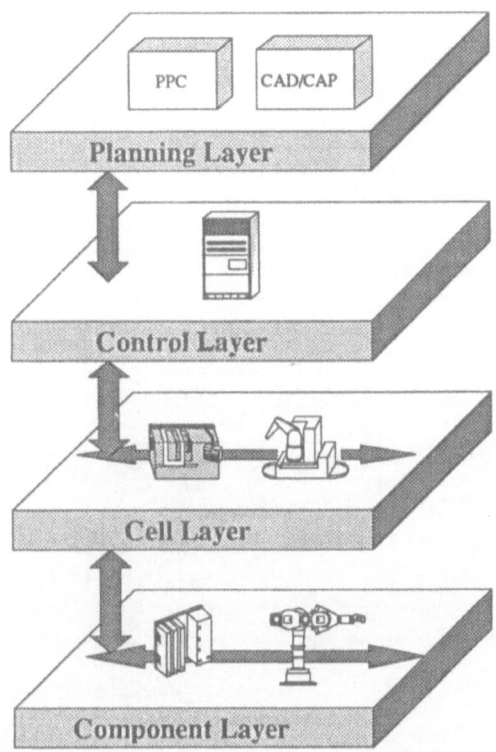

Figure 3:   *Layers of information processing in flexible manufacturing systems*

dates. Since this layer is concerned with tasks for systems which are more or less autonomous, marketable standard components are put to use.

At the control layer, the planning and realization of the production tasks in view of given restrictions, such as fixed time stamps, take place. The control layer is intercellularly active. Planning occurs according to schedule and capacity.

The cell layers's major tasks concern the realization of the job assignments tranferred from the control layer, as well as the supervision of the production process. Sensors, computers and inspection strategies enable the automation of these tasks [GROH 88]. As a result, the synchronization of an autonomous mobile robot and a machine during the loading of a work piece could be one of the tasks occurring at the cell layer. Specific procedure patterns must be developed ("open doors of a machine" oder "close fixture after loading the workpiece").

The component layer takes over the operative planning and the execution of the directions from the cell layer. It translates/converts the actions into processing, handling, and transport sequences. As a result, the component layer of the autonomous mobile robot can carry out path planning or collision inspection in a simulation run.

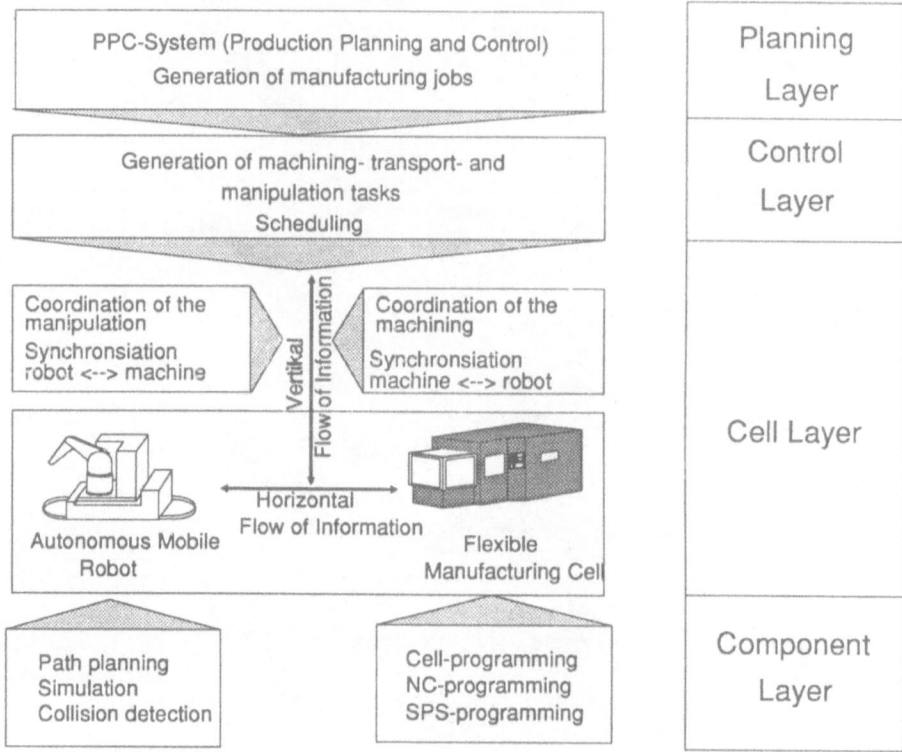

Figure 4:   *Flow of information between the layers of information processing*

2. Planning on the Control Layer

Problems and Initial Stages of Solutions

The control layer's main task is to find an optimal plan for the machines based on a specific evaluation criterion. This plan should determine an optimal distribution of the individual work steps to the working resources. This fine adjusting falls into the category of scheduling problems. This category is responsible for finding the optimal sequence of the plans, not only for the machines, but also for the autonomous mobile robots. One must also determine the optimal sequence of the job oders. A quick algorithm is necessary, since disturbances and failures of the elemental components of a flexible manufacturing system occur. After diagnosing the failure, the machines' new plans must be created in real time in order to prevent a machinery standstill for a long period of time.

A transparent and standardized form of knowledge representation is the presentation of all surrounding factors in the form of constraints. With this descriptive and explanatory formalism, one can effectively and appropriately express insecurity, deficiency and time dependency of knowledge through standardized formulated constraints. The planning problem consist of finding the right plan which considers and observes the various constraints [HUBE 90a, HUBE 90b, LAMA 90].

The planning steps

The algorithm mentioned above introduces the individual steps of the task planning at the control layer. The planning takes place in different stages in order to preserve the transparency as well as the clearly defined definitions [KUPE 89a].

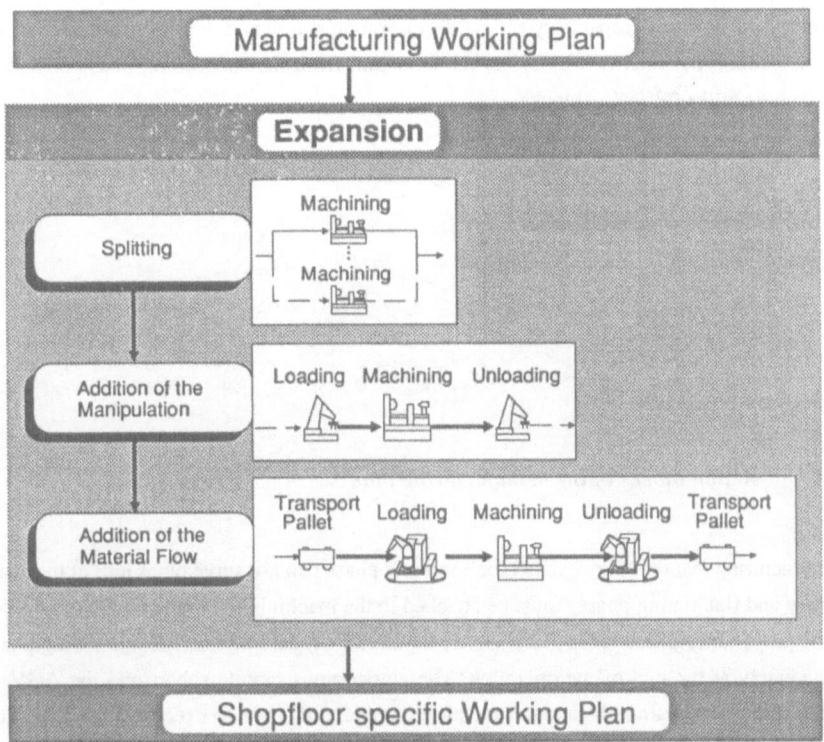

Figure 5:   The steps of network expansion

The data to be entered into the control system is comprised of manufacturing tasks extracted from the PPC-System. These jobs are converted to a network containing a linear chain of operations for the machines or machine groups. In order to integrate the autonomous mobile robots into the production flow, this network is expanded to include handling operations as well as transportation operations. Then the control system chooses out of a group of manufacturing ressources the best fitting one.

The basic idea of the expansion of the network-plan is that each operation must be prepared as well as subsequently worked over. As a result, the network runs through many times, the chain of operations is opened and new nodes are inserted [KUPE 89b].

Three expansion phases are necessary to expand the network correctly. The exit of the previous phase provides the entrance to the following phase (see figure 5). In the first step, the job orders are separated according to the maximum number of work pieces fitting on one pallet. If the lot size is larger than the size of the pallet, then as many portions are generated to fit on the pallets. The lot sizes are distributed evenly among the pallets. The loading and unloading tasks are generated next. Lastly, the operations of material flow are inserted into the network.

The network is then divided into blocks. Each block contains the machining operation and the appropriate preparing and finishing actions. In figure 6, one sees that each operation contains three inseparable components of action; the preparatory action, the executing action and the succeeding action. One sees the point of separation between two blocks at the transition from the succeeding phase of one block to the preparing phase of the following block.

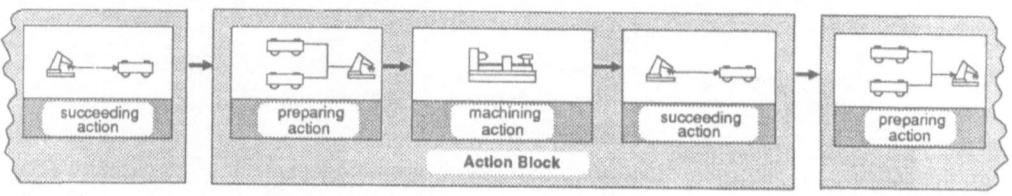

Figure 6:   Action blocks of the manufacturing plan

The manufacturing resources designated for a sepcific phase can not serve other jobs at the same time. The preparatory and fine tuning phases must be attached to the machining operations without time lapses. As a result, the preparatory phase takes place shortly before the processing and the fine tuning phase is carried out immediately at the end of the operation. The autonomous mobile robots and the AGV are released during the actual processing, the actual execution stage, and are free to execute other jobs. Subsequently, the actual fine planning regarding aspects of time and capacity takes place. An algorithm based on constraints (Constraint Satisfaction Problem = CSP) was developed for the fine planning. These constraints are basically a set of restrictions which apply when discussing the problems. The following three constraints refer to the problem assessing:

before (b):   An operation may begin once its direct predecessor has finished. This contstraint is constructed from the network plan.

meet (m):   A stricter condition is defined by this constraint. The next stage of execution in the network succeeds the previously completed operation without delay. For example; after the autonomous mobile robot places workpieces into the machine, the machine immediately begins with the processing.

not overlap (no): This is the most important constraint. Two operations may not overlap, with respect to time using the same resource.

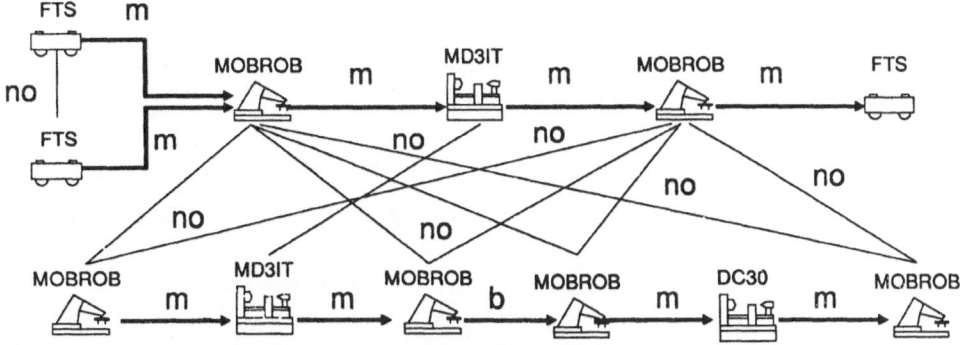

m = meets (Tasks of an action block are fixed together)
b = before (Action blocks of a task do not have to be fixed together)
no = not overlaps (No access to the same resource)

Figure 7:   *Example of a constraint network*

Figure 7 gives an example of the restriction network described with three production jobs. The constraints "b" and "m" are constructed within the margins of the network. Concisiveness with respect to time is demanded within a phase. This refers to the somewhat weaker constraint "b" for the margins between two blocks. The constraint "no" appears between the nodes which make use of the same resource.

Subsequently, the jobs are included along the time axis. Lists determining the occupancy of the machines are supplied for the representation of the contraint "no". These lists are supplied for all manufacturing resources and administer the time intervals in which the resource is engaged. The following intervals are listed:

- Failure caused by technical interferance,
- stand-still time due to maintenance, lack of operators, or occuring at certain times when the production line is shut down (for example on the weekend or at night),
- times at which the resource is occupied by other jobs and
- time intervals during which job run-throughs are tested, in order to simulate the effect on the production situation.

The operations are included in the planning on a trial basis. During this time, the time intervals are appropriately entered into the lists. Since the limitations (for example fixed time stamps) are already determined by the PPC-System, this inclusion in the planning occurs at the earliest moment possible. This occurs in blocks, in order to link the autonomous mobile robots' job orders to the machine's processing jobs. One begins with the first block of a job order. The production job runs through block by block. If overlapping at a resource occurs, the times are shifted to a later time equivalent to the amount of the overlapping. One repeats this process until a block containing the previously determined restrictions can be included in the planning. Figure 8 presents an example taken from a bar diagram showing the relation of the job orders to the resources in relation to the time. As a result, one can see the outcome and determine where the overlapping occurs.

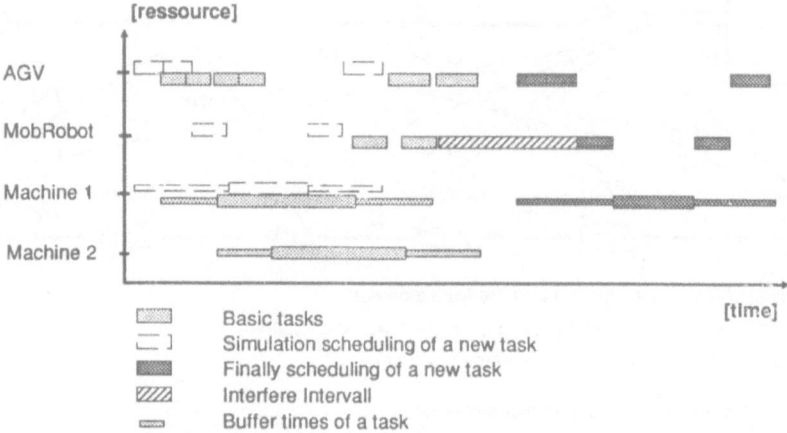

Figure 8:   *Example of a bar-diagramm*

3. Planning at the Cell Layer

After adaptation of the tasks to the autonomous mobile robots with respect to time and capacity has taken place, the tasks can be transferred from the control layer to the cell layer. During the execution of a specific task normally several autonomous units have to cooperate, e.g. a mobile robot serving a machine tool with a workpiece. The research group "Real-Time Systems and Robotics" at the "Institut für Informatik" has developed a rule-based task planner for planning and controlling task execution for autonomous units of a flexible manufacturing system. This task planner plans and executes individual tasks for autonomous units using behaviour patterns. These behaviour patterns are generalized plans which solve individual tasks. They are represented in the form of sets of rules. The task planner is able to process asynchronously occurring knowledge in order to react to asynchronous events. As a result, the task planner is able to react flexibly to exceptional situations. [FISCH 88, FISCH 89] describes the task planner's exact construction. In [FISCH 91] is described a distributed rule-based multi-agent system which was developed for the implementation of the task planner.

The Need for GIPSY

It is not always possible to carry out task planning with the help of a rule-based task planner. One reason may be that a specific autonomous unit is not connected with an appropriately efficient computer, or task execution must be very fast for reasons of efficiency. This is why, GIPSY (graphical interactive programming system) was developed. With GIPSY's help it is possible to compile task descriptions - stored in the form of sets of rules and facts in the class hierarchy of an object-oriented knowledge base -

into robot and machine programs. These programs can be directly executed by the controlling devices of the individual systems.

The User's Point of View

From the point of view of the user, GIPSY helps to create robot programs off-line on the basis of an abstract description of the task planning environment and the task to be solved. The solutions to the individual tasks depend not only on the participating autonomous units, but also on the objects with which the handling tasks are supposed to be carried out. In a flexible manufacturing system, new workpieces are defined frequently. With the help of GIPSY one can plan the necessary handling operations leading to the end product.

On the other hand it is seldom that a robot, a tool machine, a milling center, or an AGV are newly introduced. Workpieces are normally constructed with the help of a CAD system. That means, the geometrical data of a workpiece is present from the beginning. Autonomous units are normally not constructed by the factory which uses them. Therefore, their geometrical data has to be defined with the help of a CAD system in a separate working operation. Additionally, the kinematic data of all movable parts of the autonomous unit (machines, robots, etc.) must be described for GIPSY. Only when all this information is available can GIPSY plan the handling tasks of the newly incorporated autonomous unit.

How GIPSY functions

The starting point for the implementation of GIPSY was the simulation system USIS. USIS was developed at the "Institut für Werkzeugmaschinen und Betriebswissenschaften (iwb)". It is the test environment for the robot programs generated by GIPSY. The expansion of USIS to a graphical, interactive, task-oriented robot programming system takes place in several steps. During the final phase, the newly developed components from the task planner are fully integrated into USIS. That is to say, that the interaction of the user takes place using only the graphical interface of USIS. The user is able to put USIS in a state where a task-oriented interface is made available by selecting the appropriate command in the graphical interface of USIS. This means that the user can specify task which in turn are automatically planned and executed by the system. Examples of task-oriented instructions for an autonomous robot are:

    TAKE o_1024 FROM palette_1 AND PUT IT ONTO palett_2
    TAKE o_1025 FRIN palette_1 AND PUT IT INTO dc30

An example of a task-oriented instruction for a machine is:

    LOAD o_1024

For this planning step, knowledge about the task planning environment as well as about the tasks to be carried out must be stored in a knowledge base. If a user specifies a task for which there is not enough knowledge available, he is requested to give this knowledge to GIPSY. The system supports him here in as much as possible. If there is enough knowledge available to complete the task, the programs for the participating robots and machines are automatically produced during the planning and the execution of the specified task. These programs supply solutions to the tasks specified by the user in the real factory environment.

The User Interface

How the user proceeds and the system's reactions to it will now be described in more detail. To begin with, the user loads a task planning environment from a layout file. This is the usual procedure in USIS. This task planning environment determines which robots and which machines are available for the task description. It also determines which geometrical position the robots and machines have in relation to one another. Once this has been done, the user can activate the task-oriented interface by choosing the appropriate command in the graphical interface. He can then specify a task either textually or in a menu-controlled form.

These task specifications are passed on to a task planner, which first tests whether those objects listed in the task specification are available in the current scene shown on the screen. If this is not the case, the user is notified and task planning ist not started. If all of the objects needed are present, the necessary knowledge from the knowledge base is transferred to the task planner. The system allows the user to supply the missing knowledge if the knowledge is not complete. In the light of this knowledge, GIPSY plans the necessary individual steps of the solution to the task for each of the autonomous units. All subtasks are presented to the user in a simulation and may be interactively modified.

When the task has been completely planned, the user is brought back to the level of the task-oriented interface. By selecting the appropriate command in the graphical interface the user can see on the screen the whole simulation of the programs generated for a specific task. If the user is satisfied with the result, he can specify new tasks or change to the normal USIS interface by selecting the corresponding command.

The Individual Components of GIPSY

GIPSY has a divided structure (see figure 9). The higher level is composed of a rule-based task planner. At this level, the complex task-oriented instructions are broken down into simple individual steps. This division of tasks into subtasks is done by means of knowledge stored in an object-oriented knowledge base. The so-called executing components of the subordinate level receive individual actions for each autonomous unit. The geometric planning for the individual steps must take place at this level. In general, several autonomous units can be addressed. It is therefore possible that movements can be carried out simultaneously by several autonomous units. GIPSY has, for that reason, two different execution components. The task planner passes all the individual actions to be executed concurrently to the first execution component CEC (concurrent execution component). CEC transfers the separate actions individually to the execution component IEC (individual execution component).

IEC plans the trajectories for individual actions given to it. The planning results are shown on the graphic terminal, where the user can make appropriate changes. This may result in the user modifying only a point of an already complicated trajectory. In an extreme situation, the user may transform a straight-line movement from the initial point to the end, into a complicated trajectory avoiding collisions. It is easily

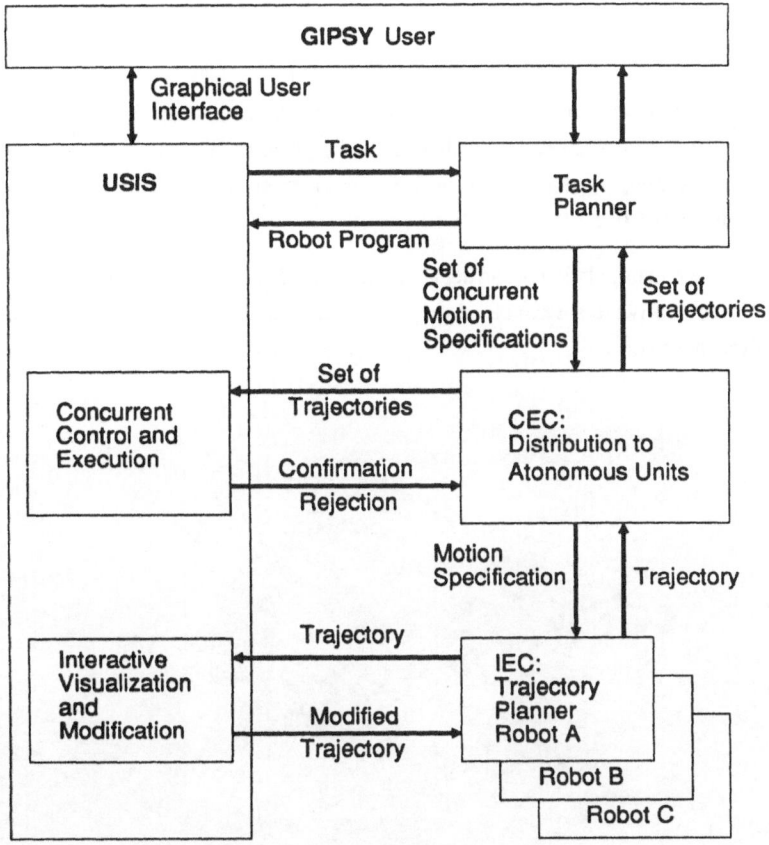

Figure 9:   *Concept of the robot programming system*

possible for the user to run through the trajectory, reach specific points in the trajectory, and modify specific points.

Once the trajectories for all of the current individual actions have been generated, CEC shows their execution and their interactions to the user. Here again it is easy for the user to modify the trajectories and their interactions for the individual autonomous units. He can switch back and forth between planning a sequence of movements for a single autonomous unit by means of IEC and executing concurrent movements in CEC.

Only after the user has indicated that the concurrent execution of the movements in CEC is correct, can the sets of planned trajectories be transferred back to the task planner. The task planner converts the trajectories into appropriate commands of robot and machine programs. At the same time, the task planner creates synchronisation commands necessary for the interaction in the form of signal handling in the robot and machine programs.

4. Planning on the component layer

The planning and simulation of manipulation processes on the component layer takes place offline. An advantage of this is that one can concurrently operate the planning times for the actual processing times. While the actual handling system is tied up in the production process, another handling procedure is planned and simulated offline.

The simulation of the manipulation occurring leans on 3D-CAD models of the individual components. The insertion of USIS, a software system based on 3D graphics, ensures a clear description of the operations which take place in the cell.

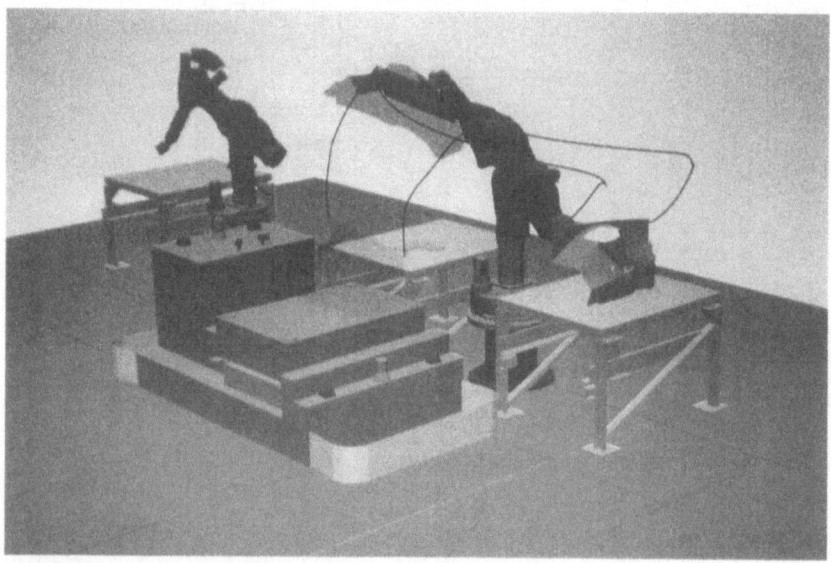

Figure 10: Simulation of robot movements in USIS

The teach functions integrated in the system extend beyond pure movement simulation. With the help of these functions, the robot programs can be made available offline. The programs generated are written in the specific robot language. As a result, a direct transfer of a robot program generated offline to an actual robot is possible without any post-processing. Communication with the actual system occurs by means of standardized protocols. In order to test the program, it is possible to follow the procedure of a manipulation in real time on the screen. A central theme of robot programming is the recognition of possible collisions the manipulators may have with their surroundings. USIS supplies several options for detecting such collisions.

A large number of sensors used to inspect the manipulation process are installed with the introduction of an autonomous mobile handling system in a flexible manufacturing cell. In order to be able to flexibly react to the changing restrictions during the simulation, a whole row of sensors was reproduced as software and integrated into the simulation system. These sensors are able to measure the positions of the mobile system regarding reference marks. They can also serve to supervize the presence of workpieces. Tactile sensors as well as distance sensors are available for the simulation of join movements.

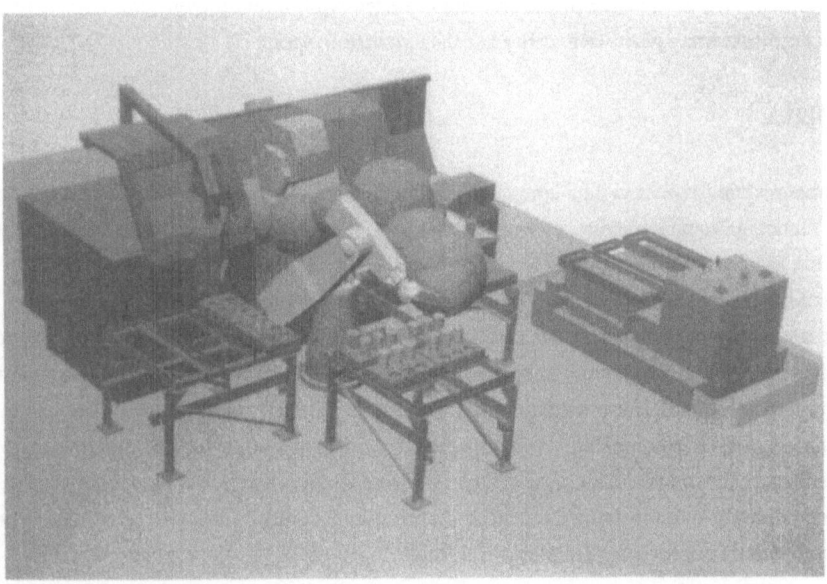

Figure 11: Automatic path planing for industrial robots

The sensors can be used not only for the simulation of manipulation operations but also for the planning of movements. Interval sensors act as the basis for the algorithm of the path planning. They are used to investigate the manipulator's system surroundings. The information attained from the sensors is used to adapt the paths generated by the system to the obstacle constellation. This should then prohibit collisions.

The very time consuming programming of industrial robots is drastically reduced by the use of such a planning tool. The direct takeover of programs generated offline increases the cell component's degree of disposition.

The consideration of physical effects, such as gravitation and friction lead to more realistic simulation results. The inclusion of these forces proves to be especially important by the simulation of gripping operations.

One of the most frequent problems concerning the changing demand profile of those jobs to be carried out by the mobile autonomous system is the variation of the grippers and their gripping sequence. Not only the simulation, but also the planning of gripping operations is important for an optimal integration of the mobile unit. The gripping planning determines the most favorable gripping position. This decision is based on several criteria, such as sliding and tipping security. Sensors are used to include the environment necessary for the automatic production of robot programs. Points along the path are adapted to the actual course with obstacles in order to prevent the manipulator from colliding with its obstacles. For difficult assembly situations, the linkage of the simulation system with a real picture processor is necessary. This compares the discrepancy between reality and the simulation model.

5. Outlook

Autonomous mobile robots can be integrated into a manufacturing process by means of the three layers of the job planning introduced above. For the most part, these systems have already been realized. In the future, the emphasis will be on overcoming disturbances autonomously. As a result, these controlling systems must be expanded to include feedback structures. This feedback plays an important role. In this context, autonomy means that in the case of a conflict, a local disturbance must be solved within the affected system. Only if the autonomous system is not capable of solving the problem, the next highest layer must be informed. These autonomous areas should be able to provide a greater availability of the system in spite of their complexity. One sees here the concept's advantage of dividing the planning into separated layers. Each individual layer can be improved itself, whereas the interactions between the other layers are confined to to the defined interface. As a result, decoupled further develpoment is possible. This is especially important for interdisciplinary reasearch.

6. Literature

[DUEL 86]   Duelen, G.; Linnemann H.; Bernhardt, R.: "Die Informationsarchitektur in datengetriebenen Fabriken", Produktionstechnisches Kolloquium, Berlin, 1986

[GROH 88]   Groha, A.: "Universelles Zellenrechnerkonzept für flexible Fertigungssysteme", Dissertation TU München, Springer Verlag, 1988

[FISCH 88] Fischer K.: "Regelbasierte Synchronisation zwischen Roboter und Maschine", Technischer Bericht, TUM-I8861, Institut für Informatik, TU München

[FISCH 89] Fischer K.: "Knowledge-Based Task Planning for Autonomous Mobile Robot Systems", Proc. of the 2nd Inter Conf. on Intelligent Autonomous Systems, Amsterdam, December 1989, pp. 761

[FISCH 91] Fischer K.: "Ein Agentensystem für eine flexible Fertigungssteuerung", Tagungsband: Prozeßrechensysteme '91, Februar 1991

[HUBE 90a] A. Huber: Wissensbasierte Überwachung und Planung in der Fertigung; Betriebliche Informations-und Kommunikationssysteme, Band14, H. Krallmann (Hrsg.), Erich Schmidt Verlag, 1990

[HUBE 90b] A. Huber, H. Krallmann: Zeitrepräsentation in der industriellen Produktionsplanung und-steuerung; in: KI 3/90, Oldenbourg Verlag, 1990

[KLIP 88]   Klippel C.: "Mobiler Roboter im Materialfluß eines flexiblen Fertigungssystems", Dissertation TU München, Springer Verlag, 1988

[KUPE 89a] Kupec T.: "Integration of Autonomous, Mobile Robots in Flexibe Manufacturing Systems" Proceedings of IAS-2, Amsterdam, 11.-14.12.1989

[KUPE 89b] Kupec T.: "Wissensbasierte Aufgabenplanung für autonome, mobile Roboter in flexiblen Fertigungssystemen", Vortrag zum 5. Fachgespräch Autonome mobile Systeme, TU München, 30.11.1989

[LAMA 90]   A. Lamatsch: Fertigungsleitsysteme; Wissensbasierte Werkstattsteuerung; Forschungsbericht, FAW Ulm, S.195-217, 1990

[LUTZ 88]   Lutz, P.: "Leitsysteme für die rechnerintegrierte Auftragsabwicklung", Dissertation TU München, Springer Verlag, 1988

[OTTA 86]   N.N.: "The Ottawa Report on Reference Models for Manufacturing Standards", Version 1.1, ISO TC 184/SC5/WG1 Dokument N51, 1986

[WEST 86]   Westkämper, E.: "Auftragsabwicklung in der computerintegrierten und automatisierten Fertigung", VDI-Bericht 611 (1986), S. 289-310

List of Authors

Badreddin, E.   Robotics Research Group   171
                       Inst. of Automatic Control,
                       CH-8092 Zürich - Switzerland

Bernhardt, R.   Fraunhofer-Institut für Produktionsanlagen und   311
                       Konstruktionstechnik (IPK)
                       Pascalstraße 8-9, D-1000 Berlin 10, Germany

Brussel Van, H.   Katholieke Universiteit Leuven,   105
                       Dept. of Mechanical Engingeering,
                       Celestijnenlaan 300B, B-3001 Leuven, Belgium

Carriker, W. F.   The Robotics Institute and Laboratory   45
                       for Automated Systems and Information Processing
                       Dept. of Electrical and Computer Engineering
                       Carnegie Mellon University
                       Pittsburgh, Pennsylvania 15213, U.S.A.

Cox, I. J.   NEC Research Institute   23
                       4 Independence Way
                       Princeton, NJ 08540, U.S.A.

Detlefsen, J.   Lehrstuhl für Mikrowellentechnik   93
                       Technische Universität München
                       P.O.Box 20 24 20, D-8000 München 2, Germany

Dierks, F.   Institute for Robotics Research (IRF)   295
                       Otto-Hahn-Str. 8, D-4600 Dortmund 50, Germany

Dillmann, R.   Institut für Prozeßrechentechnik und Robotik   279
                       Universität Karlsruhe
                       P.O.Box 69 80 D-7500 Karlsruhe, Germany

Dorn, J.	Technical University Vienna, Christian Doppler Laboratory for Expert Systems Paniglgasse 16 A-1040 Wien, Austria	219
Duelen, G.	Fraunhofer-Institut für Produktionsanlagen und Konstruktionstechnik (IPK) Pascalstraße 8-9, D-1000 Berlin 10, Germany	311
Elfes, A.	Dept. of Computer Sciences IBM T.J. Watson Center Yorktown Heights, NY 10598, U.S.A.	77
Färber, F.	Lehrstuhl für Prozeßrechner Technische Universität München P.O.Box 20 24 20, D-8000 München 2, Germany	247
Fischer, K.	Institut für Informatik Real-Time Systems and Robotics Group Technische Universität München Orlenastr. 34, D-8000 München 80, Germany	331
Freund, E.	Institute for Robotics Research (IRF) Otto-Hahn-Str. 8, D-4600 Dortmund 50, Germany	295
Freyberger, F.	Lehrstuhl für Steuerungs- und Regelungstechnik Technische Universität München P.O.Box 20 24 20, D-8000 München 2, Germany	61
Fröhlich, C.	Lehrstuhl für Steuerungs- und Regelungstechnik Technische Universität München P.O.Box 20 24 20, D-8000 München 2, Germany	61
Hallam, J.	Dept. of Artificial Intelligence University of Edinburgh 5 Forrest Hill, Edinburgh EH1 2QL, Scotland	267
Heller, J.	Inst. für Steuerungstechnik der Werkzeugmaschinen und Fertigungseinrichtungen (ISW) Universität Stuttgart Seidenstr. 36, D-7000 Stuttgart 1, Germany	201

Krogh, B. H.	The Robotics Institute and Laboratory for Automated Systems and Information Processing Dept. of Electrical and Computer Engineering Carnegie Mellon University Pittsburgh, Pennsylvania 15213, U.S.A.	45
Kupec, T.	Inst. für Werkzeugmaschinen und Betriebswissenschaften (IWB) Technische Universität München P.O.Box 20 24 20, D-8000 München, Germany	331
Lammen, B.	Institute for Robotics Research (IRF) Otto-Hahn-Str. 8, D-4600 Dortmund 50, Germany	295
Lange, M.	Lehrstuhl für Mikrowellentechnik Technische Universität München P.O.Box 20 24 20, D-8000 München 2, Germany	93
Leonard, J. J.	NEC Research Institute 4 Independence Way Princeton, NJ 08540 U.S.A.	23
Levi, P.	Institut für Informatik Image Interpretation and Artificial Intelligence Technische Universität München Orleanstr. 34, D-8000 München 80, Germany	35
Mayr, R.	Institute for Robotics Research (IRF) Otto-Hahn-Str. 8, D-4600 Dortmund 50, Germany	295
Merklinger, A.	Fraunhofer-Institute for Manufacturing Engineering and Automation (IPA) Nobelstr. 12, D-7000 Stuttgart 80, Germany	121
Milberg, J.	Inst. für Werkzeugmaschinen und Betriebswissenschaften (IWB) Technische Universität München P.O.Box 20 24 20, D-8000 München 2, Germany	331

Mori, H.	Faculty of Engineering, Yamanashi University Takeda-4, Kofu 400, Japan	135
Munkelt, O.	Institut für Informatik Image Interpretation and Artificial Intelligence Technische Universität München Orleanstr. 34, D-8000 München 80, Germany	35
Nehmzow, U.	Dept. of Artificial Intelligence University of Edinburgh 5 Forrest Hill, Edinburgh EH1 2QL, Scotland	267
Ozdes, D.	Fa. Ozdes, Alte Straße 1, D-7802 Merzhausen (Freiburg), Germany	311
Pritschow, G.	Inst. für Steuerungstechnik der Werkzeugmaschinen und Fertigungseinrichtungen (ISW) Universität Stuttgart Seidenstr. 36, D-7000 Stuttgart 1, Germany	201
Puttkamer von, E.	Computer Science Dept. University of Kaiserslautern Erwin-Schrödinger-Straße, D-6750 Kaiserslautern, Germany	187
Radig, B.	Institut für Informatik Image Interpretation and Artificial Intelligence Technische Universität München Orleanstr. 34, D-8000 München 80, Germany	35
Rožmann, M.	Lehrstuhl für Mikrowellentechnik Technische Universität München P.O.Box 20 24 20, D-8000 München 2, Germany	93
Ruß, A.	Lehrstuhl für Prozeßrechner Technische Universität München P.O.Box 20 24 20, D-8000 München 2, Germany	247

Sattler, R.	Institut für Informatik Image Interpretation and Artificial Intelligence Technische Universität München Orleanstr. 34, D-8000 München 80, Germany	35
Schmidt, G.	Lehrstuhl für Steuerungs- und Regelungstechnik Technische Universität München P.O.Box 20 24 20, D-8000 München 2, Germany	3, 61, 151
Schutter De, J.	Katholieke Universiteit Leuven, Dept. of Mechanical Engingeering, Celestijnenlaan 300B, B-3001 Leuven, Belgium	105
Schweiger, J.	Institut für Informatik Real-Time Systems and Robotics Group Technische Universität München Orleanstr. 34, D-8000 München 80, Germany	231
Siegert, H.-J.	Institut für Informatik Real-Time Systems and Robotics Group Technische Universität München Orleanstr. 34, D-8000 München 80, Germany	231, 331
Smithers, T.	Dept. of Artificial Intelligence University of Edinburgh 5 Forrest Hill, Edinburgh EH1 2QL, Scotland	267
Song, K.T.	Katholieke Universiteit Leuven, Dept. of Mechanical Engingeering, Celestijnenlaan 300B, B-3001 Leuven, Belgium	105
Stetter, R.	Inst. für Werkzeugmaschinen und Betriebswissenschaften (IWB) Technische Universität München P.O.Box 20 24 20, D-8000 München 2, Germany	331